Springer Proceedings in Earth and Environmental Sciences

The series Springer Proceedings in Earth and Environmental Sciences publishes proceedings from scholarly meetings and workshops on all topics related to Environmental and Earth Sciences and related sciences. This series constitutes a comprehensive up-to-date source of reference on a field or subfield of relevance in Earth and Environmental Sciences. In addition to an overall evaluation of the interest, scientific quality, and timeliness of each proposal at the hands of the publisher, individual contributions are all refereed to the high quality standards of leading journals in the field. Thus, this series provides the research community with well-edited, authoritative reports on developments in the most exciting areas of environmental sciences, earth sciences and related fields.

Alexei Dmitriev · Janos Lichtenberger ·
Oksana Mandrikova · Emmanuel Nahayo
Editors

Solar-Terrestrial Relations and Physics of Earthquake Precursors

Proceedings of the XIII International Conference

Springer

Editors
Alexei Dmitriev
Department of Space Science
and Engineering
National Central University
Taoyuan, Taiwan

Janos Lichtenberger
Department of Geophysics, Space Research
Group
Eötvös Loránd University
Budapest, Hungary

Oksana Mandrikova
Institute of Cosmophysical Research
and Radio Wave Propagation FEB RAS
Paratunka, Russia

Emmanuel Nahayo
South African National Space Agency
Hermanus, South Africa

ISSN 2524-342X ISSN 2524-3438 (electronic)
Springer Proceedings in Earth and Environmental Sciences
ISBN 978-3-031-50247-7 ISBN 978-3-031-50248-4 (eBook)
https://doi.org/10.1007/978-3-031-50248-4

This Springer imprint is published by the registered company Springer Nature Switzerland AG
The registered company address is: Gewerbestrasse 11, 6330 Cham, Switzerland

Paper in this product is recyclable.

Preface

Based on the plan of meetings, conferences and workshops approved by the Ministry of Science and Education of the Russian Federation, the Institute of Cosmophysical Research and Radio Wave Propagation, Far Eastern Bruch of the Russian Academy of Sciences (IKIR FEB RAS), Kamchatka, held the XIII International Conference "Solar-Terrestrial Relations and Physics of Earthquake Precursors" on September 25–29, 2023.

During the opening ceremony of the conference, Yu. V. Marapulets, the Director of IKIR FEB RAS, welcomed the guests whishing everybody fruitful work. I. A. Larionov, the Chairman of the Organizing Committee, made some opening remarks reminding of the time limit for speeches, presentation format and excursion program schedule.

Five plenary reports were presented during the first day of the conference. S.I. Svertilov opened the plenary section with the report in which he described a space project "SOZVEZDIE-270" of the Moscow University. Within the framework of the project, a group of CubeSat nanosatellites with an instrument set is developed, providing, in particular, monitoring of space environment radiation state, control over geo- and heliophysical state. By the present day, 11 CubeSat satellites have been launched. They send scientific and telemetric information. At the plenary section, Prof. Chen Tao's (National Space Science Center, Chinese Academy of Sciences) report was devoted to the analysis of atmospheric electric field from three stations (Ganzi, Guzhan and Yanzigou) located in the same seismic fracture zone near an earthquake epicenter. Possible precursors, anomalies in atmospheric electrostatic signal, were detected about 2 days before a strong earthquake at the distances of 11 km, 60 km and 299 km from the epicenter to the three sites, respectively. Prof. Xu Jiyao (National Space Science Center, Chinese Academy of Sciences) presented "Chinese Meridian" project that is a network for ground complex monitoring of space weather. Its architecture, building stages and location of the observatories, included into this network, were shown. The capabilities of this network were illustrated. It investigates solar atmosphere, interplanetary space, magnetosphere, ionosphere as well as middle and upper atmosphere of the Earth. Shevtsov B. M. in his plenary speech told about lidar observations aiming at investigating thermosphere optical properties modification, studying the solar activity impact on ionization and transparency of the middle and lower atmosphere and developing methods for lidar sounding taking into account different types of scattering at different atmospheric heights. The considered observations were compared with analogous works by foreign colleagues from China and Germany. O. V. Mandrikova presented the final report at the plenary section describing different methods for ionosphere investigation. Special attention was paid to different algorithms for the detection of anomalies in the investigated ionospheric parameters, modeling of the associated processes and data processing by neural networks.

After the plenary section, the "Atmosphere Physics" section began, where 21 reports, including 1 video speech, were presented. All reports corresponded to the section subject and were devoted to the disturbance processes of the atmosphere and its upper part, ionosphere. The section was opened by the researchers from China. Yuan Wei presented the results of investigation of plasma density increase using the observations from different equipment. Li Lei described the changes in the near-ground atmospheric electric field depending on cosmic galactic ray intensity. I. A. Mironova compared precipitated energetic electrons according to the observations in high-latitudinal atmosphere on balloons and NOAA POES satellites. In the presentation by V. P. Sivokon, impact of the dust from Geminids flux on scattering and dynamic characteristics of magnetically oriented ionospheric inhomogeneities was estimated. It was based on the analysis of electromagnetic radiation of the EISCAT stand. V. N. Marichev presented analysis of the data of annual changeability of background aerosol vertical-time structure in the stratosphere. The data were obtained at a lidar complex of atmospheric height sounding station for 2022. The reporter described stable tendency of stratospheric aerosol accumulation during cold period of the year with the maximum content in January and the decrease in spring to almost its absence during summer months. Then the presentations were made by E. I. Malkin, A. V. Timchenko, A. A. Toropov, D. M. Podorozhnyy, I. A. Pavlov, A. D. Mironov, E. A. Kazakov, D. A. Tverdyy, S. A. Tashkun and Yu. A. Kurbatov. The reports were focused on the development and presentation of the approaches, methods and observation results to solve the problems of geophysical data analysis. Experiments on recording the high-energy particles and lightning activity of different nature were described. Special attention was paid to the changes in ionospheric parameters; new models for calculation of total electron content were suggested; propagation of electromagnetic waves with the ionosphere waveguide boundary was modeled. The reports touched upon the subjects of observations of micron dust particle concentration and factors affecting its change as well as CO_2 variations during seismic events. Updates for the polar ozone concentration model were suggested, and statistics of electromagnetic radiation during volcanic activity was illustrated. Yu. V. Balabin continued the section. He illustrated the results of application of magnetosphere storm Tsyganenko model for investigation of cosmic rays. Their dynamics and spectra during the event GLE66 were shown. The section was concluded by the presentations of A. A. Cheremisin, Y. A. Kurdyaeva and A. Y. Gvozdarev. The presented reports analyzed air mass motion to determine aerosol layers motion occurred during meteorite fall. Formulas of approximation of short atmospheric waves, with respect to the vertical variable, in numerical models of atmosphere global dynamics were shown. Magnetic susceptibility of near-surface rocks was investigated during different temperatures, and expected effects in the magnetic field were estimated on models. The possibility of taking into account the obtained results in real practice of magnetic observatories was estimated. Solodchuk A. A. showed the results of the experiment on the determination of the distance to acoustic radiation sources in near-surface sedimentary rocks.

During the section "Geophysical fields and their interactions", reports devoted to a wide spectrum of investigations from magnetosphere physics to the mathematical modeling of field interactions were presented. In total, 21 presentations, including 4 video presentations, were made during the section. Interesting results on the investigations from the device, unique for our country, for recording geomagnetically induced currents in Kolenergo electric grid during an intense magnetic storm were presented in the report by I. V. Despirak. The second report by I. V. Despirak was devoted to the description of substorm features on a compressed auroral zone based on the results of PGI investigations at the magnetic station and all-sky camera in Barenburg together with the analysis of solar wind characteristics based on ACE data. It was shown that such substorms are developed during slow flow of the solar wind at southern polarity of the interplanetary magnetic field. Based on the observation results of Cluster and Double Stars satellites, Cai Chunlin (National Space Science Center, Chinese Academy of Sciences) described the results of investigations on the detection of magnetosheath filaments, which were predicted theoretically earlier and are formed by energetic ions. The presentation illustrated three episodes of observations of filament-like structures. In each case, quasiperiod fluctuations of magnetosheath energetic ions were observed. L. M. Bogomolov with co-authors described the results of investigations of back seismoelectric effect of the 2nd kind. It was noticed in the report that current density in the region of acoustic instrumentation location reached 1 μA/sqm. Similar magnitudes were observed during the experiments on seismicity modulation on test field sites in Kirgiziya and Tadzhikistan, and generated during geomagnetic storms at the regions of deep faults in cases of seismic activity modulation by geomagnetic disturbances. S. Yu. Khomutov showed some kinds of pulse noises at Novosibirsk magnetic observatory. They were recorded after the introduction of digital magnetometers in magnetic measurement practice. These pulses are assumed to be caused by the railway net working on the direct current and located 10 km from the observatory. Complexity of the problem has been noticed. It requires engagement of railway specialists to be solved correctly. I. A. Solodukhin presented the history of magnetic measurements on the territory of Belarus Republic, stages of their development and current state of Pleshchenitsy magnetic observatory operating since 1960. Geomagnetic field elements dynamics was described both in historical perspective and during the observatory work. At the present time, the observatory undergoes the stage of software updating. Cooperation with the Paratunka KGFO is planned. M.A. Mishchenko's report was devoted to the recording of seismoacoustic and seismoelectric effects in sedimentary rocks at Karymshina site, Kamchatka. Correlation between hypocentral distances and earthquake energy classes, causing this effect, was shown by rank correlation methods. N. N. Semakov analyzed the dynamics of virtual pole location and its equivalent dipole moment, calculated from Novosibirsk observatory data and cleared from diurnal variations by estimating the moving average in a 24-hour window during strong earthquakes. Bay-like disturbances for these characteristics were illustrated in some cases. Then the section was continued by a group of 5 reports by E. A. Kazakov, R. I. Parovik, L. K. Feshchenko, G. M. Vodinchar and O. V Sheremet'eva. They were united thematically as they discussed mathematical models of geospheric processes and systems (systems of space dynamo, turbulent systems of geophysical hydrodynamics, microseismic modes) describing memory or hereditarily effects. Manifestation of such effects is observed in real systems, and consideration of memory in mathematical models allows one to understand physical nature of the processes better. Two reports by V. I. Korochentsev were devoted to the development of hydroacoustic antenna complexes aimed at operating under complicated ice conditions. Both model and experimental work results were presented. The author has been participating in the conference for a number of years and presented his work in this field of hydroacoustics. I. N. Myagkova's report was devoted to the development of space weather forecast tools in NIIYaF during the latest 5 years. During one of the previous conferences at IKIR, the author already presented the capabilities of the forecast center. The report described the changes, which have been made. Periodicity of geomagnetic jerks was discussed in the report by S. A. Ryabkova. Based on the analysis of Moscow geomagnetic observatory data, it was shown that jerks form a quasi-periodical sequence with the typical period of 3–4 years. Based on these data, the next jerk can be expected in 2024–2025. The authors also suggested a hypothetic mechanism of this sequence formation. It is associated with instabilities in geodynamo action. N. S. Khaerdinov presented the results of records of unique behavior of atmospheric electric field at Baksanskaya neutrino observatory. Variations typical for thunderstorm phenomena were recorded during fair weather conditions. It was shown that telluric current was the source of the variations. It was formed as the result of tectonic activity in the form of slow oscillations of the ground surface. V. V. Bogomolov's report was devoted to the review of the developments of the scientific equipment for CubeSat satellites. V. I. Osedlo presented the results of operation of cosmic radiation complex detector launched on board the Monitor-1 spacecraft. The data were obtained both for geomagnetically calm conditions and during geomagnetic disturbances caused by helio-activity intensification. Yu. V. Poklad presented the results of the analysis of very long wave signal at Mikhnevo observatory. Signal parameter variations during the catastrophic earthquake in Turkey were detected. D. A. Tverdyi presented a mathematical model describing the process of increase in radon concentration in the near-surface layer under the conditions of activation of geo-environment stress–strain state. E. V. Liperovskaya suggested a statistical approach to the search of ionospheric precursors of earthquakes based on Japanese data. The author continued the paper series on statistical methods in geophysics. N. A. Sycheva investigated the seismicity of the earth crust in Turkey by the means of Tsallis statistics. B. M. Shevtsov considered invariant statistical model of anomalous phenomena in hereditary theory of criticality.

During the section "Physics of Earthquake Precursors", 7 reports, including 2 video speeches, were presented. Original information, obtained during the latest years and attracting the interest of the scientific community, was presented in the majority of the presented reports. V. V. Bodganov considered the earthquake preparation from the point of view of non-equilibrium thermodynamics. He formulated a hypothesis that background earthquakes with the main shock preparation process of several years have the tendency to group at large space scales approximately at the depth of a preparing great event. A. V. Pavlov told about the updated method for detection of earthquake ionospheric precursors that resulted in the improvement of forecast reliability. V. N. Sychev showed that Tsallis parameter can be used as an effective instrument for identification of seismic events. In spite of the similar magnitude of events, Tsallis parameter differs for natural seismicity and technogenic (bursts) one. E. O. Makarov described a successful forecast of earthquakes based on radon anomalies and proposed the basis for the model of earthquake precursor formation. In the video presentation, S. A. Riabova showed ionosphere response on a strong earthquake, the epicenter of which was at the distance of 1200–1500 from the site of ionosphere parameter measurements. V. A. Pilipenko considered the question on the possibility of global geomagnetic disturbance occurrences during remote earthquakes and pointed out that it was necessary to analyze detected field anomalies as a single package and together with the specialists in earthquake Physics and space weather. B. P. Komitov provided evidences that space weather phenomena may play the role of triggering mechanisms for the significant part of ground seismic and volcanic events.

In the course of the conference, 28 posters were presented and during the breaks the participants discussed them with work authors.

At the end of the conference, the reports were discussed and the conclusions were made. Yu. V. Marapulets, N. N. Semakov and L. M. Bogomolov made short speeches. The participants highlighted the following advantages of the Conference: good organization; compliance with the report subjects and the time limits; high level of scientific presentations; wide geography of the participants that indicates the topicality of the Conference themes; active participation of young researchers (30 percent of reports have young researchers as co-authors), correspondence to the standards of information resources, databases and archives. The following conclusions were made: importance and necessity to make complex, both experimental and theoretical, investigations of the processes, occurring in the lithosphere, atmosphere, ionosphere and magnetosphere and their interaction; importance of maintenance and development of the observational (experimental) bases for the researches including long continuous measurements; necessity of active involvement of theoretical groups to interpret experimental data and to analyze the obtained results; importance of participation of the scientists from other countries for future development of international cooperation.

Participants of the conference spoke for joining the efforts of scientists from different organizations on ground-space monitoring of physical fields and analysis of space weather parameters especially long-term observation standards at observatories. They supported the idea to continue the works on unification of the methods for processing and storage of the data on geophysical fields in the lithosphere, atmosphere, ionosphere as well as the works in the field of investigation of magnetic disturbances, thunderstorms with a large number of strokes including powerful strokes affecting technical systems, volcanic lightning occurrences and their interactions with geodynamic processes. The Scientific Committee approved selected reports to be published in "Springer Proceedings in Earth and Environmental Sciences".

September 2023

I. A. Larionov
Yu. A. Polozov
A. Yu. Gvozdarev
S. A. Solodchuk
G. M. Vodinchar
M. A. Mishchenko

Organization

Organizing Committee

Chairman

Igor Larionov — Institute of Cosmophysical Research and Radio Wave Propagation, Russia

Deputy Chairman

Mikhail Mishchenko — Institute of Cosmophysical Research and Radio Wave Propagation, Russia

Secretary

Nina Cherneva — Institute of Cosmophysical Research and Radio Wave Propagation, Russia

Committee members

Yurii Polozov — Institute of Cosmophysical Research and Radio Wave Propagation, Russia

Aleksandra Solodchuk — Institute of Cosmophysical Research and Radio Wave Propagation, Russia

Nikolai Popov — Institute of Cosmophysical Research and Radio Wave Propagation, Russia

Olga Filipova — Institute of Cosmophysical Research and Radio Wave Propagation, Russia

Anna Larionova — Institute of Cosmophysical Research and Radio Wave Propagation, Russia

Roman Parovik — Institute of Cosmophysical Research and Radio Wave Propagation, Russia

Lyubov Feshchenko — Institute of Cosmophysical Research and Radio Wave Propagation, Russia

Irina Gileva — Institute of Cosmophysical Research and Radio Wave Propagation, Russia

Natalia Chelnyntseva	Institute of Cosmophysical Research and Radio Wave Propagation, Russia
Valentina Semagina	Institute of Cosmophysical Research and Radio Wave Propagation, Russia
Svetlana Sapozhnikova	Institute of Cosmophysical Research and Radio Wave Propagation, Russia
Ludmila Marchenko	Institute of Cosmophysical Research and Radio Wave Propagation, Russia
Arcadyi Ivanov	Institute of Cosmophysical Research and Radio Wave Propagation, Russia
Ivan Zhidkov	Institute of Cosmophysical Research and Radio Wave Propagation, Russia
Roman Martynov	Institute of Cosmophysical Research and Radio Wave Propagation, Russia

Section Chairmen

Igor Larionov	Institute of Cosmophysical Research and Radio Propagation, Russia
Yuriy Polozov	Institute of Cosmophysical Research and Radio Wave Propagation, Russia
Gleb Vodinchar	Institute of Cosmophysical Research and Radio Wave Propagation, Russia
Aleksey Gvozdarev	Institute of Cosmophysical Research and Radio Wave Propagation, Russia
Mikhail Mishchenko	Institute of Cosmophysical Research and Radio Wave Propagation, Russia
Aleksandra Solodchuk	Institute of Cosmophysical Research and Radio Wave Propagation, Russia

Scientific Committee

János Lichtenberger	Eötvös Loránd University, Hungary
Vadim Bogdanov	Institute of Cosmophysical Research and Radio Wave Propagation, Russia
Kusumita Arora	National Geophysical Research Institute, India
Oksana Mandrikova	Institute of Cosmophysical Research and Radio Wave Propagation, Russia
Dr Emmanuel Nahayo	South African National Space Agency, South Africa

Vladimir Sivokon'	Institute of Cosmophysical Research and Radio Wave Propagation, Russia
Dr. Battuulai Tsegmed	Institute of Astronomy and Geophysics, Mongolia
Sergey Khomutov	Institute of Cosmophysical Research and Radio Wave Propagation, Russia
Alexei Dmitriev	National Central University, Taiwan
Boris Shevtsov	Institute of Cosmophysical Research and Radio Wave Propagation, Russia
Wei Xue	Harbin Engineering University, China
Yuriy Senkevich	Institute of Cosmophysical Research and Radio Wave Propagation, Russia

Contents

Atmosphere Physics

Analysis of Ionospheric Parameters During Solar Events and Magnetic Storms

Oksana Mandrikova and Nadezhda Fetisova(✉)

Institute of Cosmophysical Research and Radio Wave Propagation FEB RAS, Mirnaya Str., 7, Paratunka, 684034 Elizovskiy District, Kamchatka Region, Russia

nv.glushkova@yandex.ru

Abstract. The paper presents a detailed analysis of the critical frequency data of the ionospheric F2 layer (*foF2*) during magnetic storms in 2022 and 2023. Events of different intensity that occurred during the winter period were analyzed. To investigate the *foF2* temporal variations, we used the ionosonde data at Paratunka station (Kamchatka region), registered by IKIR FEB RAS. The analysis was performed using a generalized multicomponent model (GMCM) developed by the authors. The GMCM is based on the joint application of the wavelet transform, threshold functions, and ARIMA models. Application of the model allows us to detect anomalous changes in *foF2* variations, to study their spectral-temporal structure, and to estimate their intensity. In the analyses we performed a comparison with the interplanetary medium parameters (IMF, solar wind), and we also took into account the geomagnetic disturbance power and the nature of its occurrence (CME, CIR, events of mixed nature).The study identifies the typical features of the ionospheric parameter dynamics during disturbed periods determined by space weather factors. The analysis shows that strong and moderate magnetic storms were preceded by positive pre-storm disturbances of *foF2*.

Keywords: ionospheric disturbances · wavelet transform · autoregressive models

1 Introduction

The response of the ionosphere to changes in the geomagnetic condition (geomagnetic storms and substorms, CMEs, and solar flares, and CIR) manifests itself in the form of ionospheric storms [1–3]. Their timely detection is important to prevent adverse effects on modern technical systems and to maintain reliable radio communication (for example, [2, 3]). Ionospheric storms can include positive (increase of electron concentration) and negative (decrease of electron concentration) phases, which are caused by different physical mechanisms [1–3]. The nature of ionospheric storm (the presence of positive and (or) negative phase, its intensity, duration and spatial extent) depends on geomagnetic disturbance intensity and local time of its onset, season, solar activity, and other parameters [1, 3]. Recent studies [4–6] show that ionospheric disturbances of both signs can be observed during the preparatory phase of magnetic storms. At the same time,

A. Dmitriev et al. (Eds.): STRPEP 2023, SPEES, pp. 3–13, 2023.
https://doi.org/10.1007/978-3-031-50248-4_1

the appearance of the pre-increase effect of electron concentration is noted in numerous papers [4, 7, 8]. In recent years, many papers have been devoted to pre-storm effects in the ionosphere, however, the question of the nature of their formation remains open [4, 5, 9].

The papers of many authors are aimed at studying ionospheric processes and creating methods for estimating its condition [2, 8, 10–14]. However, the application of existing methods (e.g., empirical [10, 14], physical [11, 13], and neural networks [2, 12] approaches, etc.) is not effective enough due to the complex nonstationary structure of ionospheric parameters and the absence of complete a priori knowledge of the processes occurring in the disturbed ionosphere. The authors of this paper aim to create automated methods for analyzing ionospheric parameters and detecting ionospheric disturbances [7, 8, 15, 16]. The authors have proposed a generalized multicomponent model of the ionospheric parameter time series (GMCM) [7, 8, 16]. Model identification is based on the joint application of different wavelet transform schemes [17], threshold functions, and a class of autoregressive-integrated moving average models (ARIMA, [18]). The GMCM describes three classes of anomalies characterizing strong (class 3), moderate (class 2), and weak (class 1) ionospheric disturbances. Approbation of the model on the ionospheric critical frequency data (*foF2*) confirmed the possibility of its application to detect ionospheric disturbances of different intensity, including low-amplitude short-period ionospheric disturbances [7, 8]. Comparison of the GMCM with the median method and the International IRI model [10] showed its efficiency and prospective usefulness for analysis of the ionospheric behavior [8, 16].

In this study, we carried out a joint analysis of the *foF2* variations and space weather parameters during magnetic storms (moderate and strong magnetic storms occurring in 2022 and 2023 were considered) using the GMCM. Ionospheric anomalies were detected during analyzed events, and their parameters were estimated (intensity, duration and onset moment). The study identified features of the ionospheric parameter dynamics during disturbed periods typical for the Kamchatka region (according to the *foF2* data at the Paratunka station). On the eve of the considered magnetic storms, positive pre-storm perturbations of foF2 were observed, which were given special attention in the paper. The pre-storm disturbances were analyzed together with the parameters of interplanetary medium and magnetosphere.

2 Generalized Multicomponent Model of Ionospheric Parameters and a Method for Detecting Ionospheric Disturbances

Generalized multicomponent model of ionospheric parameter time series proposed by the authors takes the form [7, 8]:

$$f(t) = A^{REG}(t) + U(t) + e(t) = \sum_{\mu=\overline{1,T}} \sum_{k=\overline{1,N^{\mu}_{jreg}}} s^{\mu}_{jreg} b^{\mu}_{jreg}(t) + \sum_{i=\overline{1,3}} \sum_{\eta,n} P_{i,\eta}(d_{\eta,n}) \Psi_{\eta,n}(t) + e(t), \quad (1)$$

where $A^{REG}(t)$ is the regular component describes typical periodic variations of the ionospheric parameters, $U(t)$ is anomalous component describes sudden changes of the ionospheric parameters (disturbances of terrestrial and extraterrestrial nature), $e(t)$ is the noise component (hardware failures, industrial explosions, etc.).

2.1 Regular Component of the Model and a Method for Detecting Intense Ionospheric Disturbances

The regular component $A^{REG}(t)$ takes the form (the algorithm of component identification is described in the papers [7, 16]):

$$A^{REG}(t) = \sum_{\mu=\overline{1,T}} \sum_{k=\overline{1,N^{\mu}_{j^{reg}}}} s^{\mu}_{j^{reg},k} b^{\mu}_{j^{reg},k}(t) + e(t), \tag{2}$$

where $s^{\mu}_{j^{reg},k} = \sum_{l=1}^{p^{\mu}_{j^{reg}}} \gamma^{\mu}_{j^{reg},l} \omega^{\mu}_{j^{reg},k-l} - \sum_{n=1}^{h^{\mu}_{j^{reg}}} \theta^{\mu}_{j^{reg},n} a^{\mu}_{j^{reg},k-n}$ ($\mu = \overline{1,T}$ is the component number) is a μ-th component obtained using MRA [13]; $p^{\mu}_{j^{reg}}$, $\gamma^{\mu}_{j^{reg},l}$ are the order and parameters of autoregression of the μ-th component; $\omega^{\mu}_{j^{reg},k} = \nabla^{\nu^{\mu}} \delta^{\mu}_{j^{reg},k}$, ν^{μ} is the order of difference of the μ-th component, $\delta^{1}_{-m^{reg},k} = c_{-m^{reg},k}$, $\delta^{\mu}_{j^{reg},k} = d_{j^{reg},k}$, $\mu = \overline{2,T}$, T is the number of components, $c_{-m^{reg},k}$, $d_{j^{reg},k}$ are the wavelet coefficients of MRA components; $h^{\mu}_{j^{reg}}$, $\theta^{\mu}_{j^{reg},n}$ are the order and parameters of the moving average of the μ-th component; $a^{\mu}_{j^{reg},k}$ are errors of the μ-th component model; $N^{\mu}_{j^{reg}}$ is the length of the μ-th component; $b^{1}_{-m^{reg},k} = \phi_{-m^{reg},k}$; $b^{\mu}_{j^{reg},k} = \Psi_{j^{reg},k}$, $\mu = \overline{2,T}$; $\phi_{-m^{reg},k}(\mathrm{t})$ is the scaling function; $\Psi_{j^{reg},k}(\mathrm{t})$ is the wavelet function.

A method for detecting intensive ionospheric disturbances is based on the analysis of residual errors $a^{\mu}_{j^{reg},k}$ of the model component $A^{REG}(t)$ (Eq. (2)) [8]. The following condition is tested:

$$\varepsilon^{\mu}_{j^{reg}} = \sum_{q=1}^{Q_\mu} \left| a^{\mu}_{j^{reg},k+q} \right| > H_{\mu,j^{reg}}, \tag{3}$$

where $q \geq 1$ is the data lead step; Q_μ is the lead length of the μ-th component model; $a^{\mu}_{j^{reg},k+q} = s^{\mu,fact}_{j^{reg},k+q} - s^{\mu,model}_{j^{reg},k+q}$; $s^{\mu,model}_{j^{reg},k+q} = \sum_{l=1}^{p^{\mu}_{j^{reg}}} \gamma^{\mu}_{j^{reg},l} \omega^{\mu}_{j^{reg},k+q-l} - \sum_{n=1}^{h^{\mu}_{j^{reg}}} \theta^{\mu}_{j^{reg},n} a^{\mu}_{j^{reg},k+q-n}$; $H_{\mu,j^{reg}}$ is the threshold value of the μ-th component determining the presence of disturbances [8, 18]:

$$H_{\mu,j^{reg}}(Q_\mu) = u_{\xi/2} \left\{ 1 + \sum_{q=1}^{Q_\mu - 1} \left(\psi^{\mu}_{j^{reg},q} \right)^2 \right\}^{1/2} \sigma_{a^{\mu}_{j^{reg}}}, \tag{4}$$

where $u_{\xi/2}$ is the quantile of the level $(1 - \xi/2)$ of standard normal distribution; $\sigma^2_{a^{\mu}_{j^{reg}}}$ is the variance of residual errors of the μ-th component model; $\psi^{\mu}_{j^{reg},q}$ are the weighting coefficients of the μ-th component [8, 18].

2.2 Anomalous Component of the Model and a Method for Detecting Shot-Period Ionospheric Disturbances

Anomalous component $U(t)$ of the model describes sudden shot-period changes of the parameters of three classes (weak, moderate and strong disturbances) [8]:

$$U(t) = \sum_{\eta,n} P_{1,\eta}\left(d_{\eta,n}\right)\Psi_{\eta,n}(t) + \sum_{\eta,n} P_{2,\eta}\left(d_{\eta,n}\right)\Psi_{\eta,n}(t) + \sum_{\eta,n} P_{3,\eta}\left(d_{\eta,n}\right)\Psi_{\eta,n}(t), \tag{5}$$

$$P_{1,\eta}\left(d_{\eta,n}\right) = \begin{cases} 0, \textit{ if } \left|d_{\eta,n}\right| \le T_{1,\eta} or \left|d_{\eta,n}\right| > T_{2,\eta} \\ d_{\eta,n}, \textit{ if} T_{1,\eta} < \left|d_{\eta,n}\right| \le T_{2,\eta} \end{cases},$$

$$P_{2,\eta}\left(d_{\eta,n}\right) = \begin{cases} 0, \textit{ if } \left|d_{\eta,n}\right| \le T_{2,\eta} or \left|d_{\eta,n}\right| > T_{3,\eta} \\ d_{\eta,n}, \textit{ if} T_{2,\eta} < \left|d_{\eta,n}\right| \le T_{3,\eta} \end{cases},$$

$$P_{3,\eta}\left(d_{\eta,n}\right) = \begin{cases} 0, \textit{ if } \left|d_{\eta,n}\right| \le T_{3,\eta} \\ d_{\eta,n}, \textit{ if } \left|d_{\eta,n}\right| > T_{3,\eta} \end{cases},$$

where $d_{\eta,n} = \langle U, \Psi_{\eta,n}\rangle$ are the wavelet coefficients; $\left\{\Psi_{\eta,n}\right\}_{\eta,n\in Z}$ is the wavelet basis; $\left|d_{\eta,n}\right|$ is the amplitude of wavelet coefficients taken as a measure of disturbance intensity on the scale η [7], $T_{1,\eta}$, $T_{2,\eta}$, $T_{3,\eta}$ are thresholds determining disturbances of weak (class 1), moderate (class 2) and strong (class 3) intensity respectively.

To detect sudden short-period anomalous changes (sharp increases/decreases of electron density), the adaptive thresholds $P^{ad}_{i,\eta}$, $i = \overline{1,3}$ are applied (Eq. (5)) and the coefficients $d_{\eta,n}$ are taken to be equal:

$$d_{\eta,n} = \left[\begin{array}{l} \begin{cases} d^{1+}_{\eta,n}, \textit{ if } P^{ad}_{1,\eta} < \left(d_{\eta,n} - d^{med}_{\eta,n}\right) \le P^{ad}_{2,\eta} \\ 0, \textit{ if } \left|d_{\eta,n} - d^{med}_{\eta,n}\right| < P^{ad}_{1,\eta} or \left|d_{\eta,n} - d^{med}_{\eta,n}\right| > P^{ad}_{2,\eta} \\ d^{1-}_{\eta,n}, \textit{ if } -P^{ad}_{2,\eta} \le \left(d_{\eta,n} - d^{med}_{\eta,n}\right) < -P^{ad}_{1,\eta} \end{cases} \\ \begin{cases} d^{2+}_{\eta,n}, \textit{ if} P^{ad}_{2,\eta} < \left(d_{\eta,n} - d^{med}_{\eta,n}\right) \le P^{ad}_{3,\eta} \\ 0, \textit{ if } \left|d_{\eta,n} - d^{med}_{\eta,n}\right| < P^{ad}_{2,\eta} or \left|d_{\eta,n} - d^{med}_{\eta,n}\right| > P^{ad}_{3,\eta} \\ d^{2-}_{\eta,n}, \textit{ if} -P^{ad}_{3,\eta} \le \left(d_{\eta,n} - d^{med}_{\eta,n}\right) < -P^{ad}_{2,\eta} \end{cases} \\ \begin{cases} d^{3+}_{\eta,n}, \textit{ if } \left(d_{\eta,n} - d^{med}_{\eta,n}\right) > P^{ad}_{3,\eta} \\ 0, \textit{ if } \left|d_{\eta,n} - d^{med}_{\eta,n}\right| < P^{ad}_{3,\eta} \\ d^{3-}_{\eta,n}, \textit{ if } \left(d_{\eta,n} - d^{med}_{\eta,n}\right) < -P^{ad}_{3,\eta} \end{cases} \end{array}\right. \tag{6}$$

where $P^{ad}_{i,\eta} = V_i \cdot St_\eta$; V_i is the threshold coefficient; $St_\eta = \sqrt{\frac{1}{\Phi-1}\sum_{n=1}^{\Phi}\left(d_{\eta,n} - \overline{d}_{\eta,n}\right)^2}$; $\overline{d}_{\eta,n}$ and $d^{med}_{\eta,n}$ are the average value and the median estimated for each sample within the current day using a moving time window of the length Φ (taking into account the diurnal variations of ionospheric data); the coefficients $d^{i+}_{\eta,n}(d^{i-}_{\eta,n})$ determine the positive (negative) anomalies of class i.

The intensity of positive $\left(I^{i+}(n)\right)$ and negative $\left(I^{i-}(n)\right)$ ionospheric disturbances of each class i at the specific time $t = n$ is estimated as

$$I^{i+(-)}(n) = \sum\nolimits_{\eta} \left| d^{i+(-)}_{\eta,n} \right|. \tag{7}$$

3 Experimental Results and Discussion

This section presents the results of processing and analysis of the *foF2* data during magnetic storms in 2022 and 2023. Events of different intensity that occurred during the winter period were analyzed. We used hourly and 15-min *foF2* data registered by an ionosonde at Paratunka station (53.00 N, 158.70 E, Kamchatka Peninsula, IKIR FEB RAS). Identification and estimation of model parameters (Sect. 2) were performed for the *foF2* data recorded during magnetically quiet periods and taking into account seismic activity of Kamchatka region. The models were constructed separately for each season (winter, summer) and solar activity level. The process of identification and estimation of the GMCM parameters is described in detail in the paper [8].

The analysis of the *foF2* ionospheric critical frequency dynamics was carried out in conjunction with magnetospheric data and space environment parameters. Geomagnetic activity was analyzed using Dst- (https://wdc.kugi.kyoto-u.ac.jp/), Kp- (https://wdc.kugi.kyoto-u.ac.jp/) and K-indices (Paratunka station). Solar wind parameters (https://omniweb.gsfc.nasa.gov/form/dx1.html) and interplanetary magnetic field (IMF) (https://omniweb.gsfc.nasa.gov/form/dx1.html) were also analyzed. When describing events, we used space weather data (http://ipg.geospace.ru).

Figure 1 presents the results of *foF2* data processing during a moderate magnetic storm that commenced on February 10, 2022. Based on the space weather review (http://ipg.geospace.ru), an inhomogeneous accelerated flux from coronal mass ejection (CME on February 5) arrived at the end of February 9. IMF southern component fluctuations increased to Bz = −6 nT, and SWS reached 450 km/s by February 10. Analysis of the *foF2* behavior shows that a smooth increase of *foF2,* leading to the formation of a positive anomaly of moderate intensity (class 2, Fig. 1d, e), occurred at the moment of arrival of the inhomogeneous accelerated flow. The anomalous threshold was exceeded 6 h before the development of the magnetic storm and lasted about 29 h. Modeling results (Fig. 1b) confirm the anomalous changes in the *foF2* temporal variations on the eve of the event (the GMCM errors exceed the threshold values $H_{\mu,j^{reg}}$; $H_{\mu,j^{reg}}$ correspond to the 70% confidence interval). During the first hours of the magnetic storm, the anomaly showed its maximum intensity (Fig. 1e). Analysis of the spectral-temporal structure of the anomaly (Fig. 1d) shows a sharp shift of the spectrum to high frequencies during this period (scales on the vertical axis in Fig. 1d). Comparison with the interplanetary medium parameters shows a significant correlation between the ionospheric anomaly formation and solar wind data (Fig. 1h, i), indicating its possible solar nature.

During the main phase of the storm, we observed a decrease of electron concentration to the background level (Fig. 1e), which continued during the recovery phase and led to the formation of a negative ionospheric anomaly of strong intensity (class 3). At the beginning of the day on February 11, an inhomogeneous accelerated flux arrived from two coronal mass ejections (CMEs on February 8), southern component fluctuations grew to Bz = −13 nT (http://ipg.geospace.ru). In the afternoon of February 11, an inhomogeneous accelerated flux from a coronal hole arrived. Fluctuations of the southern component intensified to Bz = −16 nT, SWS (solar wind speed) began to increase and reached a maximum value of 600 km/s on February 12 (http://ipg.geospace.ru). A significant increase of disturbances in near-Earth space was accompanied by an intense decrease of electron concentration (strongly exceedance of class 3 threshold and reaching up to 3500 conventional units, Fig. 1e). At the end of the day on February 11, at the background of a sharp increase of SWS (Fig. 1 h) and a southward turn of the IMF component (Fig. 1j), a short-term positive ionospheric anomaly is also observed (Fig. 1e). According to the modeling data, during this period there was a long-term (about 3 days) change in the *foF2* temporal variations: the exceedance of $H_{\mu,j^{reg}}$ was 3.2 SD for the $f_{-3}(t)$ component (it is shown in Fig. 1b in black) and 3.8 SD for the $g_{-3}(t)$ component (it is shown in Fig. 1b in green). The moving median data (median is shown in Fig. 1a in blue) are consistent with the GMCM results.

Figure 2 shows the result of the analysis of foF2 behavior during a strong magnetic storm beginning on 26 February 2023. On the eve of the storm, a weak increase in geomagnetic activity is observed (Fig. 2f). According to http://ipg.geospace.ru, an inhomogeneous accelerated flux from a coronal mass ejection (CME on 21 February) arrived at the beginning of the day on 23 February; the IMF southern component fluctuated from Bz = −9 nT to Bz = +7 nT, and SWS increased to 610 km/s by 05.// UT. The dynamics of ionospheric data on 23 February was typical (Fig. 2d, e). At 07.// UT on 25 February, an inhomogeneous accelerated flux from a coronal mass ejection arrived, and further at 12.// UT, an inhomogeneous accelerated flux from a coronal hole arrived. SWS was 450 km/s by 07.// UT on 26 February, and the fluctuations of the southern component intensified to Bz = −9 nT. During the period from 24 to 25 February, according to the results of data processing (Fig. 2d, e), anomalous oscillatory processes of moderate intensity (class 2) appeared in the ionosphere. These anomalies reached 700 conventional units and more at some time intervals. The GMCM errors exceeded the 70% confidence interval (3.3 SD for the $f_{-3}(t)$ component, Fig. 2b). Comparison of the detected oscillatory processes with the near-Earth environment parameters indicates their possible connection with terrestrial processes. It should be noted that an earthquake with magnitude 5.4 (https://sdis.emsd.ru/main.php) registered in Kamchatka on 25 February; the moment of the earthquake is marked with a vertical arrow in Fig. 2a. An inhomogeneous accelerated flux from two coronal mass ejections (CMEs on 24 February) arrived at 07.// UT on 26 February. SWS began to increase, reaching 770 km/s by the end of the day on 26 February. During the arrival of the accelerated flux, a smooth anomalous increase of ionospheric electron concentration was observed. It exceeded the threshold of class 1 8 h before the event and led to the formation of a positive anomaly of high intensity (class 3, Fig. 2d, e). The anomalous increase in the pre-storm period is observed at the background of a sharp increase of the solar wind density (Fig. 2i) and the southward-directed component of IMF (Fig. 2h). The positive anomaly lasted for about a day and a half, and during the main phase of the storm its intensity reached very high values (more than 3000 conventional units, Fig. 2e). During the recovery phase, the positive anomaly was replaced by a negative one, which also had a very high intensity (about 4000 conventional units, Fig. 2e). A strong decrease of electron concentration led to a change in the foF2 temporal course, as indicated by strong deviations of foF2 from the median (Fig. 2a, c) and a significant increase of modelling errors (10 SD for the $f_{-3}(t)$ component and 4.3 SD for the $g_{-3}(t)$ component, Fig. 2b). Comparison of the GMCM results with median method during the storm confirms their reliability (Fig. 2a, c).

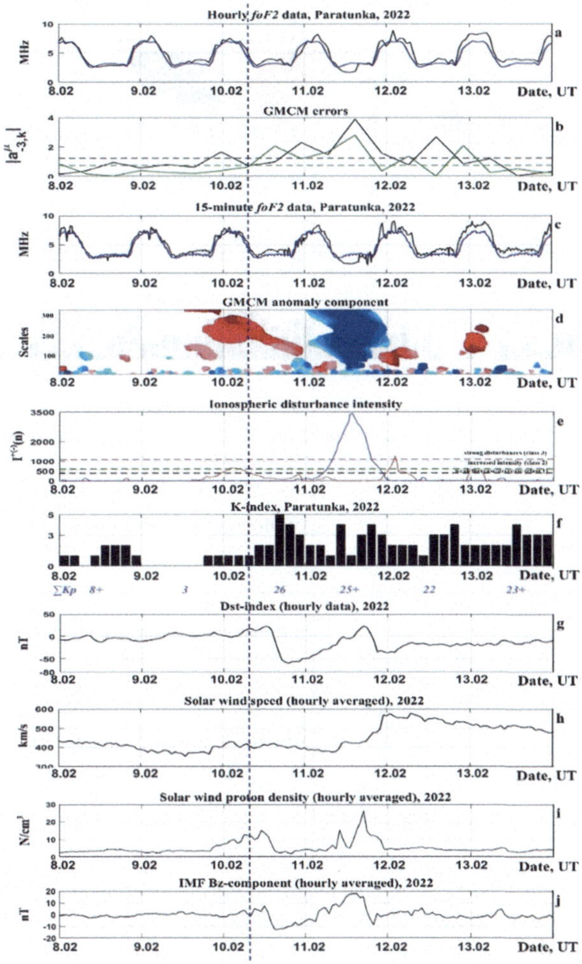

Fig. 1. Analysis of the *foF2* data during the magnetic storm on February 10, 2022: a), c) recorded *foF2* data (black), 27-day median (blue); b) errors of smoothed (black) and detailing (green) components (Eq. (3)), dashed lines show the thresholds $H_{\mu,j^{reg}}$ (Eq. (4)); d) detected different-scale positive (red) and negative (blue) ionospheric disturbances; e) intensity of detected ionospheric disturbances (Eq. (7)), dashed lines show the class thresholds. Dashed blue line shows the magnetic storm commencement.

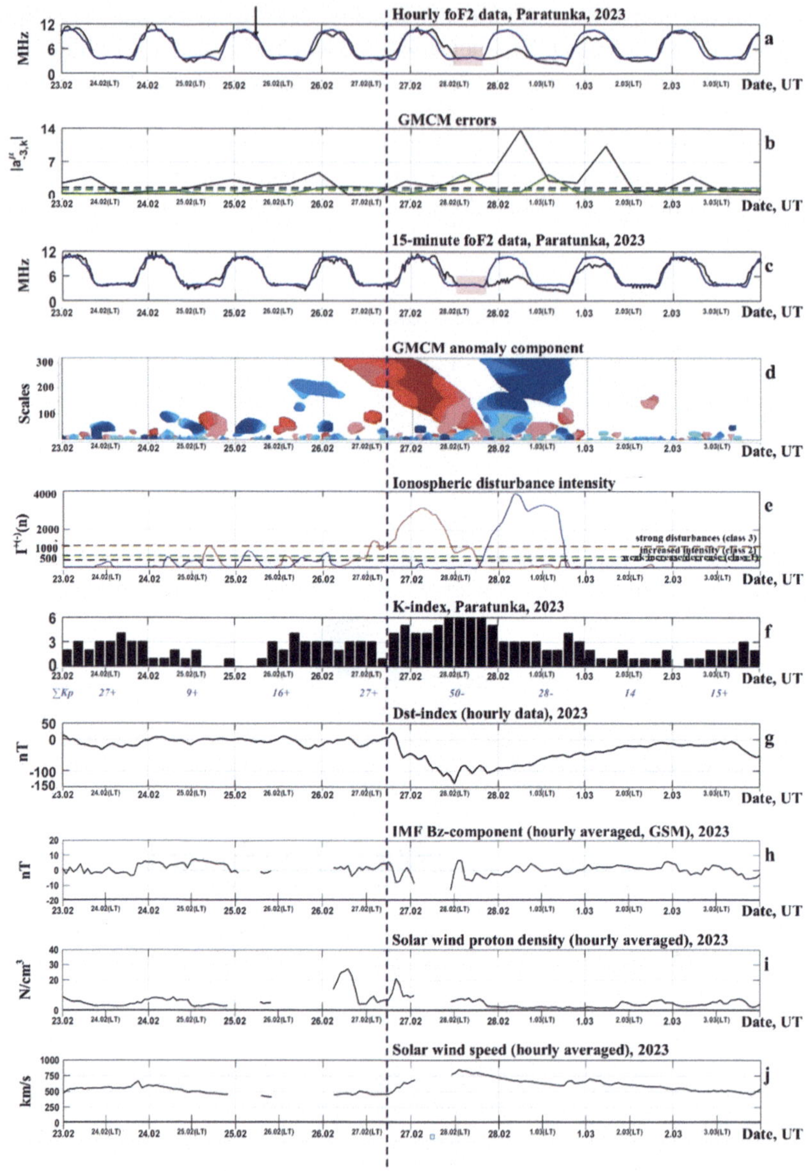

Fig. 2. Analysis of the *foF2* data during the magnetic storm on February 26, 2023: a), c) recorded *foF2* data (black), 27-day median (blue); b) errors of smoothed (black) and detailing (green) components (Eq. (3)), dashed lines show the thresholds $H_{\mu,j^{reg}}$ (Eq. (4)); d) detected different-scale positive (red) and negative (blue) ionospheric disturbances; e) intensity of detected ionospheric disturbances (Eq. (7)), dashed lines show the class thresholds. Dashed blue line shows the magnetic storm commencement.

Figure 3 presents the processing results of foF2 data recorded from 3 to 12 January 2019. During the analyzed period, as shown by the vertical arrows in Fig. 3a, two earthquakes of magnitudes 5.35 and 5.45 were recorded in Kamchatka, according to information from https://sdis.emsd.ru/main.php. In contrast to the interval of increased geomagnetic activity (marked with a dotted line, Fig. 3), the quiet periods, similar to those considered above, are accompanied by oscillatory processes with an intensity not exceeding the class 1 threshold. During the weak increase of geomagnetic activity from 4 to 7 January 2019, disturbances reaching the class 3 threshold are observed in the ionosphere. The results are consistent with the paper [19] and indicate a strong influence of solar processes on the ionospheric dynamics. Therefore, the ionospheric anomalies of solar nature, in most cases, exceed in intensity the disturbances caused by other factors.

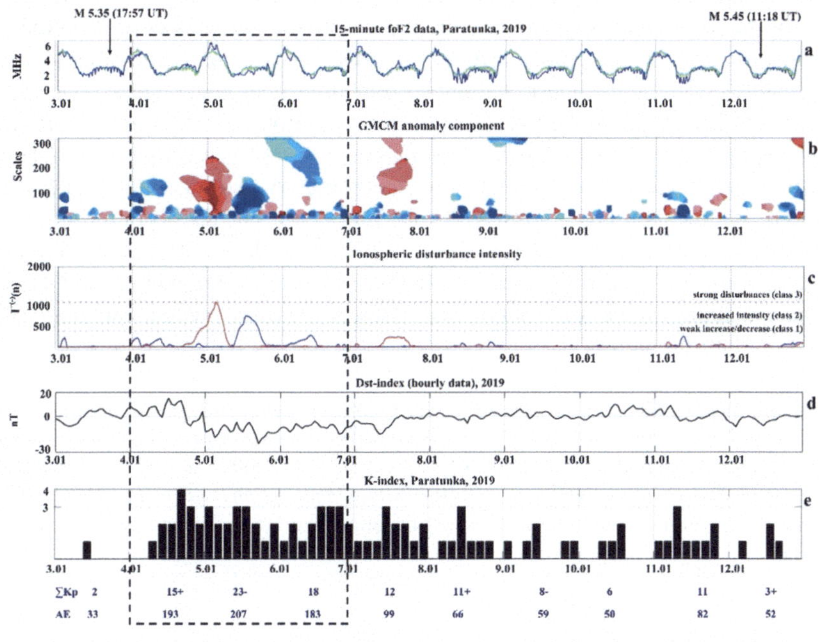

Fig. 3. Analysis of the *foF2* data during the period from 3 to 12 January, 2019

Analyzing the *foF2* behavior during the events described above, as well as those presented in the papers [7, 8, 15], we can note a general picture of the response of the ionosphere over Kamchatka to geomagnetic storms. On the eve of magnetic storms, we observed a smooth long-term anomalous increase in electron concentration. This increase was occurred at the background of a quiet or weakly disturbed magnetic field. The anomalous threshold in intensity was exceeded 3–12 h before the onset of weak magnetic storms caused by CIR and 1–2 days before the onset of moderate and strong storms caused by CME or mixed nature (CME+CIR). During the main phase of storms, the dynamics of ionospheric parameters changed, leading to a significant decrease in electron concentration at the recovery phase. A detailed analysis of pre-storm ionospheric anomalies and comparison with the parameters of interplanetary medium and magnetosphere indicate their possible connection with solar processes. The formation of anomalies is observed during quiet or weakly disturbed geomagnetic field, correlates with southward rotations of the IMF component and increases in solar wind density. The results of the study are consistent with the papers [5, 6, 9, 16]. However, clarifying the nature and determining the mechanisms of anomaly formation on the eve of magnetic storms require further comprehensive studies involving more statistics, as well as expanding the set of analyzed environmental parameters and methods of data study.

4 Conclusions

The presented results indicate the complex dynamics of the ionospheric process during increased solar activity and magnetic storms. Using the GMCM, we performed a detailed analysis of the *foF2* behavior, allowing us to estimate the parameters of detected ionospheric disturbances (onset moments, duration and intensity of anomalies) and to identify typical features of the ionospheric parameter dynamics over Kamchatka. According to the authors, an important applied result of the investigation is the fact of high frequency of the pre-increase effect in the ionosphere on the eve of storms. Comparison of detected pre-storm anomalies with the parameters of interplanetary medium and magnetosphere allows us to assume their connection with solar processes. The results of the study agree with the papers [5, 6, 9, 19]. The analysis of events of different intensity confirmed the efficiency of the GMCM for studying the ionospheric data dynamics and detecting anomalies, including weak ionospheric disturbances that may precede magnetic storms and serve as their predictors.

Acknowledgements. The work was carried out within the framework of the State task on the topic (2021–2023) "Physical processes in the system of near space and geospheres under solar and lithospheric impact", registration number AAAA-A21-121011290003-0. The work was realized by the means of the Common Use Center "North-Eastern Heliogeophysical Center" CKP_558279, USU 351757. The authors are grateful to the organizations recording the data which were applied in the paper.

References

1. Danilov, A.D.: Ionospheric F-region response to geomagnetic disturbances. Adv. Space Res. **52**(3), 343–366 (2013)
2. Nakamura, M., Maruyama, T., Shidama, Y.: Using a neural network to make operational forecasts of ionospheric variations and storms at Kokubunji, Japan. J. Natl. Inst. Inf. Commun. Technol. **56**, 391–406 (2009)
3. Tsagouri, I.: Space weather effects on the earth's upper atmosphere: short report on ionospheric storm effects at middle latitudes. Atmosphere **13**(2), 346 (2022). https://doi.org/10.3390/atmos13020346
4. Kumar, V.V., Parkinson, M.L.: A global scale picture of ionospheric peak electron density changes during geomagnetic storms. Space Weather **15**, 637–652 (2017). https://doi.org/10.1002/2016SW001573
5. Danilov, A.D.: Prestorm ionospheric disturbances: precursors or Q-disturbances? Adv. Space Res. **69**(1), 159–167 (2022). https://doi.org/10.1016/j.asr.2021.09.027

6. Danilov, A.D.: Detailed analysis of the behavior of the F2-layer critical frequency prior to geomagnetic storms. Adv. Space Res. **70**(4), 1078–1088 (2022). https://doi.org/10.1016/j.asr.2022.05.025
7. Mandrikova, O., Polozov, Yu., Fetisova, N., Zalyaev, T.: Analysis of the dynamics of ionospheric parameters during periods of increased solar activity and magnetic storms. J. Atmos. Solar-Terr. Phys. **181**, 116–126 (2018). https://doi.org/10.1016/j.jastp.2018.10.019
8. Mandrikova, O., Fetisova, N.: Modeling and analysis of ionospheric parameters based on multicomponent model. J. Atmos. Solar-Terr. Phys. **208,** 105399 (2020). https://doi.org/10.1016/j.jastp.2020.105399
9. Joshua, B.W., Adeniyi, J.O., Amory-Mazaudier, C., Adebiyi, S.J.: On the pre-magnetic storm signatures in NmF2 in some equatorial, low and mid-latitude stations. J. Geophys. Res. Space Phys. **126**, e2021JA029459 (2021). https://doi.org/10.1029/2021JA029459
10. Bilitza, D., Reinisch, B.W.: International reference ionosphere 2007: improvements and new parameters. Adv. Space Res. **42** (2007). https://doi.org/10.1016/j.asr.2007.07.048
11. Solomentsev, D.V., Titov, A.A., Khattatov, B.V.: Three-dimensional assimilation model of the ionosphere for the European region. Geomagn. Aeron. **53**(1), 73–84 (2013). https://doi.org/10.1134/S0016793212060114
12. Tebabal, A., Radicella, S.M., Nigussie, M., Damtie, B., Nava, B., Yizengaw, E.: Local TEC modelling and forecasting using neural networks. J. Atmos. Sol. Terr. Phys. **172**, 143–151 (2018). https://doi.org/10.1016/j.jastp.2018.03.004
13. Knyazeva, M.A., Namgaladze, A.A., Beloushko, K.E.: Field-aligned currents influence on the ionospheric electric fields: modification of the upper atmosphere model. Russ. J. Phys. Chem. **9**(5), 758–763 (2015). https://doi.org/10.1134/S1990793115050206
14. Shubin, V.N., Karpachev, A.T., Telegin, V.A., Tchybulya, K.G.: Global model SMF2 of the F2-layer maximum height. Geomagn. Aeron. **55**(5), 609–622 (2015). https://doi.org/10.1134/S001679321505014X
15. Polozov, Yu., Fetisova, N.: Estimation of ionosphere state in Aurora online data analysis system. In: E3S Web of Conferences, vol. 127, 01003 (2019). https://doi.org/10.1051/e3sconf/201912701003
16. Mandrikova, O.V., Fetisova, N.V., Polozov, Y.A., Solovev, I.S., Kupriyanov, M.S.: Method for modeling of the components of ionospheric parameter time variations and detection of anomalies in the ionosphere. Earth Planets Space **67**(1), 131–146 (2015). https://doi.org/10.1186/s40623-015-0301-4
17. Mallat, S.: A Wavelet Tour of Signal Processing, 3rd edn. Academic Press, London (2008)
18. Box, G., Jenkins, G.: Time Series Analysis: Forecasting and Control. Holden Day, San Francisco (1970)
19. Konstantinova, A.V., Danilov, A.D.: Ionospheric precursors of magnetic storms. 2. Analysis of Slough station data. Geomagn. Aeron. **60**, 311–317 (2020) https://doi.org/10.1134/S001679322003010X

Estimation of Characteristics of Atmospheric Ionization Rates Taking into Account Different Energies of Precipitating Electrons

I. Mironova(✉) and A. Mironov

Saint Petersburg State University, Universitetskaya Emb., 7-9, St. Petersburg 199034, Russia
i.a.mironova@spbu.ru, st095857@student.spbu.ru

Abstract. Energetic electron precipitating into the Earth's atmosphere can significantly change chemical compositions and ozone of the upper and middle atmosphere. Understanding the atmospheric impacts of energetic electron particle deposition remains challenging and needed quantifying the response atmosphere to altitudinal profiles of ionization rates. Usually for computation ionization rates produced at altitudes low thermosphere and mesosphere takes into account precipitation electrons with energies between 30 keV up to several MeV based on data of MEPED instrument of NOAA POES satellites. However electrons from several keV up to tens of keV can also play important role for ionization of these altitudes. In this paper, we retrieve the ionization rates in the atmosphere caused by precipitation of energetic electrons taking into account energy from several keV up to several MeV and show importance of whole electron energy range for ionization rates of the upper atmosphere.

Keywords: Energetic electron precipitation · electron spectra · ionization rates induced by electron flux with energy range of several keV up to several MeV

1 Introduction

The number of ions per second or in other words, the ionization rate caused by the deposition of energetic species leads to the formation of odd nitrogen NOx (N, NO, NO_2) and hydrogen HOx (H, OH, HO_2) groups, which are important catalysts and participate in ozone loss reactions from the lower thermosphere to the middle stratosphere [1, 2]. Estimation of ozone losses under energetic electron forcing in polar atmosphere can reach from 20% up to 80% in the mesosphere during periods of strong geomagnetic storms [3, 4]. To assess the impact of energetic electrons on the ozone layer, various chemical-climate models are now increasingly used, covering the altitudinal range from the ground to the lower thermosphere [e.g., 3, 5]. Where an important source of atmospheric ionization are energetic electrons with energies from about 30 keV to about 1 meV obtained from data from the MEPED instrument of the NOAA POES satellites. However, electrons of not only these energies take part in the ionization of the lower thermosphere to the middle stratosphere; therefore, for accurate assessments of the destruction of the ozone layer, it is important to take into account the entire range of precipitating electrons from energies of several keV to several MeV.

A. Dmitriev et al. (Eds.): STRPEP 2023, SPEES, pp. 14–19, 2023.
https://doi.org/10.1007/978-3-031-50248-4_2

The analytically calculated formation of ions in the Earth's atmosphere by energetic primary electrons in the energy range from 1 keV to 300 keV, taking into account auroral excitation, has a long history [6–15]. The works [8–10] proposed parameterization models for assessing the impact of medium-energy electrons (30–1000 keV) on the ionosphere. The collection of articles [11–15] presents a new scheme for calculating the formation of ions oriented towards energetic electrons in the energy range extended to tens of MeV, with angular distributions of incidence of primary electrons and secondary bremsstrahlung. If information about the electron spectra of the auroral belt and the radiation belt can be obtained from a satellite, then the bremsstrahlung radiation of energetic electrons can be measured using balloon instruments.

The purpose of this research to estimate characterized ionization rates based on various assumptions including various electron energy ranges. In this paper, we reconstruct atmospheric ionization rates caused by energetic electron precipitation, accounting for energies from a few keV to a few MeV, and show the importance of the entire range of electron energies for the ionization rates of the upper atmosphere.

2 Ionization Rates Taking into Account Electrons with Initial Energy from Several of keV Up to Several MeV

The computation of ionization rates induced by energetic particles requires knowledge of the parameterization of ion production via ionization yield functions [6–15] and energy spectra of particle flux. In this study, we used modified ionization yield functions for mono-energetic electrons with initial energy from several keV to several MeV. Both direct ionization by primary electrons as well as the secondary Bremsstrahlung electromagnetic emissions are considered in this model [14].

Figure 1 presents ionization yield function of monoenergetic electron precipitation taking into account energies of electrons from several keV till several MeV. Table 1 and Fig. 2 show four examples of electron spectra that can be observed at different levels of geomagnetic disturbances.

Each example in Table 1 characterizes an electron flux that includes energies from 3 keV to 1 meV. The difference in the experiments is due to the intensity of the electron flux, which can vary greatly depending on geomagnetic disturbances. These different electronic spectra can be observed both using a variety of instruments, from satellites to balloon observations, and during different levels of disturbance.

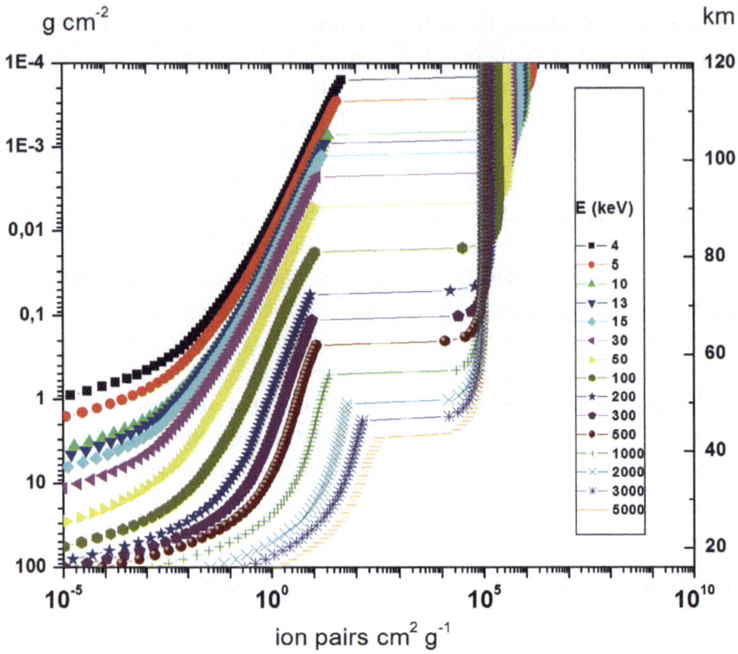

Fig. 1. Ionization yield function of monoenergetic electron precipitation taking into account energies of electrons from several keV till several MeV.

Table 1. Examples of spectra covering various energy ranges of energetic electron precipitation from 3 keV up to 1 meV. Flux intensity J in $cm^{-2}s^{-1}sr^{-1}keV$.

Examples	1	2	3	4
E (keV)	J	J	J	J
3 keV	1E9	1E8	1E7	1E6
10 keV	1E8	1E7	1E6	1E5
30 keV	1E7	1E6	1E5	1E4
100 keV	1E6	1E5	1E4	1E3
300 keV	1E5	1E4	1E3	1E2
1000 keV	1E4	1E3	1E2	1E1

Example 1 can characterize strong geomagnetic disturbances where Example 4 can characterize quite geomagnetic conditions.

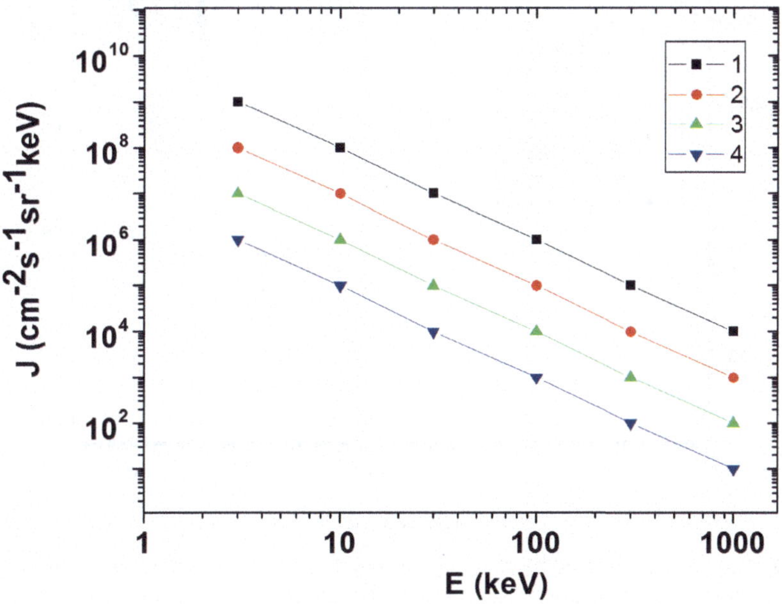

Fig. 2. Examples of spectra covering various energy of electron precipitation into atmosphere, see explanation in the Table 1. Spectra presented in are fitted by power-low distribution.

Using scheme for calculating ionization rates [14] with yield functions presented in Fig. 1 and spectra shown in Table 1 and Fig. 2 it is possible to obtain the ionization rates presented at Fig. 3. Open pattern of Fig. 3 present ionization rates that includes all energies from 3 keV to 1 meV where shaded pattern of Fig. 3 shows ionization rates based only on energies from 30 keV to 1 meV, which often used from MEPED instrument of the NOAA POES satellites. Here one can see the same behavior for all ionization rates at altitudinal band from about 16 km up to about 100 km. There are differences in ionization rates at the altitudes higher then 100 km. At altitudinal band from 100 km till 120 km ionization rates differ by an order of magnitude. This means that electrons with energies between 3 keV and 30 keV will only be important to ionization rates and atmospheric chemistry above 100 km.

Comparing ionization rates in different experiments, one can see that for each experiment ionization rates differ by two orders of magnitude for all altitudes from the stratosphere to the lower thermosphere.

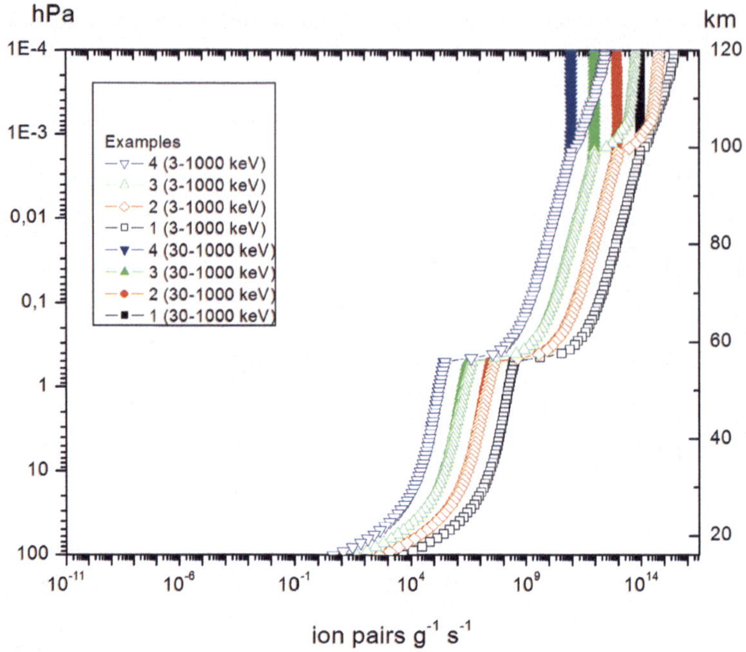

Fig. 3. Ionization rates induced by electron precipitation. Open pattern presents ionization rates that include flux with all energies from 3 keV to 1 MeV. Shaded pattern shows ionization rates are based only on energies from 30 keV to 1 MeV.

3 Conclusions

In this work, we reconstructed the rates of atmospheric ionization caused by the precipitation of energetic electrons, which can occur during quiet and fairly strong geomagnetic disturbances. We have looked at a couple of cases where all energies from 3 keV to 1 meV are used to calculate ionization rates and where we only considered energies from 30 keV to 1 meV. Ionization rates in the altitude range from about 16 km to about 100 km have the same behaviour regardless of the energies of the particles in the flow, but will vary greatly depending on the intensity of the flow. Ionization rates will vary greatly at altitudes above 100 km.

In conclusion, it can be said that for one-dimensional radiation-convective models, as well as for three-dimensional chemical-climate models covering the altitude range from the ground to the lower thermosphere it is necessary to take into account electron energies from several keV to several MeV.

Acknowledgements. The work was carried out at the "Laboratory for the Study of the Ozone Layer and Upper Atmosphere" with the support of the Ministry of Science and Higher Education of the Russian Federation under contract No. 075-15-2021-583. The work was done according to the state order of St. Petersburg State University.

References

1. Rozanov, E., Calisto, M., Egorova, T., Peter, T., Schmutz, W.: Influence of the precipitating energetic particles on atmospheric chemistry and climate. Surv. Geophys. **33**, 483–501 (2012)
2. Mironova, I.A., et al.: Energetic particle influence on the earth's atmosphere. Space Sci. Rev. **194**, 1–96 (2015)
3. Grankin, D., Mironova, I., Bazilevskaya, G., Rozanov, E., Egorova, T.: Atmospheric response to EEP during geomagnetic disturbances. Atmosphere **14**, 273 (2023)
4. Mironova, I., Grankin, D., Rozanov, E.: Mesospheric ozone depletion depending on different levels of geomagnetic disturbances and seasons. Atmosphere **14**, 1205 (2023)
5. Mironova, I., et al.: Exceptional middle latitude electron precipitation detected by balloon observations: implications for atmospheric composition. Atmos. Chem. Phys. **22**, 6703–6716 (2022)
6. Rees, M.: Auroral ionization and excitation by incident energetic electrons. Planet. Space Sci. **11**, 1209–1218 (1963)
7. Lazarev, V.I.: Absorption of the energy of an electron beam in the upper atmosphere. Geomagn. Aeron. **7**, 219 (1967)
8. Berger, M.J., Seltzer, S.M.: Bremsstrahlung in the atmosphere. J. Atmos. Sol.-Terr. Phys., **34**, 85–108 (1972)
9. Solomon, S.C.: Auroral particle transport using Monte Carlo and hybrid methods. J. Geophys. Res. **106**, 107–116 (2001)
10. Fang, X., et al.: Parameterization of monoenergetic electron impact ionization. Geophys. Res. Lett. **37**, L22106 (2010)
11. Artamonov, A.A., Mishev, A.L., Usoskin, I.G.: Model CRAC:EPII for atmospheric ionization due to precipitating electrons: yield function and applications. J. Geophys. Res. Space Phys. **121**, 1736–1743 (2016)
12. Artamonov, A., Mironova, I., Kovaltsov, G., Mishev, A., Plotnikov, E., Konstantinova, N.: Calculation of atmospheric ionization induced by relativistic electrons with non-vertical precipitation: update of model CRAC:EPII. Adv. Space Res. **59**, 2295–2300 (2017)
13. Xu, W., Marshall, R.A., Tobiska, W.K.: A method for calculating atmospheric radiation produced by relativistic electron precipitation. Space Weather **19**, e2021SW002735 (2021)
14. Mironova, I., Kovaltsov, G., Mishev, A., Artamonov, A.: Ionization in the earth's atmosphere due to isotropic energetic electron precipitation: ion production and primary electron spectra. Remote Sens. **13**(20), 4161 (2021)
15. Mironova I., et al.: Spectra of high energy electron precipitation and atmospheric ionization rates retrieval from balloon measurements. Sci. Total Environ. 31634991 (2019)

Dust Plasma in the Region of Active Influence on the Ionosphere

V. P. Sivokon(✉)

Institute of Cosmophysical Research and Radio Wave Propagation (IKIR) FEB RAS, Paratunka, Kamchatka Territory, Russia
vsivokon@mail.ru

Abstract. The Studies of active influence on the ionosphere by means of heating stands have been carried out for quite a long time, using considerable power and various technologies. However, the expected results, as a rule, appear to be significantly over-estimated. Most likely the reason for this discrepancy are unaccounted natural factors, for example, the presence of dust in the plasma. In the report, based on the analysis of electromagnetic radiation of the heating stand EISCAT attempts to assess the effect of dust from the Geminids meteor stream on the scattering and dynamic characteristics of artificial field-aligned irregularities of the ionosphere are made.

Keywords: ionosphere modification · dust plasma · Geminids

1 Introduction

The study of dust plasma properties has relevance for solving a number of fundamental and applied problems [1]. In particular, the paper points out that dust plasma is widely spread in space, found near artificial satellites of the Earth, in thermonuclear installations. It is investigated in laboratory conditions. At the same time, the analysis of publications devoted to the study of the properties of such plasma in the ionosphere, and even more so in the field of active influence on the ionosphere using heating stands showed their small number [2, 3], mainly theoretical.

It is assumed in [2] that in the presence of space dust in the region of active influence on the ionosphere, modulation excitation of inhomogeneities of electron and ion concentrations in the ionosphere should be expected. The paper [3] describes the mechanism of formation of inhomogeneities of electron and ion concentrations in the dusty ionosphere as a result of the development of modulation instability of electromagnetic pumping waves, with the participation of dust sound perturbations. Within the framework of the proposed method, the possible scales of inhomogeneities of the electronic concentration in the dusty ionosphere resulting from the development of modulation instability in the case of monochromatic electromagnetic waves from the HAARP heating stand for positively charged dust particles were determined as an example. However, the magnitudes of artificial inhomogeneities are small, and in some cases, one should not expect noticeable formation of inhomogeneities, given that HAARP is the most powerful of modern heating stands.

A. Dmitriev et al. (Eds.): STRPEP 2023, SPEES, pp. 20–27, 2023.
https://doi.org/10.1007/978-3-031-50248-4_3

2 Method

In order to assess the influence of space dust on the result of active influence on the ionosphere, information about its properties in the area of active influence is necessary. Space dust parameters have been measured for quite a long time, and sensors installed on spacecraft are used as tools. In the publication [4] it is concluded that the most informative and reliable are methods based on ionization, flash, and changes in the electrical conductivity of shock-compressed substances. Of the noncontact methods, only the method of inducing current by flying a charged particle through some electrode system has practical application in space conditions. Other methods are unsuitable for registration of small particles on large areas of sensitive surface or, at best, can be applied only in laboratory conditions. From the analysis of this material, we can conclude that it is impossible to obtain data on the parameters of cosmic dust, especially in the field of a powerful electromagnetic wave emitted by the heating stand. If we consider the ionosphere as a space plasma laboratory [5], then at this stage it makes sense to set up an experiment using such a natural phenomenon as a meteorite flux, when in the area of active influence, the level of space dust is guaranteed to exceed the back-ground value. Moreover, using the change in the topology of the meteorite flux in time, it is possible to obtain information about the fact of its influence on the dynamic characteristics of the processes occurring in the area of active impact.

Since heating technologies are used to solve the problem, it is necessary to explain the choice of the frequency of the heating wave, since its value is a determinant of the height at which the effect will be maximized. In turn, knowledge of the exposure height allows us to analyze the experiment using data from the ionosonde DPS-4, which is available on the EISCAT heating stand.

In experiments on modification of the ionosphere to improve the effectiveness of the impact various technologies are used, including the impact on the harmonics of the gyromagnetic frequency. The use of harmonics is explained by the fact that the gyromagnetic frequency depending on the altitude and geomagnetic latitude varies within 1.4–1.5 MHz, which does not allow to realize the antenna devices of the heating stand with effective directional properties. It is known that the gyromagnetic frequency of the ionospheric plasma is defined as

$$\omega_H = \frac{eB}{m}$$

where e, m are the charge and mass of the electron, B induction of the Earth's magnetic field. In the magnetic induction, which is a function of geomagnetic latitude and altitude [6]

$$\omega_H = 2\pi f_1 (r_0/r)^3 \sqrt{1 + 3(\sin\vartheta)^2}$$

where r_0 is the Earth radius, r is the distance from the Earth center to the point under consideration, is the geomagnetic latitude of this point, $f_1 \approx 0.8$ MHz. However, the accuracy of determining ω_H using this approach is low, so we use the IGRF-13 model of the Earth's magnetic field [7] to determine the heights corresponding to the values of the heating wave frequency. In some cases, a hybrid frequency is used, which is defined as the square root of the sum of the squares of the gyromagnetic and plasma frequencies. Unlike the case when the harmonic of the gyromagnetic frequency is used, it is difficult to unambiguously determine the height of the maximum efficiency of exposure in this case.

In the basis of realized by us experimental observations of experiments on active influence on ionospheres is based on the method [8] using spatially separated Soft Defined Radio (SDR). SDR receivers have a fundamental difference from conventional radio receivers, both in terms of implementation and access to them. In this technology, the signal coming from the antenna is sampled, and its further processing is done digitally. The undoubted advantage of the technology is remote access and control of the receiver, recording the signal in "wave" format with its subsequent download. In order to exclude the influence on the parameters of electromagnetic radiation scattered on the inhomogeneities of the ionosphere heating wave effects of ionospheric propagation SDR receivers are selected within line of sight of the area of influence on the ionosphere. Since the Doppler effect is used to estimate the dynamic characteristics of ionospheric inhomogeneities, the receivers are subject to special requirements for frequency stability, which is achieved by their preselection and the use of a special mode of their operation.

It is known that the presence of cosmic dust in the ionosphere contributes to the increase of turbulence in it [9]. Therefore, the Doppler frequency shift and scintillations index S4 [10] of the heating radiation scattered in the region of active influence on inhomogeneities of electronic concentration are used as analyzed parameters.

3 Results

In November 2021 on the website EISCAT [11] appeared information that the experiment "Radar detection of meteors from the Geminids meteor shower along with artificial heating" will be held on December 14, 2021. In the description of the application for the experiment indicated that the experimenters intend to use VHF radar and heating stand to observe meteorites from the Geminids meteor shower. It was assumed that the experiments would be conducted on December 14 from 00 to 06 UT with a heating wave frequency of 5423 kHz. It was stated that the heating would be performed during a 2 min on/off cycle. It was supposed that such an impact should increase the efficiency of VHF radars at altitudes of 90–100 km. However, if we check the belonging of the frequency 5423 kHz to the harmonics of the gyromagnetic frequency, as we have shown above, it turns out that this frequency is the fourth harmonic of the gyromagnetic frequency for a height of 241 km. The belonging of the used frequency 5423 kHz of the heating wave to the fourth harmonic of the gyromagnetic frequency is confirmed by the publication [12]. Consequently, it can be expected that in the case when the plasma frequency will be small or not determined at all, the option of using a hybrid frequency in the experiment is excluded and there is an impact on the harmonic of the gyromagnetic frequency with the area of maximum manifestation of the active effect on the ionosphere at a height of 241 km.

Based on the peculiarities of the movement of the Geminids meteor shower, it was necessary to select observation points for one of which the movement of the shower would be quasi longitudinal, and for the other quasi transverse with respect to the direction of the impact area - observation point. In this case, at the moment when the flow movement becomes transverse for one of the observation directions, the distribution of the observed velocities should change dramatically. Information about the direction of motion of the Geminids meteor shower was provided by the Moscow Planetarium. If the above conditions are met, at a certain time for one of the observation points the motion of the meteor shower would be transverse and the effect of its influence on inhomogeneities of the ionosphere, if present, in the form of the Doppler effect would be minimal. At the time of the experiment, SDR receivers in Arctic (Norway) and Kokkola (Finland) were available and optimally located. Figure 1 shows in scale the distance, azimuths of the observation points and the size of the active impact area according to the half power level of the an tenement device used on the EISCAT heating stand.

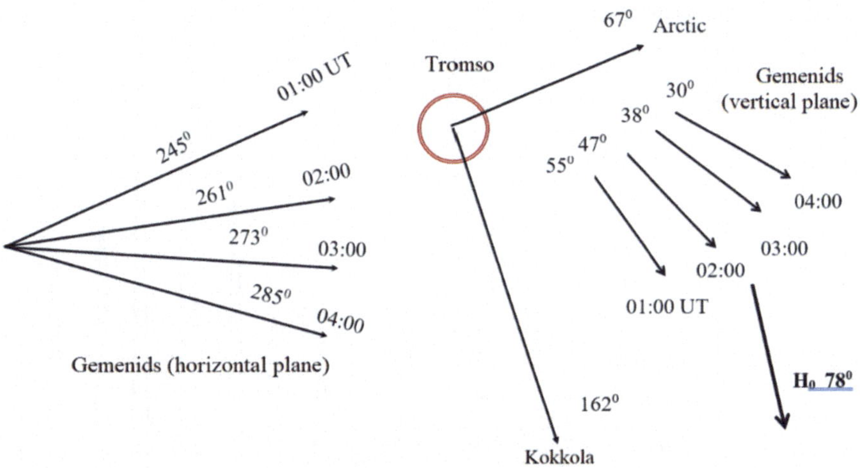

Fig. 1. Topology of the experiment.

The direction of motion of the meteoroid stream in the vertical plane for both observation points is quasi-longitudinal. In the horizontal plane at a time interval around 01.30 UT for the Tromsø - Kokkola trace should be transverse, and Tromsø - Arctic close to longitudinal motion of the meteoroid stream relative to these traces.

When conducting the experiment, it is desirable to minimize the presence of background perturbations of electron concentration, not caused by the passage of the meteorite stream. It follows that the optimal time for the experiment is the period of time with a minimum of solar activity with a quiet magnetic field of the Earth. If we refer to ionospheric [13] and magnetic data [14], it appears that these conditions at the time of the experiment were optimal. On the ionograms of station DPS-4 Tromsø, the plasma frequencies of none of the ionospheric layers are not determined. The Earth's magnetic field is quiet and the planetary magnetic index Кp = 1. In contrast to the stated schedule, the experiment was actually conducted according to the scheme 4 min heating, two minutes' pause. The record of the scattered heating radiation was subjected to spectral analysis with subsequent determination of the Doppler frequency shift distribution and, accordingly, the velocities of inhomogeneities. The distribution of their velocities at the beginning of the experiment is shown in Fig. 2. In the figure, the velocity component is plotted along the horizontal axis, and its probability is plotted along the vertical axis.

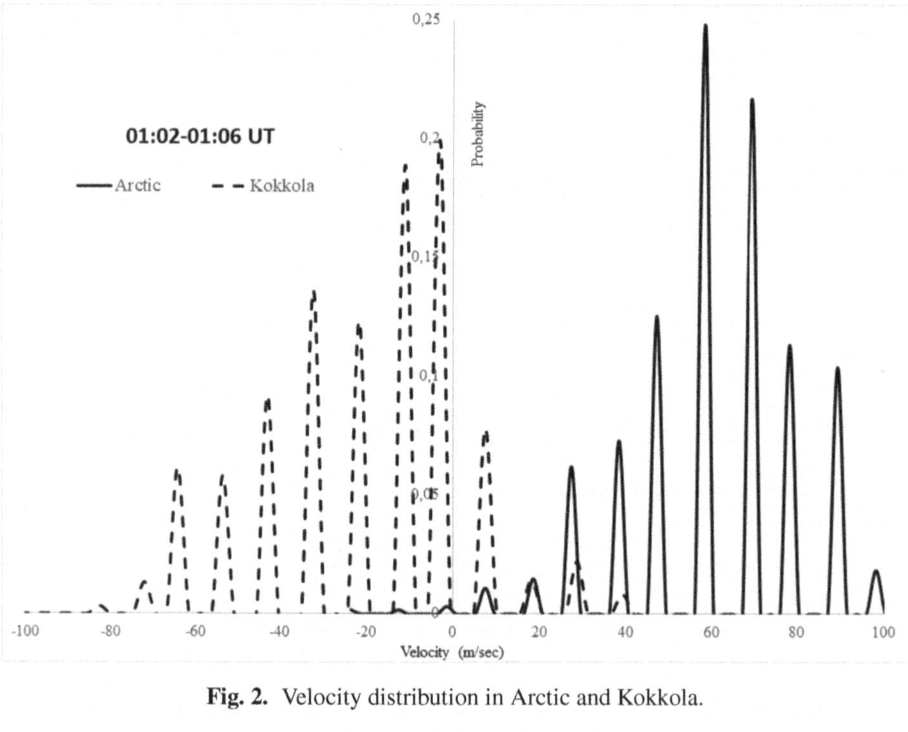

Fig. 2. Velocity distribution in Arctic and Kokkola.

The figure shows that the velocity distributions at the observation points are similar and differ in the direction of motion of inhomogeneities. A completely different picture is observed in the time interval 01:26–01:42 UT, Fig. 3.

In the time interval 01:32–01:36, a sharp change in the distribution for the Tromsø -Kokkola direction is observed, while for the Tromsø - Arctic direction, the character of the distribution has not changed. This change is most likely explained by a change in the nature of the movement of the meteor shower relative to the observation routes. In the time interval 01:32–01:36 for the Tromsø - Kokolla highway, it is longitudinal, and for the Tromsø - Arctic highway it still remains quasi-longitudinal. Consequently, our assumption about a significant influence of space dust on the dynamic characteristics of ionospheric inhomogeneities in the region of active influence on it by powerful electromagnetic radiation is confirmed experimentally.

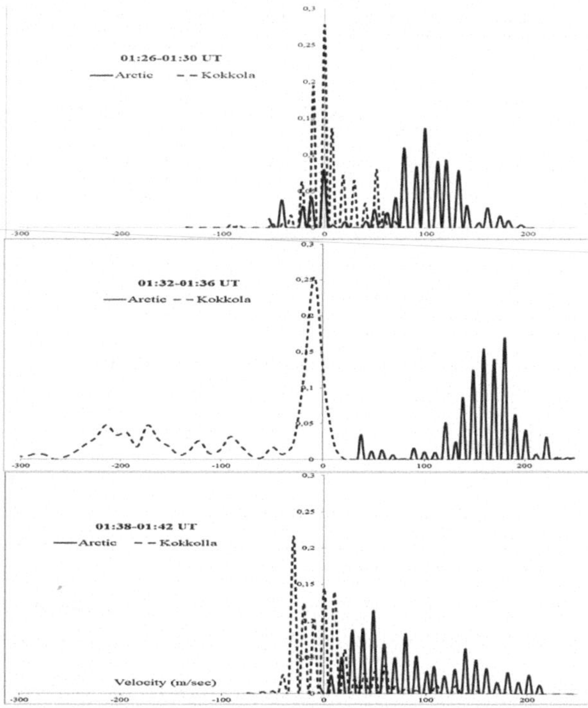

Fig. 3. Velocity distribution in the experiment 01:26–01:42 UT.

To determine the type of turbulence observed, we use an instantaneous spectrum, which showed the presence of structures with signs of vortices, Fig. 4.

Based on the shape of the spectrum, it can be assumed that radial-balloon vortices are observed in this case [15]. The same paper concludes that vortex structures can play an important role in the process of heat and particle transport and thus serve as structural elements of strong drift turbulence.

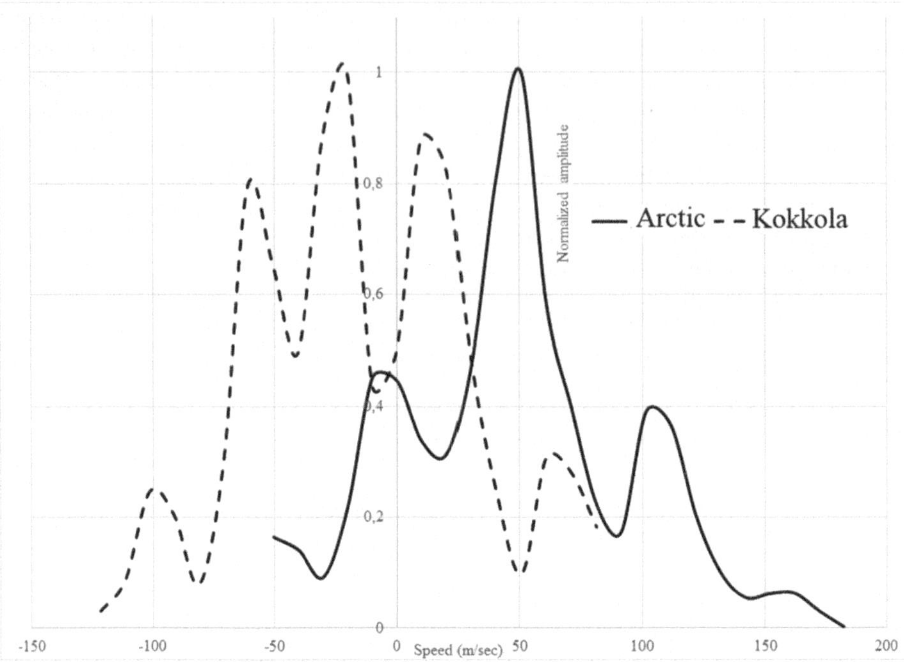

Fig. 4. Vortex structures in heating cycle 01:02–01:06 UT.

4 Concluding Remarks

1. The presence of space dust in the region of active impact on the ionosphere leads to a significant change in the dynamic characteristics of inhomogeneity's of electron concentration.
2. Turbulence in the region of impact is characterized by the presence of vortex structures, which most likely lead to additional losses of heating radiation.

The studies were carried out within the framework of the government assignment under the project AAAA-A21–121011290003-0.

References

1. Fortov, V.E., Khrapak, A.G., Khrapak, S.A., Molotkov, V.I., Petrov, O.F.: Dusty plasmas. Phys. Usp. **47**, 447–492 (2004)
2. Borisov, N.D., Kopnin, S.I., Morozova, T.I., Popel S.I.: Role of collisions with neutrals in the process of modulational excitation of dust acoustic perturbations in dusty ionosphere: physics of near-earth space. Plasma Phys. Rep. **45**(4), 346–352 (2019)
3. Kopnin, S.I., Popel, S.I., Morozova, T.I.: Modulational excitation of inhomogeneities in dusty ionospheric plasma. Plasma Phys. Rep. **41**(2), 171–177 (2015)
4. Semkin, N.D., Voronov, K.E., Piyakov, A.V., Piyakov, I.V.: Registration of a spase artificial and natural dust. Appl. Phys. **1**, 86–102 (2009)

5. Erukhimov, L.M., Genkin, L.G.: The ionosphere as a space plasma laboratory (review). Radiophys. Quantum Electron. **35**, 555–568 (1992)
6. Gershman, B.N., Ugarov, V.A.: Propagation and generation of low-frequency electromagnetic waves in the upper atmosphere. Sov. Phys. Usp. **3**, 743–764 (1961)
7. World Data Center for Geomagnetism, Kyoto. http://wdc.kugi.kyoto-u.ac.jp/igrf/point/index.html
8. Sivokon, V.P.: A new method for research on magnetically oriented ionospheric inhomogeneities using a program of radio-system determination. Geomag. Aeron. **60**(2), 236–242 (2020). https://doi.org/10.1134/S0016793220020140
9. Izvekova, Yu.N., Pope, S.I.: Nonlinear wave structures and plasma−dust effects in the earth's atmosphere. Plasma Phys. **44**(9), 747–751 (2018)
10. Sivokon, V.P.: Evaluation of the S4 scintillations index in transcontinental decameter radio lines. T-Comm: Telecommun. Transp.**17**(5), 14–24 (2023)
11. EISCAT Operations Schedule. https://portal.eiscat.se/schedule/comment.cgi?filName=20211214124973&Start=0000&End=0600
12. Borisova, T.D., et al.: Phenomena in the high-latitude ionospheric F region induced by a hf heater wave at frequencies near the fourth electron gyroharmonic. Radiophys. Quantum Electron. **57**, 1–19 (2014)
13. Digital Ionogram DataBase. https://giro.uml.edu/ionoweb/#
14. World Data Center for Geomagnetism, Kyoto. https://wdc.kugi.kyoto.u.ac.jp/ae_realtime/202112/index_20211214.html
15. Aburjania, G.D., Kharshiladze, O.A., Chargazia, K.Z.: Self-organization of IGW structures in an inhomogeneous ionosphere: 2. Nonlinear vortex structures. Geomagn. Aeron. **53**(6), 750–760 (2013)

Ozone Dynamics During Geomagnetic and Meteorological Disturbances in October 2015 and 2018

Aleksandr Timchenko[1,2(✉)], Olga Borchevkina[1,2], Yuliya Kurdaeva[1,2], and Fedor Bessarab[1,2]

[1] O3 lab, St Petersburg University, Universitetskaya Emb., St. Petersburg 199034, Russian Federation
{atimchenko,olgaborchevkina,yakurdyaeva,bessarabf}@wdizmiran.ru
[2] WD IZMIRAN, Pionerskaya st., Kaliningrad 236035, Russian Federation

Abstract. This article is devoted to the study of behavior of nighttime ozone at altitudes 80–110 km during a geomagnetic storm in October 2015 and nighttime and daytime ozone at altitudes 80–110 km during a meteorological disturbance in October 2018. The TIMED/SABER satellite data in the European mid-latitude region were used to analyze ozone dynamic. The change in the ozone concentration maximum at lower atmosphere during different heleo-geophysical factors is shown.

Keywords: TIMED SABER · SABER · ozone · ozone dynamic · geomagnetic disturbances · meteorological disturbances

1 Introduction

The investigation of variability of the mesosphere's parameters—lower termosphere during different heleo-geophysical processes is actual task of physics upper atmosphere. It is known that the process of photo-dissociation of molecular oxygen (O_2) by solar ultraviolet radiation ($\lambda < 240$ nm) effectively proceeds in the upper atmosphere of the Earth:

$$O_2 + h\nu \rightarrow O + O \tag{1}$$

which leads to the formation of increased concentrations of atomic oxygen O with a maximum at altitudes of about 90–100 km. In triple collisions involving an atom and an oxygen molecule, ozone O_3 is formed:

$$O + O_2 + M \rightarrow O_3 + M \tag{2}$$

there M is molecule of nitrogen or oxygen in the Earth's atmosphere [1] The variability of mesospheric ozone parameters is affected by a complex of photochemical and dynamic processes. Previous studies of ozone dynamics in the mesosphere and lower thermosphere showed that during geomagnetic storms there is

O. Borchevkina, Y. Kurdaeva and F. Bessarab—These authors contributed equally to this work.

A. Dmitriev et al. (Eds.): STRPEP 2023, SPEES, pp. 28–37, 2023.
https://doi.org/10.1007/978-3-031-50248-4_4

a decrease in the concentration of mesospheric O_3, this is due to an increase in the concentration of NO_x, which destroy ozone [2]. Also in the work of Hocke, [3], a study was made of the dynamics of the maximum of the tertiary ozone layer, which is located at an altitude of about 72 km, in the polar region. In this work, it is shown that during a solar proton event, the concentration of odd forms of hydrogen (OH, HO_2) increases, which leads to the loss of atomic oxygen and the disappearance of the tertiary ozone layer. Associate an increase in the concentration of NO_x and HO_x with ozone losses both in the mesosphere and in the stratosphere, in polar latitudes [3–6]. Mironova et al., 2023 [7] conducted a study of mesospheric ozone depletion depending on the season. The authors showed that the intensity of ozone destruction due to energetic electrons precipitation (EEP) is maximum (~80%) in the winter season. In spring and autumn, their effect reaches ~20%, while in summer the effect of EEP on ozone is practically absent. This is related to the intensity of solar UV. Those, in winter, during the polar night, solar ultraviolet is practically absent, and therefore EEP become the main source of NO_x and HO_x formation, which destroy ozone. Accordingly, in spring, autumn and summer, solar UV becomes larger, and therefore the influence of EEP decreases.

The aim of this work is to consider the behavior of ozone in the lower thermosphere during a geomagnetic storm in October 2015 and a meteorological storm in the Baltic Sea in October 2018.

2 Data and Selections of Event

The ozone concentration and temperature data set derived from the SABER (The Sounding of the Atmosphere using Broadband Emission Radiometry) instrument onboard the NASA TIMED (Thermosphere Ionosphere Mesosphere Energetics and Dynamics) satellite analyzed from 80 to 110 km in altitude and between 40° N and 60° N latitude, 0° E and 30° E longitude (Fig. 1) during the geomagnetic storm of October 8, 2015 and strong severe weather in this region on October 24, 2018 in night time. The TIMED satellite was launched on 7 December 2001 into a 74° inclination orbit to explore the global structure in the stratosphere, mesosphere, and lower thermosphere. The SABER instrument measured temperature and several trace species' profiles extend from 20 km to 110 km in altitude with a ~2 km vertical resolution [8]. The latitude coverage ranges from 52° in one hemisphere to 83° in the other due to the yaw cycle of ~60 days of the satellite. The local migration time is 12 min every day. The systematic temperature instrument bias of SABER is around ±5 K in the MLT region [9]. The version 2.0 Level B data set was used in this study. An advantage of the SABER temperature data analysis is that a single observed data source is used to establish the responses over a wide range of latitudes and heights. Therefore, differences in temperature variations at different latitudes and altitudes are more credibly attributed to physical mechanisms, rather than differences in measurement technique [10].

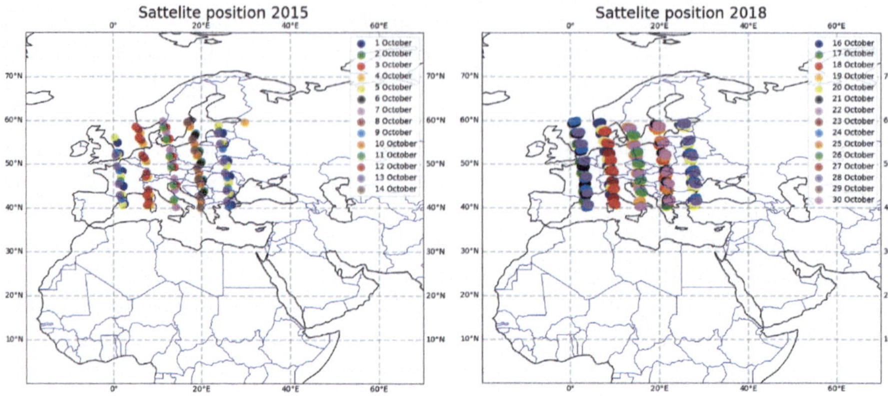

Fig. 1. The position in which the measurements of the parameters considered in this paper were made

To analyze the ozone concentration variability in the mesosphere and lower thermosphere, 2 events were selected: 1) the geomagnetic storm in October 2015. On October 7–8, 2015, the Dst-index values fell below −120 nT, the Kp index reached 7; 2) the meteorological storm in October 2018 in the Baltic Sea region. During this period, the geomagnetic situation was quiet. A detailed analysis of this storm and the impact of this event on the state of the upper atmosphere and ionosphere is presented in [11].

On Fig. 2 shows the indices of geomagnetic activity in October 2015 and 2018 [12].

3 Ozone in the Mesosphere and Lower Thermosphere

O_3 height-time variations are derived from the SABER L2A infrared 9.6 μm emission air-glow channels. Figures 3, 4 and 5 show the height-time variation of ozone concentration and neutral temperature.

The Fig. 3 shows there is an increase of the mesospheric ozone concentration during the geomagnetic storm. Such increase begins a day before the beginning of the geomagnetic storm. The temperature decreases at the same altitudes and at the same time, which corresponds to the local ozone concentration maximum. During a geomagnetic storm, the temperature decreases at 90–95 km altitudes and increases above and below this region.

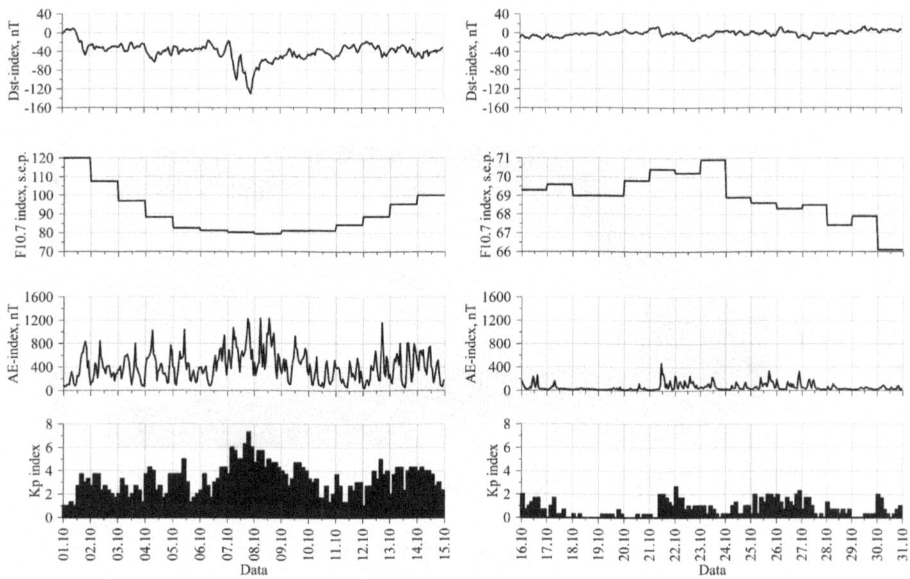

Fig. 2. Time variations in the geomagnetic and solar activity indices Dst, Kp, AE, and F10.7 during 1–10 October 2015 (left) and 16–31 October 2018 (right).

The Fig. 4 presented ozone concentration and neutral component temperatures by altitude during the October 2018 meteorological storm, during daytime. On one of the first days of the meteorological storm, heating is observed at the altitudes from 80–88 km and temperature decrease at the altitudes from 93–104 km. At the same time, the ozone concentration maximum decreases from 92 to 89 km and at the altitudes from 95 to 103 km, an area of strong decrease of ozone concentration is formed.

The Fig. 5 presented ozone concentration and neutral component temperatures by altitude during the October 2018 meteorological storm, during nighttime. During the first days of the meteorological storm, a small temperature increase from the bottom (by 10–20 K) is observed. There is also a slight decrease by altitude of ozone concentration before and during the meteorological storm and formation of two maxima of ozone concentration at a altitude of about 94 km.

To further analyze the ozone dynamics in the mesosphere during a geomagnetic storm and a meteorological storm, we have constructed plots of the change in the ozone concentration maximum and the change in the height of the ozone concentration maximum with time (Fig. 6).

Figure 6 clearly shows that on the first day of the geomagnetic storm (marked in red) there is a sharp increase of the ozone concentration maximum and a smooth decrease on the subsequent days. The increase of the ozone concentration maximum altitude occurs smoothly and with a small delay, and after reaching the highest altitude, a sharp decrease is observed. The simultaneous increase of the maximum height and the increase of the maximum concentration of mesospheric ozone indicate the dynamic nature of such changes rather than photo-chemical.

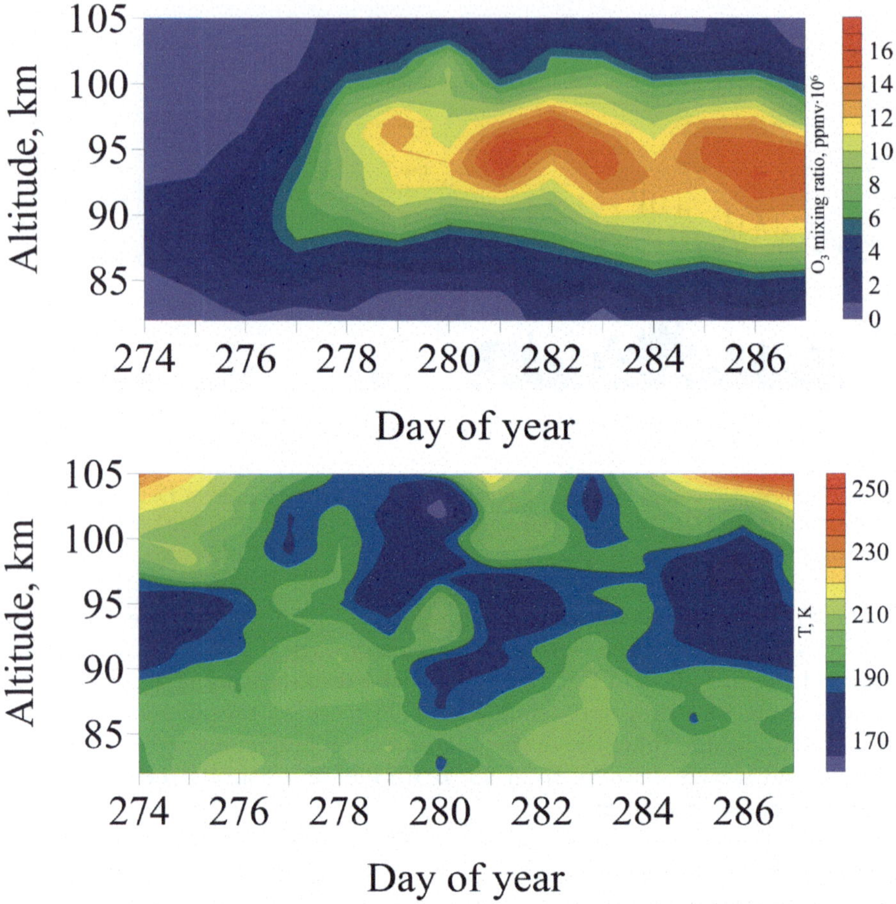

Fig. 3. The height-time variation of ozone concentration (up) and neutral temperature (down) in October 2015

The day before and the day after the 2018 weather storm (day marked in red), there is a sharp increase in the ozone concentration maximum, when on the day of the weather storm itself, the ozone concentration maximum is at a local minimum. The altitude changes smoothly during this period and reaches a minimum during the meteorological storm. Such specific behavior of these parameters suggests the convection nature of such changes.

4 Discussion

As shown above, the variation of ozone concentration at different altitudes from different heliogeophysical sources differs significantly [13–17]. This appears to be due to the different effects of different heliogeophysical sources on the properties and characteristics of the middle atmosphere. For example, [18] proved that during periods of geomagnetic storms at altitudes of 75 km there is an effect of increasing nitrogen oxides and probably chlorine, which leads to a more pronounced destruction of the ozone layer. However, the behavior of ozone at different altitudes may differ significantly. The physical mechanism of ozone changes at altitudes of 90–100 km during geomagnetic storms remains incompletely understood. Most likely, this behavior of ozone concentration is related to the change of NO_x during geomagnetic storms. The physical interpretation of this phenomenon requires a separate and detailed study.

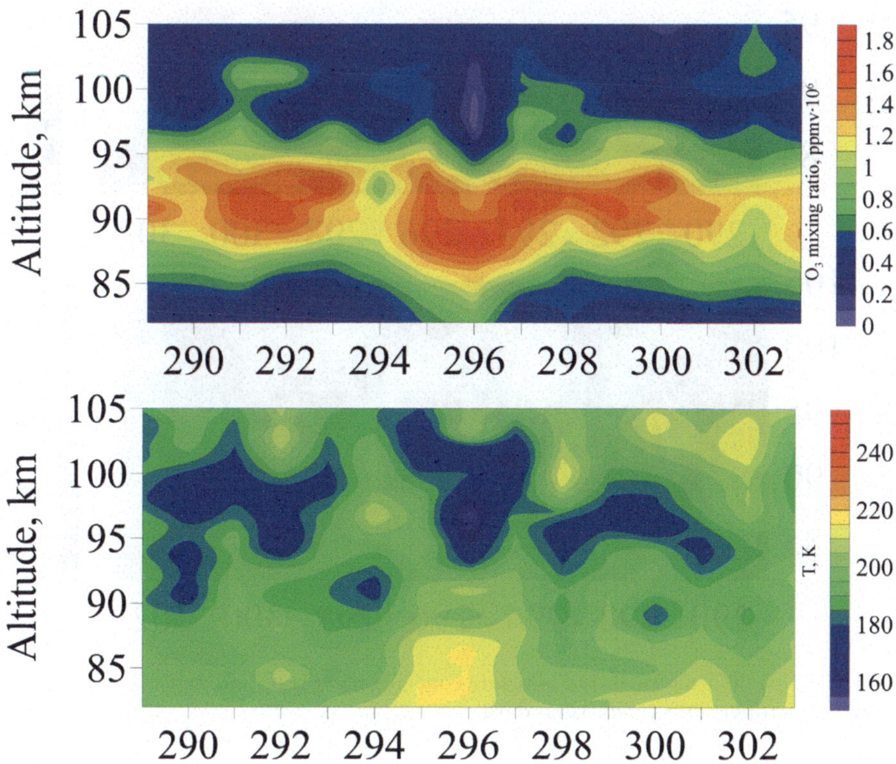

Fig. 4. The height-time variation of ozone concentration (up) and neutral temperature (down) in October 2018. Daytime data

The characteristics and properties of the mesosphere and lower thermosphere are significantly influenced by turbulence and atmospheric waves. Turbulent formations play a significant role in the processes of atmospheric heating, redistribution of chemical components of atmospheric aerosol, in the destruction and weakening of internal gravity waves (IGWs), and in the dissipation of free atmosphere jet streams [19]. Internal gravity waves (IGWs) propagating from the region of convective processes in the lower atmosphere, in particular, a weather storm, can amplify turbulent processes at the altitudes of the upper mesosphere and lower thermosphere. The amplitude of the IGW increases as the atmospheric density decreases with height. When a certain threshold value is reached [20,21], the wave becomes unstable and breaking, forming turbulent regions "moving" with it.

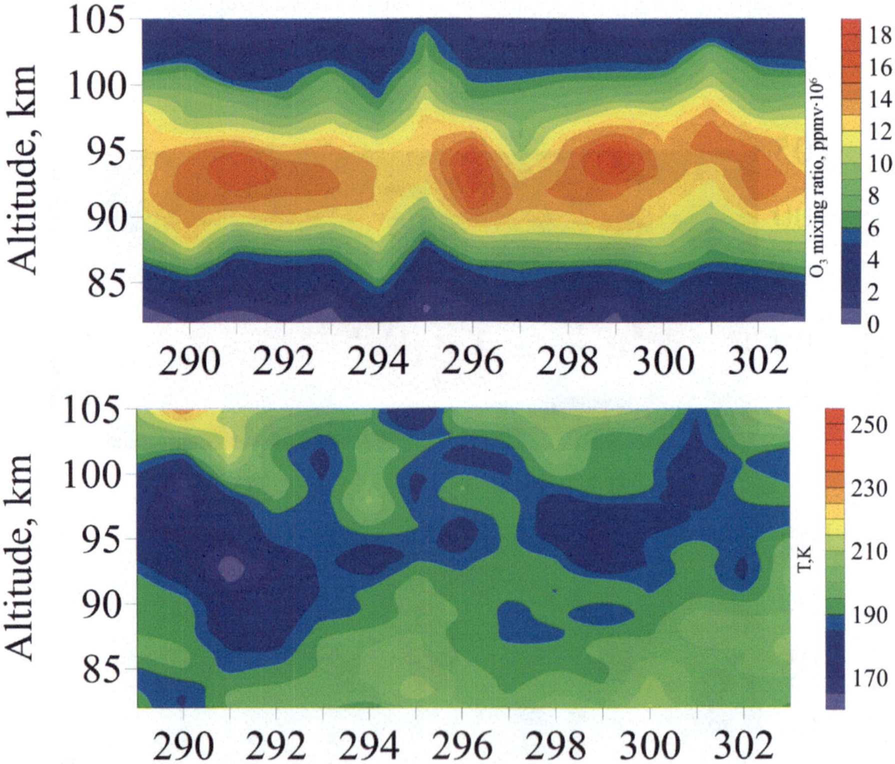

Fig. 5. The height-time variation of ozone concentration (up) and neutral temperature (down) in October 2018. Nighttime data

In works [22–24], a review of methods is given and the results of experimental studies of turbulent motions and determination of turbulence parameters at altitudes of 80–110 km are presented. In particular, in [24] new results of the study of atmospheric dynamics are presented, which complement the previously obtained data on turbulence and dynamics of the lower ionosphere. The turbopause level in the 88–110 km altitude interval is subject to both fast and slow changes. The altitudinal and temporal variations of the parameters of the neutral component have convincingly demonstrated a significant influence of wave processes on them. The changes of the parameters in time occur with a periodicity characteristic of the IGW.

Vertical transport by turbulent mixing and horizontal transport by winds redistribute constituents such as atomic oxygen, hydroxyl and ozone. Thus, long-term change in trace constituents cannot be fully explained in isolation from studies of corresponding change in temperature and neutral dynamics [23]. Vlasov et al. [22] suggest the development of localized turbulence, which can develop as a result of strong winds with mean wind shear in narrow turbulence layers bounded by boundaries with undisturbed gas.

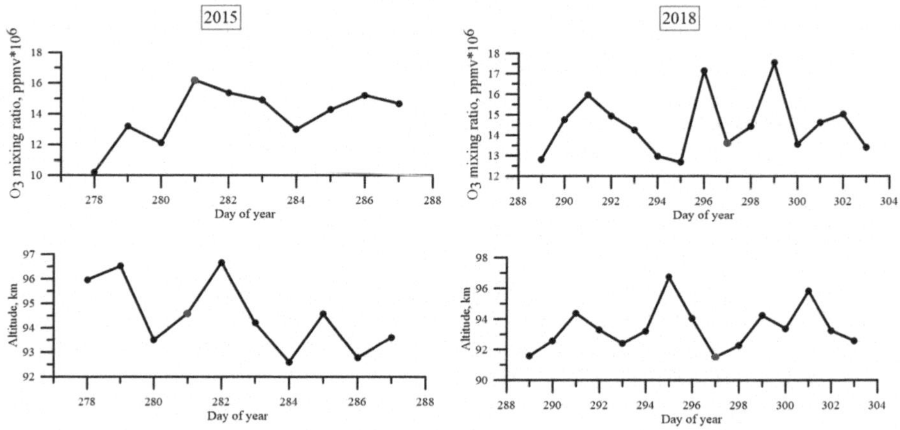

Fig. 6. Changes in the ozone concentration maximum and the height of the ozone concentration maximum from time to time in October 2015 (left) and in October 2018 (right).

Gabriel [25] observe strong effect of ozone-gravity wave coupling in the upper stratosphere/lower mesosphere (USLM) what leads to a stronger increase in gravity wave (GW) amplitudes with height during daytime than during nighttime and estimate the sensitivity of the amplification of the GW amplitudes in the upper mesosphere on changes in the ozone background.

5 Conclusion

In this work, was considered the height-time ozone variations in the lower thermosphere during a geomagnetic storm in October 2015 and a meteorological storm in the Baltic Sea in October 2018. The mid-latitude region in Europe (40–60° N, 0–30° E) was considered. Ozone concentration height-time variations were taken from experimental data of the SABER instrument on the TIMED satellite.

During the geomagnetic storm on October 7–8, 2015, there is an increase in the mesospheric ozone concentration. Such increase begins a day before the beginning of the geomagnetic storm. The temperature decreases at the same altitudes and at the same time corresponding to the local ozone concentration maximums.

According to the daytime data, at the altitudes of about 85 km, there is a significant heating of the neutral atmosphere during a meteorological storm. At the same time, a region of decreased temperature is formed at the altitudes of 95–100 km. At nighttime, a simultaneous increase in temperature and decrease in ozone concentration are observed at an altitude of 85 km. At the same time there is an increase of ozone concentration at an altitude of 100–105 km.

Propagating from the meteorological storm region, the IGW in the lower thermosphere becomes unstable and breaking, forming local areas of increased turbulence that affect the temperature and redistribution of chemical constituents at mesospheric and lower thermospheric altitudes.

Acknowledgement. The work was supported by the MegaGrant of the Ministry of Education and Science of Russia, agreement No. 075-15-2021-583.

References

1. Antonenko, O.V., Kirillov, A.S., Kulikov, Y.N.: The influence of zonal and seasonal variations of atomic oxygen on the radiation intensity and excited oxygen concentration at the heights of nightglow of the Earth. Works Kola Sci. Cent. RAS **10**(8(5)), 65–76 (2019)
2. Daae, M., Espy, P., Nesse Tyssøy, H., Newnham, D., Stadsnes, J., Søraas, F.: The effect of energetic electron precipitation on middle mesospheric night-time ozone during and after a moderate geomagnetic storm. Geophys. Res. Lett. **39**(21) (2012)
3. Hocke, K.: Response of the middle atmosphere to the geomagnetic storm of November 2004. J. Atmos. Solar Terr. Phys. **154**, 86–91 (2017)
4. Marsh, D., Smith, A., Brasseur, G., Kaufmann, M., Grossmann, K.: The existence of a tertiary ozone maximum in the high-latitude middle mesosphere. Geophys. Res. Lett. **28**(24), 4531–4534 (2001)
5. Seppälä, A., et al.: Destruction of the tertiary ozone maximum during a solar proton event. Geophys. Res. Lett. **33**(7) (2006)
6. Rodger, C.J., et al.: Radiation belt electron precipitation due to geomagnetic storms: significance to middle atmosphere ozone chemistry. J. Geophys. Res.: Space Phys. **115**(A11) (2010)
7. Mironova, I., Grankin, D., Rozanov, E.: Mesospheric ozone depletion depending on different levels of geomagnetic disturbances and seasons. Atmosphere **14**(8), 1205 (2023)
8. Russell III, J.M., Mlynczak, M.G., Gordley, L.L., Tansock Jr., J.J., Esplin, R.W.: Overview of the saber experiment and preliminary calibration results. In: Optical Spectroscopic Techniques and Instrumentation for Atmospheric and Space Research III, vol. 3756, pp. 277–288. SPIE (1999)
9. Remsberg, E.E., et al.: Assessment of the quality of the version 1.07 temperature-versus-pressure profiles of the middle atmosphere from timed/saber. J. Geophys. Res.: Atmos. **113**(D17) (2008)

10. Forbes, J.M., Zhang, X., Marsh, D.R.: Solar cycle dependence of middle atmosphere temperatures. J. Geophys. Res.: Atmos. **119**(16), 9615–9625 (2014)
11. Borchevkina, O., Karpov, I., Karpov, M., Korenkova, N., Vlasov, V., Leshchenko, V.: Impact of meteorological storms on the e-region of the ionosphere in 2017–2018. Solar-Terr. Phys. **6**(4), 74–79 (2020)
12. https://omniweb.gsfc.nasa.gov/form/dx1.html
13. Crutzen, P.J.: Tropospheric ozone: an overview. In: Isaksen, I.S.A. (ed.) Tropospheric Ozone. NATO ASI Series, vol. 227, pp. 3–32. Springer, Dordrecht (1988). https://doi.org/10.1007/978-94-009-2913-5_1
14. Vaughan, G., Price, J.: On the relation between total ozone and meteorology. Q. J. R. Meteorol. Soc. **117**(502), 1281–1298 (1991)
15. Oltmans, S.J., et al.: Summer and spring ozone profiles over the North Atlantic from ozonesonde measurements. J. Geophys. Res.: Atmos. **101**(D22), 29179–29200 (1996)
16. Laštovička, J., Mlch, P.: Is ozone affected by geomagnetic storms? Adv. Space Res. **24**(5), 631–640 (1999)
17. Kilifarska, N., Peqini, K.: Impact of for bush decreases and geomagnetic storms on the atmospheric ozone profiles. Earth Space Sci. **10**(7), 2023–002954 (2023)
18. Grankin, D., Mironova, I., Bazilevskaya, G., Rozanov, E., Egorova, T.: Atmospheric response to EEP during geomagnetic disturbances. Atmosphere **14**(2), 273 (2023)
19. Belyaev, A.: The mechanism of formation of mesoscale pulsating jet streams in the earth's atmosphere. Izv. Atmos. Ocean. Phys. **58**(3), 295–301 (2022)
20. Walterscheid, R., Schubert, G.: Nonlinear evolution of an upward propagating gravity wave: Overturning, convection, transience and turbulence. J. Atmos. Sci. **47**(1), 101–125 (1990)
21. Fritts, D.C., Bizon, C., Werne, J.A., Meyer, C.K.: Layering accompanying turbulence generation due to shear instability and gravity-wave breaking. J. Geophys. Res.: Atmos. **108**(D8) (2003)
22. Vlasov, M., Kelley, M.: Specific features of eddy turbulence in the turbopause region. In: Annales Geophysicae, vol. 32, pp. 431–442. Copernicus Publications Göttingen, Germany (2014)
23. Hall, C.M., Holmen, S.E., Meek, C.E., Manson, A.H., Nozawa, S.: Change in turbopause altitude at 52 and 70° n. Atmos. Chem. Phys. **16**(4), 2299–2308 (2016)
24. Bakhmetieva, N., Grigoriev, G., Vinogradov, G., Zhemyakov, I., Kalinina, E., Pershin, A.: Parameters of atmospheric turbulence and the dynamics of the lower ionosphere in studies at the SURA facility. Geomag. Aeron. **61**, 871–887 (2021)
25. Gabriel, A.: Ozone-gravity wave interaction in the upper stratosphere/lower mesosphere. Atmos. Chem. Phys. **22**(16), 10425–10441 (2022)

Observations of the Meteoric Aerosol in the Stratosphere Above Tomsk in August 2013

I. I. Romanchenko[1], A. A. Cheremisin[1(✉)], P. V. Novikov[1,2], V. N. Marichev[3], and D. A. Bochkovsky[3]

[1] Laboratory of Dispersed Systems, V.V. Voevodsky Institute of Chemical Kinetics and Combustion SB RAS, Institutskaya st., 3, Novosibirsk 630090, Russia
aacheremisin@gmail.com

[2] Department of Train Traffic Support Systems, Krasnoyarsk Institute of Railway Transport, Lado Ketskhoveli st., 89, Krasnoyarsk 660028, Russia

[3] Laboratory of Lidar Methods, V.E. Zuev Institute of Atmospheric Optics SB RAS, Academician Zuev sq, 1, Tomsk 634055, Russia
{marichev,moto}@iao.ru

Abstract. According to the data obtained in lidar measurements of the backscattering ratio in the atmosphere above Tomsk, aerosol was recorded at the stratospheric altitudes in August 2013. The most intense aerosol scattering peak was detected at the altitude of 29 km on August 22. The meteoric origin of this aerosol is assumed in the work. Backward trajectories of air masses were calculated, and the data on stratospheric aerosol recorded by the space lidar CALIOP and by the satellite limb sensor OMPS-LP were involved, which allowed us to determine approximate location of the region in which the meteor body burned in the atmosphere: a region of the Sea of Okhotsk near Kamchatka peninsula.

Keywords: aerosol · stratosphere · lidar · backward trajectories · satellite sounding

1 Introduction

Aerosols in the stratosphere are essential for radiation exchange processes in the Earth's atmosphere. Stratospheric aerosol is not so thoroughly studied as the tropospheric aerosol. Carbon aerosol is detected in the stratosphere as a result of pyrocumulonimbus injection arising during forest fire events [1], for example, in Canada [2] and in Siberia [3], and volcanic aerosol is detected after eruptions [4], e.g. that of Raikoke volcano [5]. Polar stratospheric clouds are often observed in the stratosphere in winter [6].

Another phenomenon resulting in the formation of aerosol in the stratosphere and mesosphere is meteors. While coarse meteors fall on the Earth, aerosol originating from these meteors is most frequently formed in the upper atmosphere [7] and less

I. I. Romanchenko, P. V. Novikov, V. N. Marichev and D. A. Bochkovsky—These authors contributed equally to this work.

A. Dmitriev et al. (Eds.): STRPEP 2023, SPEES, pp. 38–46, 2023.
https://doi.org/10.1007/978-3-031-50248-4_5

commonly in the middle and lower stratosphere [8]. The most thoroughly studies case of meteor aerosol formation in the past had been the fall of meteorite near Chelyabinsk on February 15, 2013 [9]. A dust ring, formed at an altitude of 25–45 km after the meteorite fall, filled the stratosphere of the Northern hemisphere over 50–70°N latitudes [10]. After the Chelyabinsk meteorite, unique information has been obtained with the help of land-based lidar instrumentation (Obninsk, Moscow, Yakutsk) on the aerosol layers propagating in the stratosphere; the data on the trajectories of air masses in the stratosphere were involved [11].

According to the data of lidar measurements of the backscattering ratio in the stratosphere above Tomsk in August 2013, intense peaks of aerosol scattering were detected. The most intense peak was observed at the altitude of 29 km. In the present work, the assumption concerning the meteoric origin of the recorded stratospheric aerosol is analyzed. For this purpose, the data obtained in the calculation of backward trajectories of air masses, the satellite data from the limb sensor OMPS-LP and from the space lidar CALIOP have been also considered.

2 Experimental Data and the Procedure of Trajectory Analysis

The starting point of the investigation was in the land-based measurements of the aerosol backscattering ratio. The measurements were made using the lidar station situated in Tomsk at the Institute of Atmospheric Optics SB RAS. The aerosol backscattering ratio was calculated from measurement results: $R(H) = (\beta_{\pi m}(H)+\beta_{\pi a}(H))/\beta_{\pi m}(H)$, where $\beta_{\pi m}(H)$, $\beta_{\pi a}(H)$ are the volumetric coefficients of molecular and aerosol backward scattering at the altitude H, respectively. In the presence of aerosol at a definite altitude $R(H) > 1$, while in its absence $R(H) \approx 1$ to the accuracy within measurement error. The major characteristics of this lidar station were reported in [12].

An essential part of the investigation is the use of OMPS-LP instrument (Ozon Mapping Suite Profiler – Limb Profiler), which is intended for recording the profiles of ozone and aerosols by measuring scattered solar radiation within the spectral range of 290–1000 nm. In each survey point, limb sensor OMPS-LP records the profile of aerosol extinction at the altitudes from 0.5 to 40.5 km above sea level [13]. The limb sensor measures three limb profiles spaced approximately 250 km in the cross-track direction. In particular, this instrument was used to study the dust belt that was formed after the explosion of the Chelyabinsk meteorite [10], and to study several aerosol clouds left by smaller bolides and rocket launchings [14].

The data from the space lidar CALIOP at the CALIPSO satellite on the altitudinal distribution of aerosol in the atmosphere were also involved in the study. This lidar uses three receiving channels, one at 1064 nm and the other two at 532 nm with orthogonal polarization [15].

Aerosol transport was analyzed on the basis of the method that includes tracing the backward isentropic trajectories, implemented using the software developed by us for plotting air mass trajectories. In these calculations, we used the data from UK Met Office on wind velocities in the Northern hemisphere [16]. The procedure showed excellent performance in the calculations of aerosol transport after volcanic eruptions and after the Chelyabinsk meteorite fall [11].

3 Results of the Lidar Observations of Aerosol Above Tomsk in August 2013

Aerosol layers at the stratospheric altitudes were recorded above Tomsk in August 2013 (Fig. 1). The thickest aerosol layer was observed on August 22 at the altitude of 29 km. The value of backscattering ratio R reached 24.

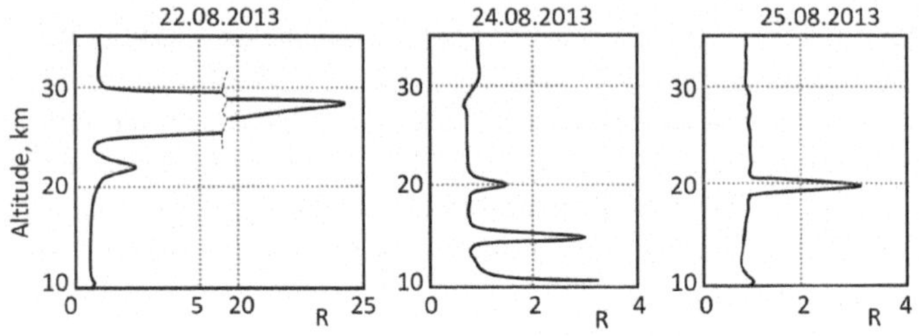

Fig. 1. The altitudinal profiles of the backscattering ratio R(H), obtained on the basis of the results of lidar measurements above Tomsk in August 2013.

These layers were formed while Perseids meteoroid flux was passing, which is observed every year since about July 17 till August 24 [17]. The peak intensity of this flux falls within August 11–13. It was assumed that the observed aerosol layers were formed as a result of combustion of meteor bodies in the stratosphere.

As a rule, the meteor bodies of this flux burn at the altitudes from 60 to 120 km above the Earth's surface. It may be assumed that the observed aerosol layers appeared at the stratospheric altitudes as a result of sedimentation of this aerosol of meteoric origin. Attention should be paid to the fact that the observed aerosol layers are very narrow, which does not agree with the assumption that they originated as a result of aerosol particles sedimentation from high altitudes. However, after the fall of the Chelyabinsk meteorite on February 15, 2013, narrow aerosol layers were detected in Obninsk, Moscow and Yakutsk [11] at an altitude above 30 km. The origin of the observed layers of meteoric aerosol was confirmed by the results of air mass trajectory analysis. The aerosol cloud formed during meteorite burning was decomposed into narrow layers as a result of wind distribution in the stratosphere.

In addition, sporadic meteors that do not belong to any meteoric fluxes are observed every year in the atmosphere. In the lower part of their trajectories, they reach an altitude of 25–35 km [18]. The existing monitoring systems in the USA and Canada [19, 20], as well as some other countries, mainly contain the data on the brightest meteors recorded in these countries is populated areas. Thus, the falls of middle-sized meteor bodies in sparsely populated areas may remain unnoticed.

In the present work, we made an attempt to reveal the origin only of the most intense aerosol peak that was observed at an altitude of 29 km above Tomsk.

4 Results of Trajectory Analysis Involving the Satellite Data

The backward isentropic trajectories of air masses were calculated for 16 days before these air masses arrived to Tomsk at het altitude of about 29 km on August 22. The trajectory corresponding to the maximum of aerosol peak showed in Fig. 1 for August 22 is presented in Fig. 2. Points on this trajectory mark 12:00 UTC moments of each day. Air circulation in the stratosphere over West Siberia is slow, so a fragment of the trajectory above Tomsk is shown in the insert on an enlarged scale. During 16 days, air masses at an altitude of about 29 km had passes from Chukotka to Tomsk since August 6 to 22 within the 55–65°N latitudes.

Then the satellite data were engaged for the analysis of aerosol filling. Along the obtained trajectory of air mass motion, the tracts of Suomi NPP satellite were found, crossing this trajectory within the meso-scale accuracy in the points where a survey point and the moving air mass occurred simultaneously. Therefore, the aerosol, carried by these air masses and later recorded at the altitude of 29 km above Tomsk on August 22, could be detected from the satellite. These tracks are shown in the Figure as bold segments, with numbers near the segments indicating the dates when the track crossed the air mass trajectory. Near Tomsk, the satellite tracks synchronously crossed the trajectory of air mass motion several times: on August 21, 22, 23, and 25. These tracks turned out to be spatially close to each other, so they are shown in Fig. 2 as one continuous line marked with the dates 21–25. The search for synchronous crossings between air mass trajectory

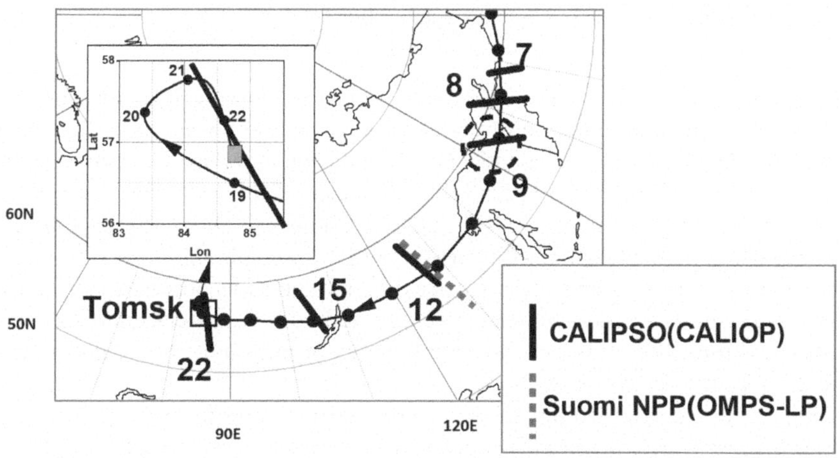

Fig. 2. The trajectory of air mass motion (a thin continuous line with points) that reached Tomsk on August 22, 2013 at the altitude of 29 km, and its synchronous intercrossings with the tracks of satellites: Suomi NPP (bond continuous lines), and CALIPSO (a grey dash line). Points on the trajectory indicate air mass shifts with a step of 24 h (corresponding to 12:00 UTC of each day). Near satellite tracks, the dates of their crossing by the air mass trajectory are indicated. A fragment of the trajectory near Tomsk is shown on the enlarged scale in the insert. Tomsk is marked as a grey square. A dash line in the region of Kamchatka encircles the site where the meteorite was assumed to fall.

and CALIPSO satellite tracks was carried out similarly. One track that crossed the air mass trajectory synchronously on August 12 has been found.

According to the data from the OMPS-LP instrument, mounted on Suomi NPP, the vertical profiles of aerosol extinction ratio (at 869 nm) in the points of nearly synchronous crossing with the air mass trajectory near Tomsk since August 21 till August 25 are presented in Fig. 3. No data from OMPS-LP on August 24 are available.

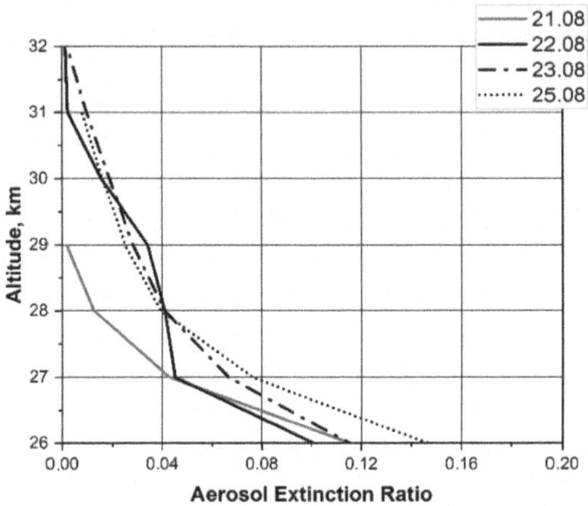

Fig. 3. The sequence of vertical profiles of the aerosol extinction ratio in the stratosphere above Tomsk in August 2013 according to the data from OMPS-LP.

According to the data, on the 21th of August, a day before lidar measurements, the aerosol content at the altitude of about 29 km was the lowest over the available observation data. On August 22, a noticeable broad peak of the aerosol extinction ratio appears around the altitude of 29 km. The appearance of this peak corresponds to the data of lidar observations in Tomsk (Fig. 1). The features of OMPS-LP probing cause broadening of the aerosol peak in comparison with the data of lidar observations. Later on, on August 23 and 25, increased aerosol extinction ratio was observed at the altitude of about 29 km near Tomsk. This may be connected with the extension of the aerosol layer that had travelled to Tomsk, and the features of atmospheric circulation at this altitude near Tomsk during those days.

In Fig. 4, the vertical profiles of the aerosol extinction ratio in the points of synchronous intercrossing with air mass trajectory on August 15, 12, 8, and 7 are shown it the similar manner. These profiles are given separately for each day as continuous black lines. For comparison, grey lines depict the vertical profiles of aerosol extinction ratio in the same spatial points but a day before the aerosol-carrying air masses reached Tomsk.

According to these data, the peak at the vertical profiles of aerosol extinction ratio with a maximum at 29–30 km appeared on about August 9, while no peak was observed on August 7 and 8. According to the formulated assumption, the meteor was to fall within this time interval and produce the observed aerosol. In Fig. 2, the assumed region of meteorite fall and the formation of recorded aerosol around August 9, 2013 is encircled with a dash line: it is above the Sea of Okhotsk to the north-west from Kamchatka.

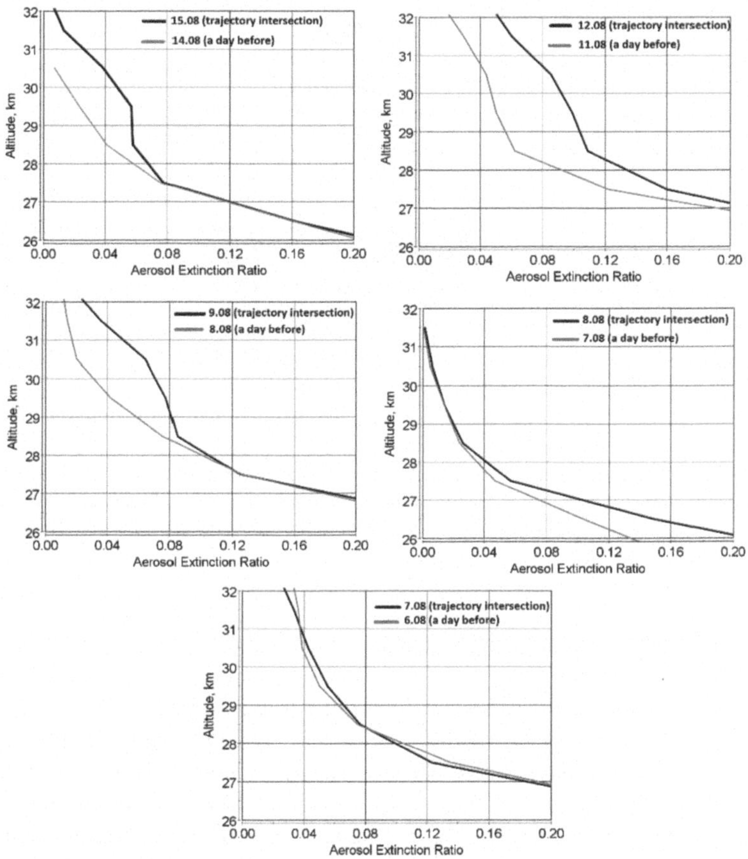

Fig. 4. A sequence of the vertical profiles of aerosol extinction ratio according to the data from OMPS-LP in the points of synchronous intersections of satellite tracks with the air mass trajectory towards Tomsk in August 2013, shown in Fig. 2 as a thin continuous line with points, and at the same geographic points a day before.

The presence of aerosol at the altitude of 29 km on August 12, in the point of air mass trajectory intersection with the track of CALIPSO satellite, is also indicated by the data on the aerosol filling of the stratosphere, obtained from the satellite lidar CALIOP. Aerosol was detected by the limb sensor of OMP S-LP at the same site and at the same time at the altitude of 29 km (Fig. 5).

The question concerning possible artificial origin of the aerosol cloud moving to Tomsk at the altitude of 29 km was also considered in the investigation.

It is known that aerosol at the altitude of 35–65 km may arise as a consequence of rocket launching [21]. There were three rocket launching evens in August 2013 that could lead to the formation of aerosol at the altitude indicated above: the rockets were launched from Tanegashima Space Center on August 3 at 19:48 UTC, from Cape Canaveral on August 8 at 00:29 UTC, and from Yasny cosmodrome on August 22 at 14:39 UTC [22]. The direct isoentropy trajectories were calculating using the data of UKMO and

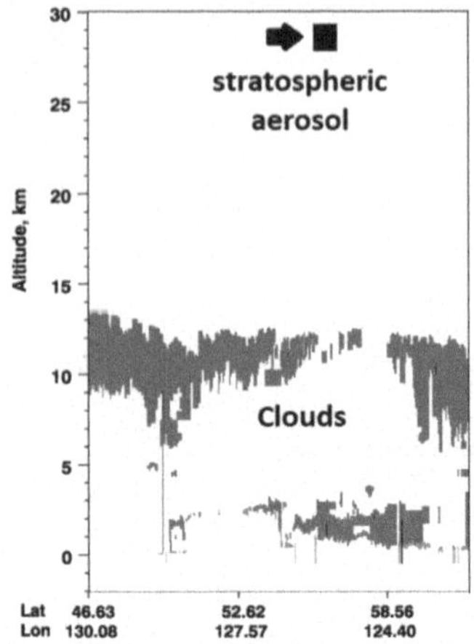

Fig. 5. A vertical section of the aerosol filling on the atmosphere along the track of CALIPSO, shown with a grey dash line in Fig. 2, according to the data from space lidar CALIOP. The arrow indicates aerosol at the altitude of 29 km, in the point where CALIPSO track crosses the air mass trajectory.

ECMWF. The former two launching sites are situated at the latitudes of 28–30°, and in summer 2013, in both cases, the air masses were moving at the altitude of 25–31 km from launching sites to the west almost in parallel to the equator within a band of 25–35° N and could not be source of the aerosol observed on August 22 above Tomsk. Yasny cosmodrome is situated at 51°N, 56°E and, as evidenced by calculation results, air masses at the altitude of 25–35 km could not reach Tomsk to the moment of lidar observation. So, aerosol from rocket launching bears no relation to stratospheric aerosol detected above Tomsk on August 22.

5 Conclusion

According to the data of lidar measurements of the backscattering ratio, aerosol was recorded at the stratospheric altitude in the atmosphere above Tomsk in August 2013. The most intense peak of aerosol scattering was detected at the altitude of 29 km on August 22.

The assumption concerning the meteoric origin of this aerosol is considered in the work. Isoentropic backward trajectories of air mass motion were calculated using the original software package and the data on wind velocities from the UK Met Office. In addition, the data on the stratospheric aerosol obtained from the space lidar CALIOP and the satellite limb sensor OMPS-LP were involved, which allowed us to determine approximate region of meteor body combustion in the atmosphere.

The peak of aerosol scattering, observed on August 22, 2013 above Tomsk at the altitude of 29 km, is most probably due to the aerosol formed as a result of meteor body fall in the region of the Sea of Okhotsk near Kamchatka.

Acknowledgements. We are grateful to the UKMO and the CEDA for the access to the data on http://badc.nerc.ac.uk/browse/badc/ukmo-assim.

We are grateful to the NASA for the access to the data of OMPS-LP and CALIPSO.

References

1. Fromm, M., Lindsey, D.T., Servranckx, R., et al.: The untold story of pyrocumulonimbus. J. Bull. Amer. Meteorol. Soc. **91**(9), 1193–1210 (2010). https://doi.org/10.1175/2010BAMS3004.1
2. Ansmann, A., Baars, H., Chudnovsky, A., et al.: Extreme levels of Canadian wildfire smoke in the stratosphere over central Europe on 21–22 August 2017. J. Atmos. Chem. Phys. **18**, 11831–11845 (2018). https://doi.org/10.5194/acp-18-11831-2018
3. Cheremisin, A.A., Marichev, V.N., Bochkovskii, D.A., et al.: Stratospheric aerosol of siberian forest fires according to lidar observations in Tomsk in August 2019. J. Atmos. Ocean Opt. **35**, 57–64 (2022). https://doi.org/10.1134/S1024856022010043
4. Cheremisin, A.A., Marichev, V.N., Novikov, P.V.: Lidar observations of volcanic aerosol content in the atmosphere over Tomsk. J. Russ. Meteorol. Hydrol. **36**, 600–607 (2011). https://doi.org/10.3103/S1068373911090056
5. Grebennikov, V.S., Zubachev, D.S., Korshunov, V.A., et al.: Observations of stratospheric aerosol at rosgidromet lidar stations after the eruption of the Raikoke Volcano in June 2019. J. Atmos. Ocean Opt. **33**, 519–523 (2020). https://doi.org/10.1134/S1024856020050097
6. Cheremisin, A.A., Marichev, V.N., Novikov, P.V.: Transport of polar stratospheric clouds from the arctic to Tomsk in January 2010. J. Atmos Ocean Opt. **26**, 492–498 (2013). https://doi.org/10.1134/S1024856013060031
7. Kruchinenko, V.G., Kozak, P.N., Taranukha, Y.G., et al.: Meteoroids as a source of aerosol in the upper atmosphere. J. Atmos. Ocean Opt. **24**, 261–270 (2011). https://doi.org/10.1134/S1024856011030110
8. Schneider, J., Weige, R., Klimach, T., et al.: Aircraft-based observation of meteoric material in lower-stratospheric aerosol particles between 15 and 68° N. J. ACP. **21**(2), 989–1013 (2021). https://doi.org/10.5194/acp-21-989-2021
9. Lobanovsky, Y.I.: Chelyabinsk meteoroid: critique of sources and proving of conclusions. Eng. J.: Sci. Innov. **7**(55) (2016). https://doi.org/10.18698/2308-6033-2016-7-1497
10. Gorkavyi, N., Rault, D.F., Newman, P.A., et al.: New stratospheric dust belt due to the Chelyabinsk bolide. J. Geophys. Res. Lett. **40**(17), 4728–4733 (2013). https://doi.org/10.1002/grl.50788
11. Ivanov, V.N., Zubachev, D.S., Korshunov, V.A., et al.: Lidar observations of stratospheric aerosol traces of Chelyabinsk meteorite. Opt. Atmos. Okeana [in Russian] **27**(02), 117–122 (2014)
12. Marichev, V.N., Bochkovsky, D.A.: Modernization of the lidar complex of the station of high-altitude sounding of the atmosphere IAO SB RAS. J. Bull. KRAUNTS. Phys. Math. Sci. **5**(25), 17–23 (2018). https://doi.org/10.18454/2079-6641-2018-25-5-17-23
13. Taha, G., Loughman, R., Zhu, T., et al.: OMPS LP version 2.0 multi-wavelength aerosol extinction coefficient retrieval algorithm. J. Atmos. Meas. Tech. **14**, 1015–1036 (2021). https://doi.org/10.5194/amt-14-1015-2021
14. Gorkavyi, N., DeLand, M., Bhartia, P.K.: Limb profiler of OMPS/NPP and registration of plumes from bolides and rockets. In: GSFC/NASA. Kaufman Symposium (2016)
15. CALIOP Algorithm Theoretical Basis Document. Calibration and Level 1 Data Products. https://www-calipso.larc.nasa.gov/resources/pdfs/PC-SCI-201v1.0.pdf. Accessed 15 May 2023

16. Swinbank, R., O'Neill, A., Lorenc, A.C., et al.: Met office stratospheric assimilated data. NCAS British (2006)
17. Meteor Data Center of the International Astronomical Union. Perseids. https://www.ta3.sk/IAUC22DB/MDC2022/Roje/pojedynczy_obiekt.php?lporz=00054&kodstrumienia=00007&colecimy=0&kodmin=00001&kodmax=01222&lpmin=00001&lpmax=01711&sortowanie=0. Accessed 28 Aug 2023
18. Moreno-Ibáñez, M., Gritsevich, M., Trigo-Rodríguez, J.M.: New methodology to determine the terminal height of a fireball. Icarus **250**, 544–552 (2015). https://doi.org/10.1016/j.icarus.2014.12.027
19. Center for Near Earth Object Studies. Fireballs. https://cneos.jpl.nasa.gov/fireballs/intro.html. Accessed 28 Aug 2023
20. Report a Fireball. https://fireballs.imo.net/members/imo/report_intro. Accessed 28 Aug 2023
21. Gorkavyi, N.N.: Spaceborne limb observations of artificial aerosol clouds. Cosmic Res. **58**, 86–91 (2020). https://doi.org/10.1134/S0010952520020045
22. The United Nations Office for Outer Space Affairs (UNOOSA) – Online Index of Objects Launched into Outer Space. https://www.unoosa.org/oosa/osoindex/index.jspx?lf_id=. Accessed 28 Aug 2023

Impact of Solar Activity on the Optical Properties of the Thermosphere

Boris Shevtsov(✉)

Far Eastern Branch of the Russian Academy of Sciences, Institute of Cosmophysical Research and Radio Wave Propagation, Mirnaya, Paratunka 684034, Kamchatka, Russia
bshev@ikir.ru

Abstract. Mechanisms of the influence of solar activity on optical characteristics of the thermosphere and the possibility of studying them, using lidar observations and monitoring the glow of the night sky, are considered. The need for an integrated approach to remote observations is substantiated by the example of the dynamics of spectral lines of optical emission over Kamchatka, which arises as a result of changes in the state of the thermosphere, and which can be used for remote lidar sensing in adaptive mode.

Keywords: optical characteristics of the thermosphere · dynamics of optical emission spectra · complex remote sensing methods

1 Introduction

One of the reasons for climate change is associated with the influence of solar activity on optical properties of the atmosphere via the increase in ionization, which causes the increase in the scattering coefficient of the atmosphere. In the upper layers, this is due to the spectral properties of excited ions, and in the lower layers, with the formation of aerosols at condensation centers, which are created by ions. Both of these mechanisms have been studied using lidar observations in the middle and lower layers of the atmosphere for a relatively long time, and in the upper layers only in recent years [1–5].

Lidar studies of the upper atmosphere are methodically related to the observations of a thin layer of metal vapors in the altitude range of 100–120 km using resonant scattering [6–10]. The proposal to create a thermospheric He lidar [11], despite the difficulties with the IR receiver, was implemented [3]. At the same time, three thermospheric O^+, N^+ and Ca^+ lidars have already been created. This circumstance is in favor of ion spectral lines when choosing a probing target. And this choice remains a big problem.

The fact that excited ions have a wider spectral line than their neutrals can be understood if we compare them according to the hydrogen-like model. The ion has twice the charge and four times the energy levels. Ions move under the action

A. Dmitriev et al. (Eds.): STRPEP 2023, SPEES, pp. 47–53, 2023.
https://doi.org/10.1007/978-3-031-50248-4_6

of electric fields, receive additional energy and transfer it to neutrals in collisions. It is obvious that ions and neutrals differ in their excitation mechanisms, it means that they interact differently with radiation. And this should be taken into account in lidar observations.

With increasing solar activity, sun thermodynamic state deviates even more from equilibrium, and its thermodynamic distribution from exponential, which manifests itself in the increase in short-wave radiation, which, acting on the earth's thermosphere, increases the excitation of its components, and this leads to the increase in the scattering coefficient.

In the middle and lower atmosphere, ionization, promoting the formation of aerosol, also causes the increase in scattering. In this scheme of the influence of solar activity on the atmosphere, the decisive role is played by ultraviolet and x-ray radiation, which is converted into longer wavelengths through the process of ionization and recombination.

We must not forget that there are other factors associated with volcanic activity, fires, dust storms and cyclones, but their contribution must be assessed integrally, then it will be quite comparable with the constantly acting low level influence of solar activity. The anthropogenic factor can be estimated at a hundredth of a percent of energy production in relation to solar heat flow. This is also comparable to the influence of solar activity. From the point of view of a systems approach, all factors can be interconnected, and their synchronization can lead to catastrophic consequences.

The excited state of the thermosphere is determined by the population of energy levels and the set of spectral lines of plasma components, and the differences in the population of levels determine the resonant conditions for the radiation propagation. Therefore, to carry out lidar observations, it is necessary to monitor optical emission spectra, the characteristics of which will be considered using the example of the results of spectral analysis of the night sky glow over Kamchatka.

2 Glow Spectra of the Night Sky

The results of spectral analysis of the night sky emission are presented in Fig. 1 and 2 (local time). Data were obtained over Kamchatka on February 28, 2012 using a SP-2500i spectrophotometer with a PicoStar HR 12 picosecond camera and a telescope aimed at the zenith with a mirror diameter of 60 cm and a focal length of 2 m, the coordinates are 52°58′23″N, 158°14′50″E, in a quiet geomagnetic environment, all K-indices were equal to 1. The observation time was 1 h 50 min with two signal accumulation sessions of 20 min each, at the beginning and at the end of this period.

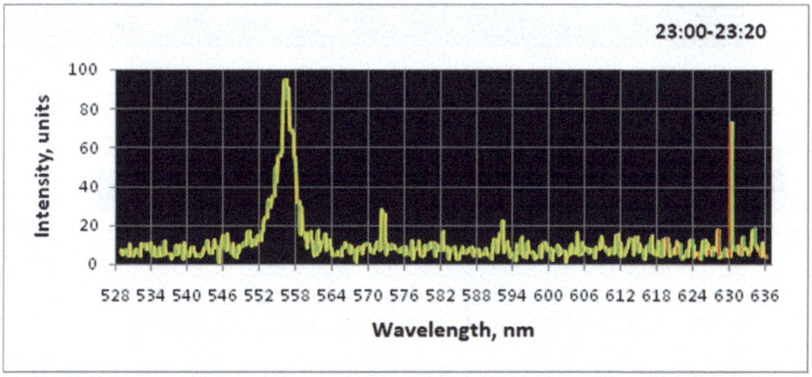

Fig. 1. Luminescence specter of the night sky on February 28, 2012, at the beginning of observations.

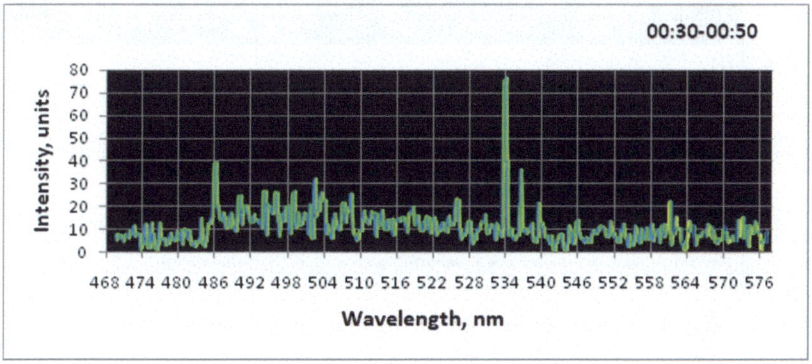

Fig. 2. Luminescence specter of the night sky on February 28, 2012, at the end of observations.

We should note the significant changes in the spectral characteristics of the glow of the night thermosphere over an observation period of 1 h and 50 min. The beginning and end of the observations are presented in Fig. 1 and 2 by different groups of spectral lines. The results of the search in the NIST Atomic Spectra Database for the yellow and red lines in Fig. 1 are shown in Fig. 3.

Grotrian diagrams of these lines are presented in Fig. 7a. In addition to the oxygen lines Fig. 1 illustrates yellow nitrogen lines N 571.3 and 572.9 nm, their data are shown in Fig. 4a, Grotrian diagrams are presented in Fig. 7b.

	Ritz Wavelength Air (nm)	Rel. Int. (?)	A_{ki} (s^{-1})	Acc.	E_i (cm^{-1})		E_k (cm^{-1})	Lower Level Conf., Term, J	Upper Level Conf., Term, J	Type	TP Ref.	Line Ref.
O I	555.4832		5.83e+05	C+	88 630.587	-	106 627.934	$2s^22p^3(^4S^\circ)3p$ ^3P 1	$2s^22p^3(^4S^\circ)7s$ $^3S^\circ$ 1		T5713LS	
O I	555.5004		9.71e+05	C+	88 631.146	-	106 627.934	$2s^22p^3(^4S^\circ)3p$ ^3P 2	$2s^22p^3(^4S^\circ)7s$ $^3S^\circ$ 1		T5713LS	
O I	555.5053		1.94e+05	C+	88 631.303	-	106 627.934	$2s^22p^3(^4S^\circ)3p$ ^3P 0	$2s^22p^3(^4S^\circ)7s$ $^3S^\circ$ 1		T5713LS	
O I 557.734	557.7339	120	1.26e+00	B+	15 867.862	-	33 792.583	$2s^22p^4$ ^1D 2	$2s^22p^4$ ^1S 0	E2	T4539,T5081	L3760

	Observed Wavelength Air (nm)	Ritz Wavelength Air (nm)	Rel. Int. (?)	A_{ki} (s^{-1})	Acc.	E_i (cm^{-1})		E_k (cm^{-1})	Lower Level Conf., Term, J	Upper Level Conf., Term, J	Type	TP Ref.	Line Ref.
O I		630.0304		5.63e-03	B+	0.000	-	15 867.862	$2s^22p^4$ ^3P 2	$2s^22p^4$ ^1D 2	M1	T4539,T5081	
O I		630.0304		2.11e-05	C+	0.000	-	15 867.862	$2s^22p^4$ ^3P 2	$2s^22p^4$ ^1D 2	E2	T4539,T5081	

Fig. 3. Yellow and red lines of atomic oxygen.

	Ritz Wavelength Air (nm)	Rel. Int. (?)	A_{ki} (s^{-1})	Acc.	E_i (cm^{-1})		E_k (cm^{-1})	Lower Level Conf., Term, J	Upper Level Conf., Term, J	Type	TP Ref.	Line Ref.
a												
N I	571.314		8.94e+04	D+	96 750.840	-	114 249.5	$2s^22p^2(^3P)3p$ $^4S^\circ$ $^3/_2$	$2s^22p^2(^3P)6d$ ^4P $^1/_2$		T5723LS	
N I	572.935		8.87e+04	D+	96 750.840	-	114 200.0	$2s^22p^2(^3P)3p$ $^4S^\circ$ $^3/_2$	$2s^22p^2(^3P)6d$ ^4P $^3/_2$		T5723LS	

	Ritz Wavelength Air (nm)	Rel. Int. (?)	A_{ki} (s^{-1})	Acc.	E_i (cm^{-1})		E_k (cm^{-1})	Lower Level Conf., Term, J	Upper Level Conf., Term, J	Type	TP Ref.	Line Ref.
b												
N I	534.4049		6.10e+04	D+	88 107.260	-	106 814.459	$2s2p^4$ ^4P $^5/_2$	$2s^22p^2(^3P)4p$ $^4D^\circ$ $^5/_2$		u18,LS	
N I	534.649		9.21e+04	D+	96 750.840	-	115 449.5	$2s^22p^2(^3P)3p$ $^4S^\circ$ $^3/_2$	$2s^22p^2(^3P)9s$ ^4P $^1/_2$		T5723LS	
N I	535.4388		1.01e+04	D	88 107.260	-	106 778.337	$2s2p^4$ ^4P $^5/_2$	$2s^22p^2(^3P)4p$ $^4D^\circ$ $^3/_2$		u18,LS	
N I	535.6623		1.41e+05	D+	88 151.170	-	106 814.459	$2s2p^4$ ^4P $^3/_2$	$2s^22p^2(^3P)4p$ $^4D^\circ$ $^5/_2$		u18,LS	
N I	536.7010		1.07e+05	D+	88 151.170	-	106 778.337	$2s2p^4$ ^4P $^3/_2$	$2s^22p^2(^3P)4p$ $^4D^\circ$ $^3/_2$		u18,LS	
N I	537.2606		8.34e+04	D+	88 170.570	-	106 778.337	$2s2p^4$ ^4P $^1/_2$	$2s^22p^2(^3P)4p$ $^4D^\circ$ $^3/_2$		u18,LS	
N I	537.2665		3.33e+04	D	88 151.170	-	106 758.731	$2s2p^4$ ^4P $^3/_2$	$2s^22p^2(^3P)4p$ $^4D^\circ$ $^1/_2$		u18,LS	
N I	537.8273		1.66e+05	D+	88 170.570	-	106 758.731	$2s2p^4$ ^4P $^1/_2$	$2s^22p^2(^3P)4p$ $^4D^\circ$ $^1/_2$		u18,LS	

Fig. 4. Yellow (a) and green (b) lines of atomic nitrogen.

Atomic nitrogen was highlighted at the beginning of observations with yellow lines, and at the end it was marked with green lines, data is in Fig. 4, diagram is in Fig. 7b. At the end of the observations, lines of atomic oxygen ion (Fig. 5) and atomic hydrogen (Fig. 6) appear.

	Ritz Wavelength Air (nm)	Rel. Int. (?)	A_{ki} (s^{-1})	Acc.	E_i (cm^{-1})		E_k (cm^{-1})	Lower Level Conf., Term, J	Upper Level Conf., Term, J	Type	TP Ref.	Line Ref.
O II	539.8300				195 710.47	-	214 229.671	$2s2p^4$ ^2S $^1/_2$	$2s^22p^2(^3P)3p$ $^2P^\circ$ $^3/_2$			L11267
O II	539.8682		2.49e+04	C	214 229.671	-	232 747.562	$2s^22p^2(^3P)3p$ $^2P^\circ$ $^3/_2$	$2s^22p^2(^3P)3d$ ^4D $^5/_2$		T6172	
O II	539.9143		4.06e+04	C	214 229.671	-	232 745.981	$2s^22p^2(^3P)3p$ $^2P^\circ$ $^3/_2$	$2s^22p^2(^3P)3d$ ^4D $^3/_2$		T6172	

Fig. 5. Atomic oxygen ion lines.

During the observations, the spectral lines shifted from the yellow-red region to the blue-green region, indicating changes in the state of the thermosphere. This should be taken into account during lidar observations.

Atomic nitrogen appeared both at the beginning and at the end of the observations, but in the second case it was in a shorter wavelength range of radiation.

H I.	Unc. (nm)	Ritz Wavelength Air (nm)	Unc. (nm)	Rel. Int. (?)	A_{ki} (s^{-1})	Acc.	E_i (cm^{-1})		E_k (cm^{-1})	Lower Level Conf., Term, J			Upper Level Conf., Term, J			Type	TP Ref.	Line Ref.
		486.1278624	0.0000024		1.7188e+07	AAA	82 258.9191133	-	102 823.894250	2p	$^2P^\circ$	$^1/_2$	4d	^2D	$^3/_2$		T7771	
486.1283363	0.0000024	486.1283363	0.0000024				82 258.9543992821	-	102 823.9094871	2s	^2S	$^1/_2$	4d	^2D	$^5/_2$	E2		L9496
		486.12841	0.00014															c63
486.1286949	0.0000024	486.1286949	0.0000024		9.6680e+06	AAA	82 258.9543992821	-	102 823.8943175	2s	^2S	$^1/_2$	4p	$^2P^\circ$	$^3/_2$		T7771	L9477
		486.1288370	0.0000024		8.5941e+05	AAA	82 258.9191133	-	102 823.8530211	2p	$^2P^\circ$	$^1/_2$	4s	^2S	$^1/_2$		T7771	
486.1296711	0.0000024	486.1296711	0.0000024				82 258.9543992821	-	102 823.8530211	2s	^2S	$^1/_2$	4s	^2S	$^1/_2$	M1		L9496
486.1297761	0.0000024	486.1297761	0.0000024		9.6682e+06	AAA	82 258.9543992821	-	102 823.8485825	2s	^2S	$^1/_2$	4p	$^2P^\circ$	$^1/_2$		T7771	L9477
		486.13250	0.00014															c64
486.135	0.005	486.1333	0.0003	180000	8.4193e+06	AAA	82 259.158	-	102 823.904	2			4				T8637	L7439c30
		486.1361516	0.0000024		2.0625e+07	AAA	82 259.2850014	-	102 823.9094871	2p	$^2P^\circ$	$^3/_2$	4d	^2D	$^5/_2$		T7771	
		486.13622	0.00005															c65
		486.1365118	0.0000024		3.4375e+06	AAA	82 259.2850014	-	102 823.894250	2p	$^2P^\circ$	$^3/_2$	4d	^2D	$^3/_2$		T7771	
		486.1374864	0.0000024		1.7190e+06	AAA	82 259.2850014	-	102 823.8530211	2p	$^2P^\circ$	$^3/_2$	4s	^2S	$^1/_2$		T7771	

Fig. 6. Atomic hydrogen lines.

The shift of spectral lines N to the short-wavelength part of the spectrum and the appearance of lines H and O^+ instead of lines O indicates movement of the excitation region of the thermosphere upwards that is quite natural for midnight, when the sun goes far below the horizon.

At the end of the observations, a rise in the spectrum continuum occurred between the H 486 nm and N 534 nm lines. Moreover, it increased in the short-wavelength part of the spectrum and ended at the hydrogen line. The spectrum interval between the H 486 nm and N 534 nm lines is an emission line cluster. This spectral range and the one to the left of the hydrogen blue line H 486 nm are of interest for studies of the resonant propagation of both solar and laser radiation. This is of particular interest when the atmosphere is strongly influenced by solar activity.

3 Discussion

The scattering coefficients of the thermosphere in individual spectral lines are very small. They increase during strong disturbances. It is interesting to estimate the integral impact of solar activity by spectrum, height and strength of a disturbance. This may be the subject of further lidar observations. In this regard, it is promising to use sufficiently excited levels with transition lines in the blue region of the spectrum. Photodetectors have high sensitivity in this area.

There are not many spectral lines that have natural line width of at least 100 MHz and are suitable for use in resonance scattering. With a narrow resonance line, vibrations increase slowly, and collisions of atoms interfere with this. In case of sudden disturbances, the loss of the resonant signal is possible. And this effect should also be taken into account.

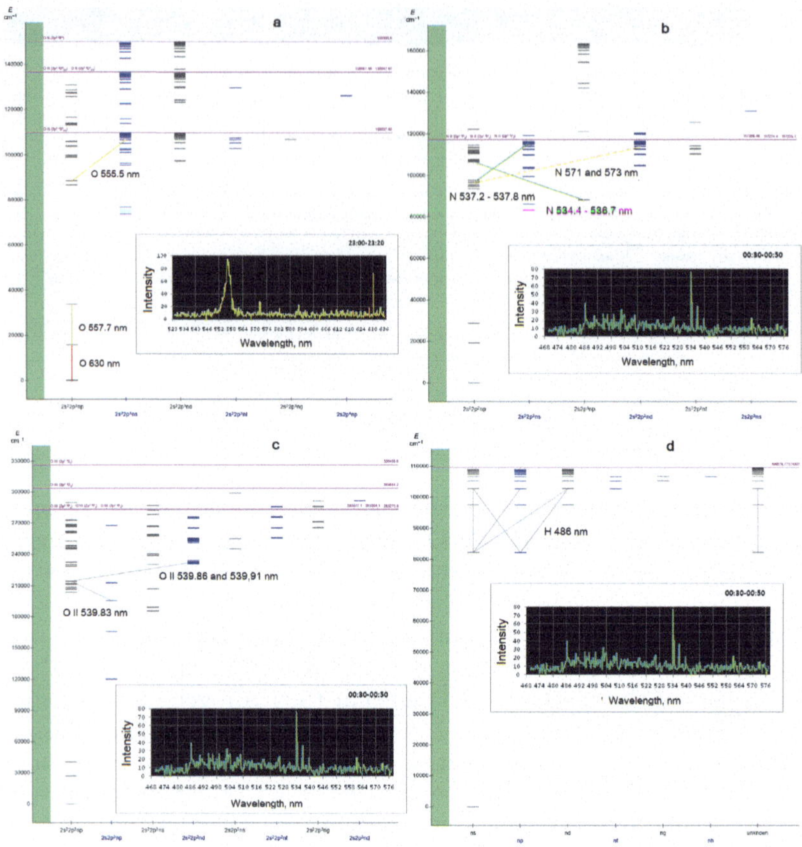

Fig. 7. Grotrian diagrams of atomic oxygen (a), atomic nitrogen (b), atomic oxygen ion (c) and atomic hydrogen (d).

4 Conclusion

Based on the results of spectral observations, it was shown that thermosphere is significantly restructured even under calm conditions, and this should be taken into account during lidar observations, which must be carried out in adaptive mode.

Acknowledgments. The work was carried out within the framework of realization of the State task № AAAA-A21-121011290003-0.

References

1. Bychkov, V.V., Nepomnyashchiy, Y.A., Perezhogin, A.S., Shevtsov, B.M.: Lidar returns from the upper atmosphere of Kamchatka for 2008 to 2014 observations. EPS **66**, 150 (2014)

2. Bychkov, V.V., Perezhogin, A.S., Seredkin, I.N., Shevtsov B. M.: Appearance of light-scattering layers in the thermosphere of Kamchatka during the autumn of 2017. In: Proceedings of the SPIE, vol. 10833, p. 10833A4 (2018)
3. Kaifler, B., Geach, C., Büdenbender, H.C., Mezger, A., Rapp, M.: Measurements of metastable helium in earth's atmosphere by resonance lidar. Nat. Commun. **13**, 6042 (2022)
4. Shevtsov, B.M., Perezhogin, A.N., Seredkin, I.N.: Atmospheric optical characteristics in the area of 30–400 km. Remote Sens. **14**, 6108 (2022). https://doi.org/10.3390/rs14236108
5. Lifang, D., et al.: The all-solid-state narrowband lidar developed by optical parametric oscillator/amplifier (OPO/OPA) technology for simultaneous detection of the Ca and Ca^+ layers. Remote Sens. **15**(18), 4566 (2023). https://doi.org/10.3390/rs15184566
6. Raizada, S., Tepley, C.A., Aponte, N., Cabassa, E.: Characteristics of neutral calcium and Ca^+ near the mesopause, and their relationship with sporadic ion/electron layers at Arecibo. Geophys. Res. Lett. **38**(9), L09103 (2011). https://doi.org/10.1029/2011GL047327
7. Raizada, S., et al.: New lidar observations of Ca^+ in the mesosphere and lower thermosphere over arecibo. Geophys. Res. Lett. **47**(5), e2020GL08711 (2020) https://doi.org/10.1029/2020GL087113
8. Granier, G., Jégou, J.P., Mégie, G.: Resonant lidar detection of Ca and Ca^+ in the upper atmosphere. Geophys. Res. Lett. **12**(10), 655–658 (1985)
9. Gardner, C.S., Kane, T.J., Senft, D.C., Qian, J., Papen, G.C.: Simultaneous observations of sporadic E, Na, Fe, and Ca^+ layers at Urbana, Illinois: three case studies. J. Geophys. Res. **98**, 16865–16874 (1993)
10. Qian, J., Gardner, C.: Simultaneous lidar measurements of mesospheric Ca, Na, and temperature profiles at Urbana, Illinois. J. Geophys. Res. **100**, 7453–7461 (1995)
11. Gerrard, A.J., Kane, T.J., Meisel, D.D., Thayer, J.P., Kerr, R.B.: Investigation of a resonance lidar for measurement of thermospheric metastable helium. JASTP **59**, 2023–2035 (1997)

Relation Between Sprites and Whistlers Based on AWDANET and WWLLN Data

Evgeniy Malkin[1(✉)], Evgeniy Kazakov[1], Nina Cherneva[1], Janos Lichtenberger[2,3], Dmitriy Sannikov[1], and David Koronczay[3]

[1] Laboratory of Electromagnetic Propogation, Institute of Cosmophysical Research and Radio Wave Propagation FEB RAS, Mirnaya, Paratunka 684034, Kamchatka, Russia
{malkin,kazakov,nina}@ikir.ru

[2] Department of Geophysics and Space Sciences, Eotvos University, Parmany Peter setany 1/A, Budapest 1117, Hungary

[3] HUN-REN-ELTE Space Research Group, Parmany Peter setany 1/A, Budapest 1117, Hungary
lityi@sas.elte.hu, david.koronczay@ttk.elte.hu

Abstract. Based on the data of the global Automatic Whistler Detector and Analyzer Network (AWDANet), World Wide Lightning Location Network (WWLLN) and the VLF direction finder of IKIR FEB RAS, statistical analysis was carried out to detect whistler generation sources. Stroke pairs with repetition interval of less than 100 ms and the distance between them of less them 40 km were selected from the WWLLN data base. The obtained time series showed high correlation degree with the time series of the recorded whistlers in AWDANet data.

Keywords: lighting stroke · whistler · sprite · ionosphere

1 Introduction

Lightning strokes are the main natural source of pulse radiation in the very low frequency range (30 Hz–30 kHz). These radio pulses propagate at long distances with low attenuation in the Earth-ionosphere waveguide and have the established name, atmospherics. Atmospherics can also propagate into the ionosphere and magnetosphere. According to the paper [1], electromagnetic radiation undergoes frequency dispersion in the result of propagation through the ionospheric and magnetospheric channel. At first, signal high-frequency components arrive, then lower frequencies arrive with increasing delay. Such signals are called whistlers (whistling atmospherics).

It is known that lightning strokes and whistlers have casual relationship. Whistlers, recorded at a certain point on the Earth surface, are initiated by

E. Kazakov, N. Cherneva, J. Lichtenberger, D. Sannikov and D. Koronczay—These authors contributed equally to this work.

A. Dmitriev et al. (Eds.): STRPEP 2023, SPEES, pp. 54–61, 2023.
https://doi.org/10.1007/978-3-031-50248-4_7

lightning strokes in magnetically conjugated regions [2]. About 20% of cloud-ground (CG) positive strokes cause sprites [3].

Sprites, as a class of separate lighting strokes, were discovered not long ago. First documentary confirmations of their existence were received with highly-sensitive video cameras in 1989 within the framework of "SKYFLASH" program realized by Minnesota University.

Horizontal dimensions of a sprite lie in the range from several kilometers to 50 km. Vertical dimensions reach the values of 90 km with the intensive illumination region at the height of about 60–70 km. As long as oxygen concentration at this height is insignificant, nitrogen is responsible for the maximum radiation intensity in the spectrum that determines red color of the stroke in a sprite. Photometric measurements show that sprite illumination duration is about 3 ms. Mainly, it occurs after an initiating stroke of a CG positive lightning [3].

Apparently, all cloud-ionosphere (CI) strokes require a CG stroke trigger, and they occur about 3 ms after this trigger. With few exceptions, the trigger CGs are identified by the National Lightning Detection Network (NLDN) as positive strokes, which may be situated up to 50 km lateral distance from the CI central region. Triggers and CIs are sometimes multiple, several occurring within 1 s.

This fact was confirmed by many papers, for example [4].

Sprite triggering CG stroke occurs at the ionosphere lower boundary, that is why the possibility to initiate a whistler should be higher than that of other types of lightning strokes.

2 Atmospherics Initiated by Sprites

The delay times between the initiating strokes and sprites triggering CG strokes reaches 70 ms and the distances are up to 40 km. It was mentioned in the paper [3] that in 96% of cases, the second stroke in a pair of CG positive strokes is a sprite. Based on these statements, paired strokes, corresponding to the described criteria, are distinguished in WWLLN data. We can assert that these strokes have initiating strokes followed by sprites triggering CG strokes. To test the truth of the latter statement, paired strokes, detected in WWLLN data base, and the pulses, recorded by the VLF direction finder of IKIR FEB RAS, were analyzed [5].

The criterion for selection of paired events in WWLLN data were the remoteness of lightning strokes from the VLF direction finder location. The distance should be less than 1000 km. The time interval, for which the comparison was carried out, was six years, from 01 January 2015 to 31 December 2021. We selected 27511 paired strokes over the whole period of comparison, 82% (22519) from them were positive stroke pairs (+CG, +CG). An example of such pairs, recorded at the VLF direction finder electric component output, is illustrated in Fig. 1.

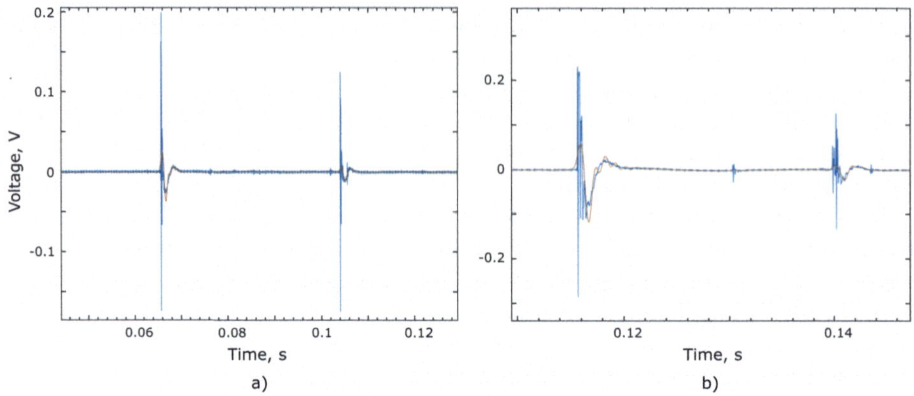

Fig. 1. Examples of paired strokes selected from WWLLN data base and recorded by the VLF direction finder

It was interesting that paired negative strokes (–CG, –CG) were observed only in 7% of cases and negative stroke - positive stroke (–CG, +CG) pairs were observed in 9%. Positive stroke - negative stroke (+CG, –CG) were not observed. In about 2% of cases, no atmospheric strokes or only one was observed at the VLF direction finder output. We can state from the analysis that paired strokes, selected from the WWLLN data base and corresponding to the criteria of time delay up to 70 ms, spaced by not more than 40 km, are really pairs of cloud–ground (+CG) positive strokes and a sprite triggering CG stroke.

3 Whistlers and Sprites

To show the presence of statistical relation between whistlers and sprites triggering CG strokes, we used the data obtained at Karymshina station ($53.0° N$, $158.7° E$), for the period from 14 July 2015 till 27 September 2021 and at Dunedin station ($45.7° S$, $170.5° E$) for the period from 01 January 2017 till 31 December 2020. The stations are the nodes of AWDANet. Lightning activity parameters were taken from the data base of the WWLLN network. At Karymshina station, $1.3 \cdot 10^6$ whistlers were recorded during the period under analysis. The number of lightning strokes, occurred during that period, was $1,28 \cdot 10^9$, $2,8 \cdot 10^8$ from them corresponded to the criteria of paired strokes. At Dunedin station, $2.8 \cdot 10^6$ whistlers were recorded. The total number of strokes for that period was $8.4 \cdot 10^8$, the number of paired strokes was $1.9 \cdot 10^8$.

The relation between whistlers and lightning strokes can be illustrated by constructing delay time distributions. This relation is calculated as a difference between whistler recording times and the times of lightning stroke occurrences. If events are not associated with each other, the distribution should be uniform. Figure 2a shows the obtained distribution, from which the maximum of $\sim +1.1 \pm 0.2$ s is clear. It indicated the fact that whistlers are really associated with atmospherics of casual lightnings. Having marked all the locations of recorded lightnings, included into the distribution, on the map, we obtain that the lightnings are spread all over the Earth, but the certain structure of their surface density is evident (Fig. 3).

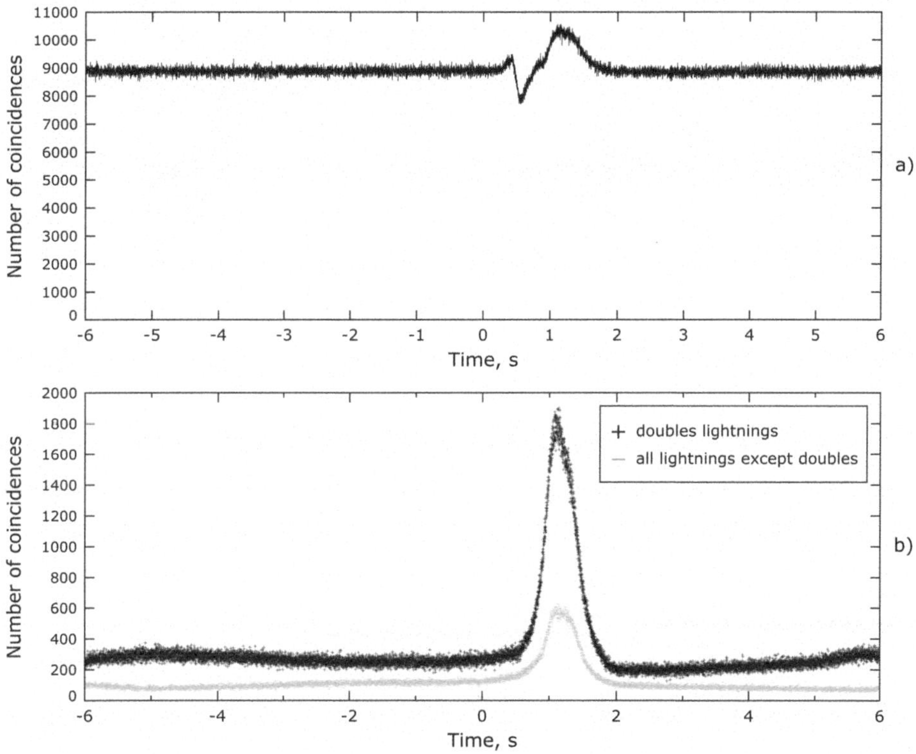

Fig. 2. Distribution of the number of coincidences in delay times between whistler recording times and lightning stroke occurrences times. The series is centered with respect to the whistler recording times (a). Distribution of delay times: + between whistler recording times and recording times of the sprites triggering CG strokes; – between whistler recording times and the recording times of all the strokes but sprites triggering CG strokes (b).

Based on the evident structure of lightning stroke distribution density, shown in Fig. 3, we can join all the locations of recorded lightnings into 11 clearly distinguished clusters. Using the k-means clustering method (the most appropriate simulation with the least sum of distances is chosen), the region centers of whistler possible sources are obtained. In order to define whistler sources unambiguously, we estimated the delay time distributions for separate regions illustrated in Fig. 4. Graph numbers coincide with the region center numbers in Fig. 3.

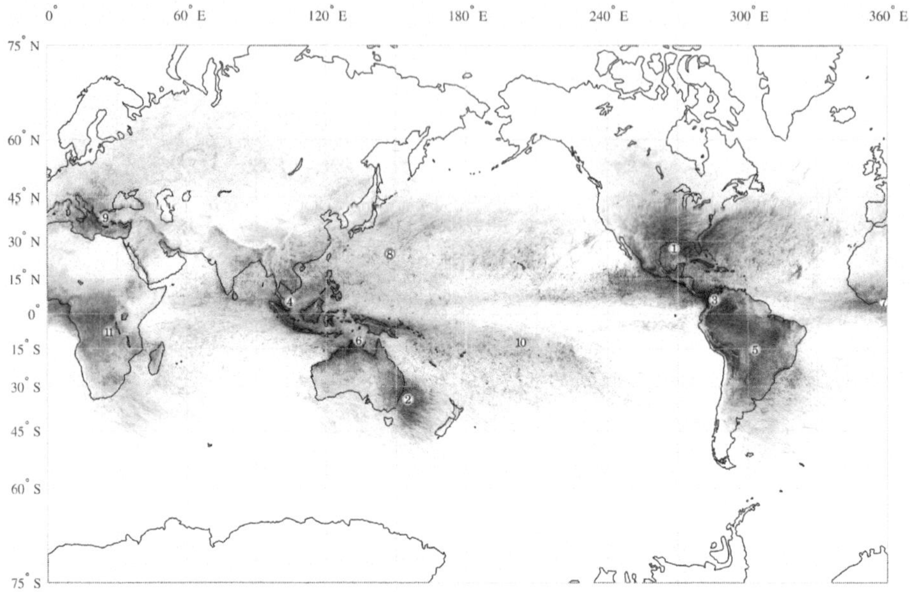

Fig. 3. Map of lightning stroke density distribution, the numbers denote the centers of the clusters entering the maximum of the distribution illustrated in Fig. 2a.

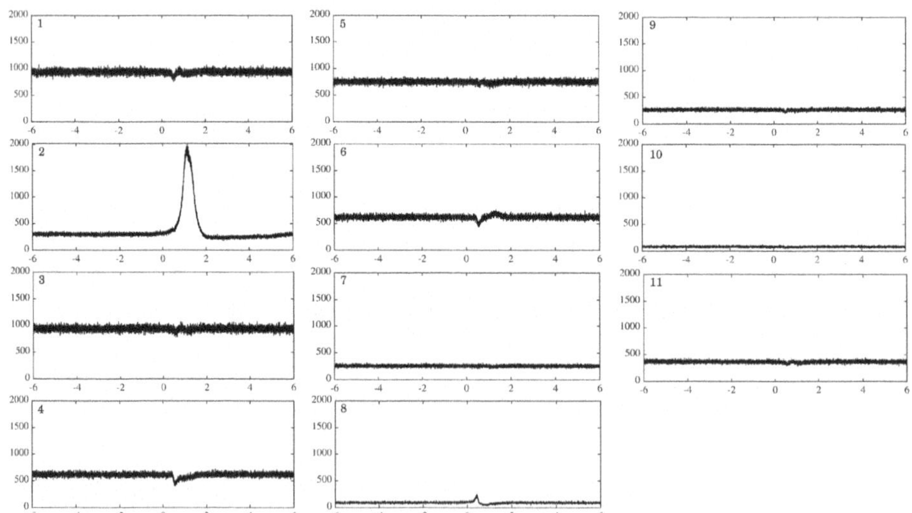

Fig. 4. Distribution of the number of coincidences in delay times between the whistler recording times and the times of lightning strokes occurred in the regions located at the distances of not more than 1400 km from its centers. Graph numbers correspond to the region center numbers shown in Fig. 3.

It is clear from Fig. 4 that the clear maximum in the distribution is maintained only for region 2, which coincides with the magnetically conjugated region for Karymshina station. Thus, we repeated the results published in the paper [2]. All further samplings of initiating lightning strokes are limited by the magnetically conjugated region.

As for the mechanism of whistler formation, it is generally accepted that a powerful lightning stroke generates a wide-band pulse capable of penetrating through the ionosphere into the magnetosphere where it turns to a whistler when dispersing as the result of its propagation along the magnetospheric channel [1].

Whistler excitation by an electromagnetic wave is one of the evident mechanisms. The second one consists in the direct change of electron density by the basis of a magnetospheric channel. This mechanism can be realized as the result of lightning stroke impact at the basis of a magnetospheric channel. Sprites, being strokes, occurring between the upper edge of a lightning cloud and the ionosphere lower boundary, neutralize negative extra charge of a lightning cloud [6]. In this case, an atmospheric generated by a sprite has positive VLF pulse (Fig. 1).

As it was shown above, there is a clear sign to determine sprites in WWLLN data. Based on that, we compared two distributions of delay times:

- between whistler recording times and recording times of the sprites triggering CG strokes (second stroke in a pair);
- between whistler recording times and the recording times of all the strokes but sprites triggering CG strokes.

The both distributions are illustrated in Fig. 2b.

If we compare the delay time distributions obtained by taking into account all the strokes (Fig. 4.2) and obtained by taking into account only sprites triggering CG strokes (Fig. 2b), occurred in the magnetically conjugated region, we obtain that the difference is less than 10% in absolute magnitude. The second distribution, shown in Fig. 2b, was obtained not taking into the account sprites triggering CG strokes (second strokes in pairs). That confirms the assumption that just second strokes of paired lightnings are the causes of whistler occurrences. Thus, based on the papers [2,3,7], it is possible to state with high confidence that sprites triggering CG strokes are whistler sources.

We made analogous estimations for AWDANet data, obtained at Dunedin observation site ($45.7°S$, $170.5°E$). The results are shown in Fig. 5 where the time delay distribution more clearly demonstrates the statement that just second strokes of paired lightnings are the whistler sources [8].

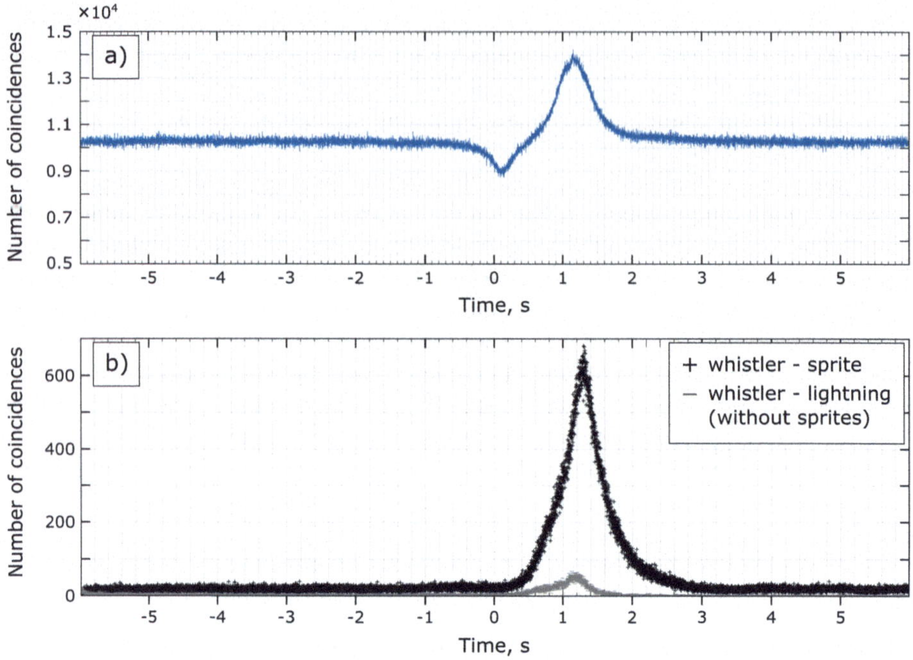

Fig. 5. a) Distribution of delay times between whistler recording times and lightning stroke occurrences times. The series is centered with respect to whistler recording times (Dunedin). b) Distribution of delay times: + between whistler recording times and recording times of the sprites triggering CG strokes; – between whistler recording times and the recording times of all the strokes but sprites (Dunedin).

4 Conclusions

Statistical analysis was carried out to detect the sources of whistler generation based on the data of the global Automatic Whistler Detector and Analyzer Network (AWDANet), World Wide Lightning Location Network (WWLLN) and the VLF direction finder of IKIR FEB RAS. As the result of the work we obtained:

- distribution of the number of coincidences in the delay times between whistler recording times and the lightning stroke occurrence times for Karymshina and Dunedin stations;
- distribution of the number of coincidences in the delay times between whistler recording times and recording times of the sprites triggering CG strokes for Karymshina and Dunadin stations;
- statistical data describing paired strokes recorded by VLF direction finder.

On the basis of the papers [6,7] and analysis of paired strokes, recorded by the VLF direction finder, we can state with high confidence degree that paired strokes are the positive cloud-ground strokes and ionosphere-cloud strokes. In this case, the second stroke is formed by a noncompensated negative cloud charge, which remained after +CG, thus the second stroke in this pair can be considered as sprite. Comparative analysis of the obtained distributions was carried out for all the strokes occurred in the magnetically conjugated region as well as taking into account only the strokes associated with sprites triggering CG strokes. The analysis showed that there is statistical relation between the two phenomena, in particular, the triggering lightning strokes of sprites accompany whistlers.

Acknowledgments. The work was carried out within the framework of realization of the State Task № AAAA-A21-121011290003-0.

References

1. Storey, L.R.O.: An investigation of whistling atmospherics. Philos. Trans. Roy. Soc. London Ser. A Math. Phys. Sci. **246**(908), 113–141 (1953)
2. Koronczay, D., et al.: The source regions of whistlers. J. Geophys. Res.: SpacePhys. **124**, 5082–5096 (2019). https://doi.org/10.1029/2019JA026559
3. Lyons, W.A.: Sprite observations above the U.S. high plains in relation to their parent thunderstorm systems. Geophys. Res. **101**, 29641–29652 (1996)
4. Rodger, C.J.: Red sprites, upward lightning, and VLF perturbations. Rev. Geophys. **37**(3), 317–336 (1999)
5. Druzhin, G.I., Pukhov, V.M., Sannikov, D.V., Malkin, E.I.: VLF-lightning direction finder. Vest. KRAUNC Fiziko-Mat. Nauki **27**(2), 95–104 (2019)
6. Winckler, J.R.: The cloud-ionosphere discharge: a newly observed thunderstorm phenomenon. Proc. Nat. Acad. Sci. **94**(20), 10512–10519 (1997)
7. Reising, S.C., Inan, U.S., Bell, T.F.: ELF sferic energy as a proxy indicator for sprite occurrence. Geophys. Res. Lett. **26**(7), 987–990 (1999)
8. Malkin, E.I., Kazakov, E.A., Sannikov, D.V., Cherneva, N.V., Marchenko, L.S., Drugin, G.I.: Statistical relationship between whistlers and sprites according to AWDANet and WWLLN data. Vest. KRAUNC Fiziko-Mat. Nauki **41**(4), 178–190 (2022). https://doi.org/10.26117/2079-6641-2022-41-4-178-190

Analysis of Electromagnetic Radiation During Shiveluch and Bezymyanniy Volcano Eruptions from 2017 to 2023

Evgeniy Malkin[1(✉)], Veronika Cherneva[2], Dmitriy Makhlai[2], Nina Cherneva[1], Dmitriy Sannikov[1], Gennadiy Druzhin[1], Rinat Akbashev[3], and Robert Holzworth[4]

[1] Laboratory of Electromagnetic Propogation, Institute of Cosmophysical Research and Radio Wave Propagation FEB RAS, Mirnaya, Paratunka 684034, Kamchatka Region, Russia
{malkin,nina,drug}@ikir.ru

[2] Faculty of Software Engineering and Computer Engineering, ITMO University, Kronverksky Pr., St. Petersburg 197101, Russia

[3] Investigations on Lithosphere Outgassing Laboratory, Kamchatka Branch of the Geophysical Service of Russian Academy of Sciences, Piip Boulevard, Petropavlovsk-Kamchatsky 683006, Kamchatka, Russia
arr@emsd.ru

[4] Earth and Space Sciences, University of Washington, 1707 NE Grant Lane, Seattle, WA 98195, USA
bobholz@uw.edu

Abstract. The paper presents the analysis of monitoring of natural pulse electromagnetic radiation (PER) in VLF range (3–30 kHz) accompanying dirty thunderstorms, which occurred during Shiveluch and Bezymianniy volcano eruptions. Lightning activity in ash-gas clouds during a volcano eruption is traced by radio technical monitoring devices located in the region of Karymshina river. Satellite data confirm the information on the motion of eruptive clouds accompanied by successive lightning strokes tracing the trajectory of their propagation. More than 40 eruptions of Shiveluch volcano were analyzed during the investigations over the period of data collection from 2017 to 2023. More than 75% from the eruptions caused lightning generation whereas during the Bezymianniy volcano eruptions over the indicated period, all the events were accompanied by PER increases typical for dirty thunderstorms. It was demonstrated that two-stage scenario of event development was characteristic for all the cases. The first stage of PER increase is accompanied by eruptive column formation and it depends on burst power and ash ejection height. The second stage depends on eruption power and eruptive cloud interaction during the process of its propagation with meteorological clouds (that can be indicated by the increase of the number of cloud-cloud strokes). The mechanism of thunderstorm formation

V. Cherneva, D. Makhlai, N. Cherneva, D. Sannikov, G. Druzhin, R. Akbashev, R. Holzworth—These authors contributed equally to this work.

A. Dmitriev et al. (Eds.): STRPEP 2023, SPEES, pp. 62–70, 2023.
https://doi.org/10.1007/978-3-031-50248-4_8

in eruptive clouds, explaining their two-stage regime, was shown. We also give preliminary description of Bezymianny and Shiveluch volcanoes events occurred in April 2023 and accompanied by ash ejections of the height up to 20 and 30 km, respectively.

Keywords: Lightning discharge · pulse electromagnetic radiation · remote methods · satellite images

1 Introduction

Investigation of electrostatic electricity in eruptive clouds by remote methods is a promising direction for obtaining the information on eruptive cloud development and the processes of their electrification. Charge separation, causing lightning occurrences, is observed not only in meteorological, rain or hailstorm clouds but also in eruptive clouds (EC) during volcano eruptions with high lightning activity ("dirty thunderstorms") [1–7]. Monitoring of EC propagation from "dirty thunderstorms", carried out by different methods, makes it possible to take preventive actions to decrease consequences. Reliable remote methods of observation of volcano explosive activity are seismic, infrasound methods and monitoring of lightning activity. The first one provides the information on explosive process intensity and duration. The second one indicates the degree of its nonstationarity and the intensity of ash ejection in the atmosphere. The third method traces ash cloud motion based on lighting stroke data. In the paper, we focus on the first and the third methods and apply satellite data for analysis reliability.

Shiveluch (56°47′ N, 157°56′ E) and Bezymianniy (55.98° N, 160.59° E) volcanoes of the Klyuchevskaya Volcano Group have been the most active volcanos from 2016 to 2023.

2 Instrumentation and Observation Method

A direction finder of very low frequencies (VLF direction finder, $f \approx$0.5–60 kHz) is used to record pulse electromagnetic radiation (PER) at the Institute of Cosmophysical Research and Radio Wave Propagation (IKIR) FEB RAS (Karymshina). The hardware-software complex for PER recording is described in the paper [8].

Besides the VLF direction finder, installed at Karymshina River observation site, IKIR FEB RAS records PER from volcanic lightning with the time accuracy up to several microseconds by a segment of the World Wide Lightning Location Network (WWLLN, http://wwlln.net), located in Paratunka [9]. It provides excellent results for the monitoring of eruptive cloud propagation in cases of optimal arrangement of the network sites. The network data were used to check the presence of lightning activity in the region [10]. The VLF direction finder of IKIR FEB RAS, based on one-point direction finding method, is used in this work as the main source of information. This method consists in the determination of the azimuth of transverse electromagnetic (TEM) wave arrival from fields components recorded separately. Base station is an antenna system including two mutually perpendicular frame magnetic antennas (oriented in north-southern and east-western directions) and one pole electric antenna, a system for signal transmission and processing. The frame 100-coil antennas have the dimension of 4 by 8 m, and the six-meter pole antenna is installed on a building roof. Electromagnetic pulses in the range of 0.5–60 kHz, exceeding the threshold level of 1 V/m, are recorded. The VLF direction finder allows us to record electromagnetic radiation from regional lightning and to locate the azimuth of its arrival.

3 Analysis of Electromagnetic Radiation During Shiveluch and Bezymianniy Volcano Eruptions

A large data volume has been collected for the period from 2016 up to the present time. The data show the efficiency of remote observation methods during volcano eruptions. Preliminary review of the Shiveluch and Bezymianniy volcano eruptions, accompanied by PER, for that period was presented in the paper [1].

3.1 Examples of Records of Lightning Strokes Accompanying Explosive Eruptions of Kamchatka Volcanos Until 2022

Based on the data of Kamchatka Bruch of the Federal Research Center "Geophysical Survey, Russian Academy of Sciences" (KB GS RAS), http://www.emsd.ru/~ssl/monitoring/main.htm, a catalogue of the strongest events on Shiveluch and Bezymianniy volcanos was composed for the period of 2016–2022. Sixty-nine eruptions of Shiveluch volcano and thirteen eruptions of Bezymianniy volcano were selected, 75% from which caused the formation of "dirty thunderstorms", radiating electromagnetic pulses recorded by the direction finder.

For example, during the eruption of Shiveluch volcano at 22:31 on 16 December 2016, based on KB GS RAS data, the eruptive cloud height, estimated from seismic signal intensity, was 5.6 km.

In HIMAWARI–8 satellite images (http://rammb.cira.colostate.edu/), the eruptive cloud moved under wind effect with the azimuth of ~75° and the velocity of ~17 m/s that corresponded to the wind azimuth and velocity in the height interval of 6.5–8.0 km according to atmosphere stratification [1].

Based on VLF direction finder data, eruptive cloud formation was accompanied by PER from the azimuth to Shiveluch volcano of 25,6° ± 10°.

The most interesting event of Shiveluch volcano eruption, from the point of view of PER monitoring, occurred at 16:26 on 14 June 2017, when EC at the height of ~12 km moved near the KLY and KZY sites (http://www.emsd.ru/~ssl/monitoring/main.htm).

It is clear from the HIMAWARI-8 satellite images, that 34 min after the eruption beginning, almost round EC with the diameter of ~70 km was formed at the height of 9 km.

According to wind stratification of balloon sounding data, the EC began to move to Klyuchi settlement with the velocity of 12 m/s. EC formation and propagation was recorded by a video camera at KLY. Almost one hour later, the

EC covered the KLY settlement where 100 g/m^2 of ash fell. Then it continued to move in the direction of Klyuchevskoy volcano. At 21:33 the eruptive cloud reached KZY, where very weak fall of finely dispersed ash was observed [1,4].

During the eruptions of Bezymianniy volcano on 20 January and 15 March 2019, the VLF direction finder recorded PER from the azimuth of 23.6° ± 0.4° to the volcano. The pulse radiation intensity of ~20 pule/min was close to the strongest events of Shiveluch volcano.

3.2 Mechanism of Lightning Formation in Eruptive Clouds

Two-stage development of PER is characteristic for all "dirty thunderstorms" recorded by the VLF direction finder of IKIR FEB RAS. It was earlier detected for the Augustine and Redoubt volcanos eruptions in Alaska [11]. The mechanism, explaining the two-stage process on the example of Kamchatka explosive eruptions, is described in the papers [1–7].

In order to illustrate the two-stage mechanism of thunderstorm formation in eruptive clouds, we consider three events of Shiveluch and Bezymianniy volcanoes.

Table 1 shows conditional characteristics of the two stages for the three most intensive "dirty thunderstorms", occurred during Shiveluch volcano eruption.

Table 1. Parameters of the three most vivid events during the Shiveluch volcano eruptions.

date	time	time 1 phase, mim	max counting rate, pulse/min	pulse number pulse	time 2 phase, min	max counting rate, pulse/min	pulse number pulse	WWLLN
11.05.2017	18:24	4	16	47	28	14	183	4
23.07.2017	17:43	4	3	10	26	18	205	0
21.10.2019	12:17	5	18	45	–	–	–	0

Table 2 illustrates the parameters of three events during Bezymianniy volcano explosions. The differential peculiarity of the Bezymianniy volcano eruption on 15 March 2022 was the burst at 12:50 that was reflected in the dynamics of "dirty thunderstorm" development in the form of two-stage process with clearly defined fist (I) and (II) stages. Owing to the atmosphere inhomogeneous stratification at the time of the eruption, the eruptive cloud became stretched with observed azimuthal angle size of 22.5° relative to the Karymshina River site. Analysis of

Table 2. Parameters of the three most vivid events during the Bezymianniy volcano eruptions.

date	time	time 1 phase, mim	max counting rate, pulse/min	pulse number pulse	time 2 phase, min	max counting rate, pulse/min	pulse number pulse	WWLLN
20.12.2017	03:39	3	5	9	54	37	781	7
15.03.2019	12:17	–	–	–	40	83	1138	11
15.03.2022	18:24	4	11	32	50	21	159	4

VLF direction finder data confirmed the presence of a thunderstorm source in the azimuthal angles from 8.2° to 27°.

According to the paper [12], bipolar particle charging depend on their size during triboelectrization. In this case, negative charge is characteristic for small particles and positive charge is typical for larger ones. During the gas-ash injection, due to the hydrodynamic resistance and sedimentation, fractional separation of ash particles takes place. This process causes clustering of larger and, consequently, positively charged particles at the bottom of an eruptive column. In its turn, that should initiate positive polarity lightning occurrences. After EC development, due to great amount of water (up to 7% of ejected material mass and up to 97% of gas weight content), classical mechanisms of charge separation, typical for meteorological clouds, should work. The data, obtained during Shiveluch volcano eruption, is illustrated in Table 1.

It is clear that the majority of the first phase strokes have positive polarity, that means that the initial dipole moment of a radiating dipole is directed upward (39 cases for the event on 11 May, 2017, 7 cases on 23 July, 2017 and 37 cases on 21 October, 2019). During the second stage, the greater part of charges have negative polarity, 98% from the total number of recorded charges in average. That, in its turn, indicates opposite direction of the initial dipole moment and, as a sequence, inversion of the EC electric field.

Azimuthal distribution of the recorded lightning events and their polarity during Bezymianniy volcano eruption on 15 March 2022, demonstrates clearly the general pattern of filed orientation.

Thus, the volcanic thunderstorm first phase, lasting for 5–7 min, accompanies eruptive column formation. High density of positive charges at the bottom part of the column is determined by fractional separation of ash particles (large+, small–). The pause between the phases is determined by charge separation in the eruptive cloud upper part. Separation into phases of a volcanic thunderstorm was observed in 43 cases that is 75% of the analyzed events. According to the paper [1], aerosols in the floating zone carry the positive charge and, as a sequence, generate a field opposite in the direction to the field in the eruptive column. During a long eruption, aerosol number increases. Thus, the volumetric charge value rises that results in the generated field inversion. The atmosphere naturally has over 1000 ions per cm^3 at these altitudes, with near equal numbers of positive and negative ions, maintained by cosmic ray flux, however the trigger for dirty thunderstorm formation is the charged tefra that was illustrated in the paper [2].

3.3 Remote Methods for Observations of Shiveluch and Bezymianniy Volcano Eruptions in April 2023

The two volcanoes became active in April 2023. Based on the data of volcano video observation of the KB GS RAS, a series of surface events, lasting for 3–6 min, was occurring on Bezymianniy volcano from 00:11 till 5:00 UTC on 7 April 2023. They were accompanied by steam-gas emission with ash content and ash ejections to the height up to 3000–4500 m above the sea level. According to the atmosphere stratification, the ash-gas plume propagated to the south-east.

At 5:34 UTC a paroxysmal explosive eruption began. The eruptive cloud raised to the maximum height of 14 km above the sea level at 5:42 UTC. Half an hour later, from 06:00 untill 24:00, a series of surface events followed. According to the HIMAWARI-9 satellite data, they were accompanied by gas-ash ejections to the height of 5000 m $\pm$ 660 m above the sea level.

Based on the data of monitoring remote methods, sudden increase of seismic activity was observed at BZM seismic station. The VLF complex recorded PER increase from the direction to Bezymianniy volcano (Fig. 1). According to satellite data, the maximum height reached 20000 m (Fig. 2a).

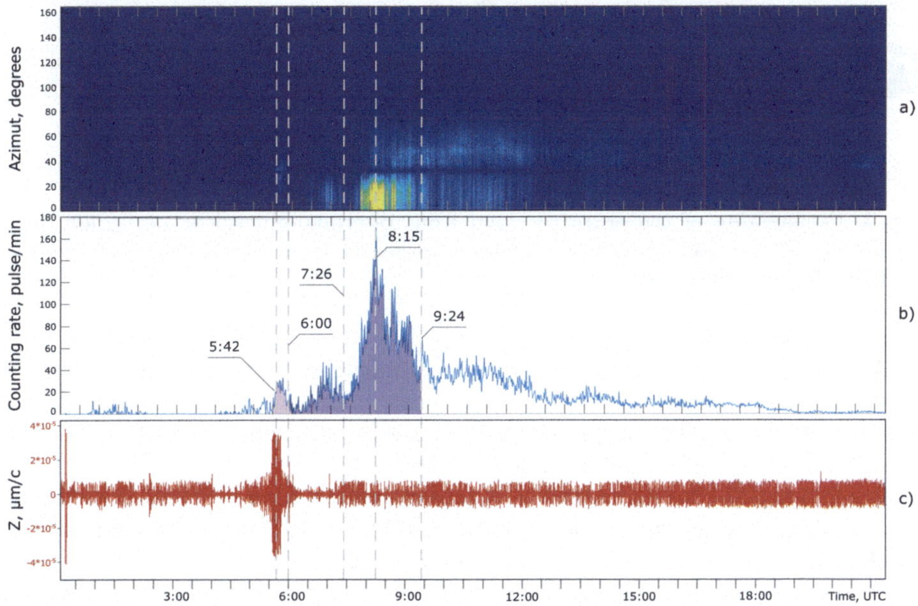

Fig. 1. Azimuthal distribution of PER direction finding in the range of 0°–60° (a). The counting rate of PER, azimuthally arrived from Bezymianniy volcano direction, during the eruptions (b). Record of explosive earthquakes at BZM seismic station during Bezymianniy volcano eruption on 7 April 2023 (c)

The two-stage process of the volcanic thunderstorm development was also observed.

Shiveluch volcano became active on 10 April. Based on HIMAWARI-9 satellite data, an aerosol cloud appeared at 10:10 UTC, raised to the height of 6 km above the sea level and propagated to the north-north-west. From 12:10 UTC on 10 April, ash ejection from the volcano in the form of a steam-gas plume began to move in the direction of the north-west (Fig. 2b).

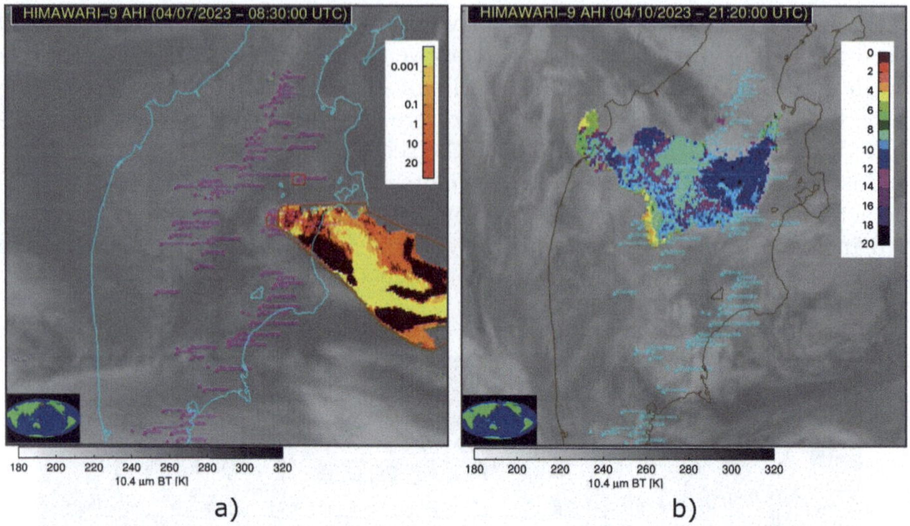

Fig. 2. Propagation of the eruptive cloud from Bezymianniy volcano eruption on 7 April 2023 (a) and from Shiveluch volcano eruption on 10 April 2023 (b) based on the HIMAWARI-9 satellite image data

At 13:07 UTC on 10 April, the explosive eruption of Shiveluch volcano began. The lava dome in its crater was significantly destructed. An ash cloud was observed at 13:20 UTC in the HIMAWARI-9 satellite image. It moved in the direction of west-north-west from the volcano.

Based on HIMAWARI-9 satellite data, the eruptive cloud raised to the height of about 10 km above the sea level at 13:20 UTC on 10 April. According to the satellite data, the maximum height reached 30000 m (Fig. 2).

Scientists from KVERT gave the volcano observatory notice for aviation (VONA) for Shiveluch volcano changing the aviation color code from orange to red at 19:29 UTC on 10 April. The explosive eruption lasted continuously for 3 days from 10 to 13 April.

A powerful cyclone, covering the whole Kamchatka peninsula, stretched the EC to the south-west, then to the north and directed it to the east from the volcano.

The dynamics of ash and aerosol cloud development for this eruption is illustrated on animations, developed on the basis of HIMAWARI-9 satellite data for the period from 08:00 UTC on 10 April to 07:50 UTC on 14 April. The eruption is described in detail in the paper [13].

Based on the remote observations by WWLLN, the first notification on the explosive eruption arrived at 13:06 UTC.

In Fig. 3, according to the VLF direction finder data, lightning activity beginning coincides with the HIMAWARI-9 satellite data. The first stage was traced from 13:10 until 13:35. The second stage of the two-stage lightning process was

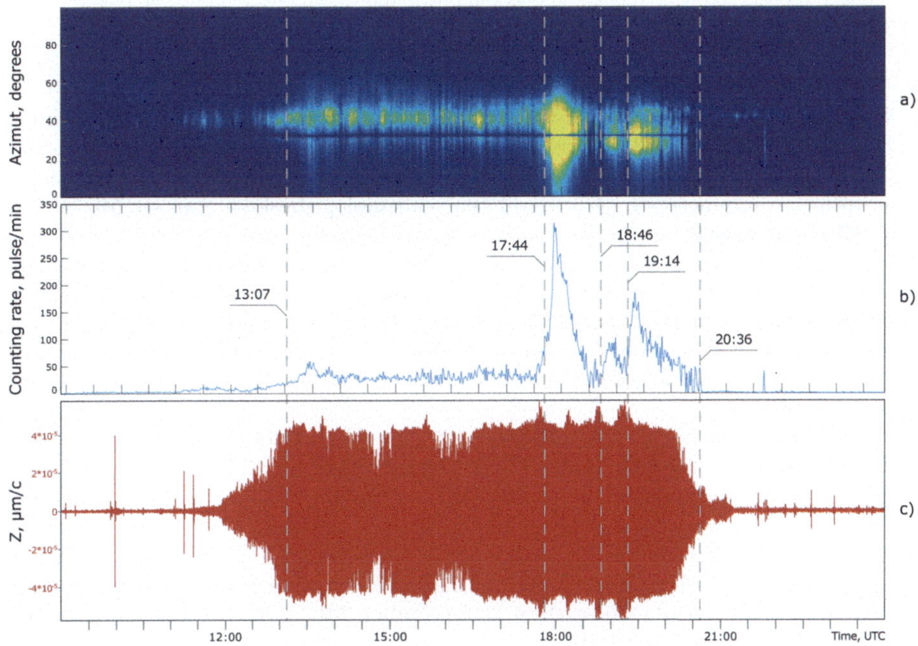

Fig. 3. Azimuthal distribution of PER direction finding in the range of 0°–60° (a). The counting rate of PER, azimuthally arrived from Shiveluch volcano direction, during the eruptions (b). Record of explosive earthquakes at BDR seismic station during Shiveluch volcano eruption on 10 April 2023 (c)

observed from 13:50 until 20:35. Sudden increase of lightning activity corresponds to the explosions at 17:44, 18:46, 19:14 recorded by the BDR seismic station. The total number of strokes during the Shiveluch volcano eruption reached 26101.

4 Conclusions

The Kamchatka volcano group is located near international aviation routes. Due to that fact, eruptions are a serious threat for air communication safety. In order to decrease the risks, online systems for eruption detection are required. Remote observation methods, such as tracing of accompanying thunderstorm sources, make it possible to receive the information at the stages of formation and fragmentation of eruptive clouds when electrification processes develop the most intensively. Lightning strokes give the information on eruptive plume formation and trace its motion during the eruption initial period.

Acknowledgments. The work was carried out within the framework of realization of the State Task № AAAA-A21-121011290003-0.

References

1. Malkin, E.I., Cherneva, V.I., Makhlai, D.O., Cherneva, N.V., Akbashev, R.R., Sannikov, D.V.: Remote methods for observations of Shiveluch and Bezymyanniy volcano eruptions (2023). https://doi.org/10.26117/2079-6641-2023-43-2-141-165
2. Rulenko, O.P., Tokarev, P.I.: Atmospheric-electric effects of large fissure tolbachik eruption in july-october 1975. Byulleten' vulkanologicheskikh stantsiy **56**, 96–102 (1979). (in Russian)
3. James, M.R., Lane, S.J., Gilbert, J.S.: Volcanic plume monitoring using atmospheric electrical potential gradients. J. Geol. Soc. Lond. **155**, 587–590 (1998). https://doi.org/10.1144/gsjgs.155.4.0587
4. Firstov, P.P., et al.: Registration of Atmospheric-Electric Effects from Volcanic Clouds on the Kamchatka Peninsula (Russia) (2020). https://doi.org/10.3390/atmos11060634
5. Malkin, E.I., Cherneva, N.V., Firstov, P.P., Druzhin, G.I., Sannikov, D.V.: Dirty thunderstorms caused by volcano explosive eruptions in Kamchatka by the data of electromagnetic radiation (2021). https://doi.org/10.1088/1755-1315/946/1/012015
6. Firstov, P.P., Akbashev, R.R., Malkin, E.I., Cherneva, N.V., Druzhin, G.I.: Atmospheric electrical effects during a strong explosive eruption of Bezymyanniy volcano (Kamchatka Peninsula, Russia) on 20 December 2017 (2021). https://doi.org/10.1088/1755-1315/840/1/012020
7. Malkin, E.I., Firstov, P.P., Cherneva, N.V., Druzhin, G.I.: Lightning Activity of Eruptive Clouds from Shiveluch Volcano (Kamchatka, Russia). Paper presented at the 13th International Conference and School Problems of Geocosmos, 24–27 March 2021 (2022). https://doi.org/10.1007/978-3-030-91467-7_34
8. Druzhin, G.I., Pukhov, V.M., Sannikov, D.V., Malkin, E.I.: VLF–direction finder to investigate natural radio radiations. Vestnik KRAUNTs. Fiziko-matematicheskie nauki **27**(2), 95–104 (2019). https://doi.org/10.26117/2079-6641-2019-27-2-95-104
9. Dowden, R.L., Brundell, J.B., Rodger, C.J.: VLF lightning location by time of group arrival (toga) at multiple sites. J. Atmosph. Solar-Terr. Phys. **64**(7), 817–830 (2002). https://doi.org/10.1016/S1364-6826(02)00085-8
10. Abarca, S.F., Corbosiero, K.L., Galarneau, T.J.J.: An evaluation of the worldwide lightning location network (wwlln) using the national lightning detection network (nldn) as ground truth. J. Geophys. Res. **115**(D18206) (2010). https://doi.org/10.1029/2009JD013411
11. Behnke, S.A., et al.: Observations of volcanic lightning during the 2009 eruption of redoubt volcano. J. Volcanol. Geotherm. Res. **259**, 214–234 (2013). https://doi.org/10.1016/j.jvolgeores.2011.12.010
12. Méndez, H.J., Cimarelli, C., Cigala, V., Kueppers, D.J.: Charge injection into the atmosphere by explosive volcanic eruptions. Earth Planet. Sci. Lett. **574**, 117162 (2021). https://doi.org/10.1016/j.epsl.2021.117162
13. Girina, O.A., et al.: Analysis of the development of the paroxysmal eruption of Sheveluch volcano on 10–13 April 2023, based on data from various satellite systems (2023). https://doi.org/10.21046/2070-7401-2023-20-2-283-291

Space Project "Modulation", a New Approach to Studying the Fluxes of Galactic Cosmic Rays in the Field of Solar Modulation Energies

V. V. Kalegaev, D. E. Karmanov, A. A. Kurganov, A. D. Panov, M. V. Podzolko, D. M. Podorozhny(✉), and A. N. Turundaevsky

Skobeltsyn Institute of Nuclear Physics, M.V. Lomonosov Moscow State University, Leninskie gory 1(2), Moscow 119991, Russian Federation
{klg,panov}@dec1.sinp.msu.ru, karmanov@silab.sinp.msu.ru, me@sx107.ru, dmpo@bk.ru

Abstract. The article discusses the "MODULATION" project, which is supposed to be implemented aboard the Russian space station ROSS. The purpose of the space mission is to study the solar modulation of galactic cosmic rays (GCR) in the heliosphere. A new approach is presented to obtain accurate long-term measurements of GCR particle fluxes, as well as solar cosmic rays (SCR), with energies from 30 to 1500 MeV/nucleon in the wide charge range $Z = 1 - 26$. The project describes a new method for measuring the energy of cosmic particles, which makes it possible to create a lightweight and compact but precision instrumentation that will allow measuring not only the energy but also the direction of arrival of cosmic rays. The project involves the creation of a data bank of GCR and SCR data for the entire solar activity cycle. Such a bank is necessary to improve numerical models of the fluxes of energetic heliospheric particles.

1 Introduction

One of the main goals of the proposed MODULATION experiment is to study the solar modulation of galactic cosmic rays (GCR) in the heliosphere. The objective of the experiment is to obtain precision measurements of protons and nuclei from He to Fe of GCR in the solar modulation energy range ≈30–600 MeV/nucleon outside the region of the Earth's magnetic field for a long time spanning periods of low and high solar activity. In the inner heliosphere, GCR fluxes with energies of less than a few GeV/nucleon are subject to "Solar modulation" – the solar wind partially "blows" less energetic GCR particles further away from the Sun. In the region of the Earth's orbit, the modulation effect is most pronounced for particles with energies less than 600 MeV/nucleon. At the same time, the GCR fluxes in the modulation energy region depend on solar activity.

Accurate measurements of GCR particle fluxes with energies less than 600 MeV/nucleon are quite difficult because of the significant impact of more

A. Dmitriev et al. (Eds.): STRPEP 2023, SPEES, pp. 71–90, 2023.
https://doi.org/10.1007/978-3-031-50248-4_9

energetic GCR particles penetrating through the spacecraft body and the side walls of the instrument. In a large number of previous experiments measuring the fluxes of energetic heliospheric particles, the "side background" screening was absent or did not work properly. Therefore, long-term precision measurements of the energy spectra of protons and nuclei from He to Fe of GCL in the energy range 30–600 MeV/nucleon with screening of the "side background" by the MODULATION instrumentation will allow us to obtain a more accurate numerical description of the shape of the GCR spectrum in the modulation energy range and its dependence on solar activity.

To date, measurements of proton and nuclei fluxes of GCR in the modulation energy range of tens – hundreds of MeV/nucleon have been obtained mainly with instruments on several U.S. high-orbit satellites and automatic interplanetary stations (AIS):

- GME/IMP-8, 1973–2002, orbit at ≈22–45 Earth radii – protons, He; ≈50–500 MeV/nucleon [1] (some channels are incorrect);
- CRIS/ACE, 1997-ongoing, L1 Sun-Earth libration point – $Z \geq 5$ nuclei; ≈50–500 MeV/nucleon [2];
- EPHIN/SOHO, 1995–2014, L1 Sun-Earth libration point – protons, He <50 MeV/nucleon; protons 0.3–1.5 GeV (modelling) [3];
- COSPIN/Ulysses, 1990–2009, elliptical heliocentric orbit – protons, He 5–320 MeV/nucleon [4].

Also, measurements of proton and He GCR fluxes with energies >80 and >200 MeV/nucleon, respectively, were performed with the PAMELA massive calorimeter/magnetic spectrometer developed jointly by Russian and foreign specialists on the Russian Resurs-DK1 spacecraft in low polar orbit in 2006–2014 [5]. In addition, single short-term measurements were performed on polar stratospheric balloons: BESS series [6, 7] and others.

On the basis of these experimental data, as well as measurements of higher energy GCR fluxes by massive instruments on stratospheric balloons and low-orbit satellites and ground-based neutron monitors, a number of numerical models of GCR have been developed. Two groups of models are available. The first are "physical" models based on the numerical solution of the equations of particle transport in the heliosphere and using data from experimental GCR measurements as boundary parameters, such as the Badhwar& O'Neill [8,9] model and others. The second one is more applied empirical models of R.A. Nymmik [10,11] and N.V. Kuznetsov [12], and foreign ones: CREME-96 [13], DLR [14] and others. At present, in particular, the Skobeltsyn Institute of Nuclear Physics of Moscow State University is developing these models using the latest data from experimental measurements: PAMELA, AMS-02, measurements of Voyager-1, 2 outside the heliosphere, and others.

However, unfortunately, measurements of GCR particle fluxes in the region of modulation energies, performed over the last few decades, have incompleteness and noticeable errors. This is partly due to objective reasons: the difficulty of debugging the operation of the equipment in space and transferring to Earth a large volume of measurement data from instruments on autonomous spacecraft.

Thus, in the part of completeness of measurements of proton and nuclei fluxes of GCR in the region of modulation energies:

- GME measurements of the IMP-8 spacecraft (1973–2002) cover the required energy range of 30–500 MeV/nucleon only for protons and He nuclei; for heavier nuclei (C, N, O, Ne, Mg, Si, Fe) only the range of 8–40 MeV/nucleon have been covered;
- CRIS ACE measurements (1997–present) cover only nuclei with $Z \geq 5$;
- proton fluxes of GCR with energies 0.3–1.5 GeV/nucleon from the SOHO EPHIN instrument – actually represent annual averages calculated from modelling of lateral particle passages in the instrument using the GEANT4 transport code;
- measurements from the PAMELA instrument of the Resurs-DK-1 spacecraft are very accurate, but cover only protons with $E > 80$ MeV and He nuclei with $E > 200$ MeV/nucleon, and in time – only the solar minimum and the growth branch of cycle 24.

A specific problem in measuring GCR fluxes with energies less than 600 MeV/nucleon is the counting of higher energy particles penetrating the walls of the instrument, the fluxes of which are significant due to the shape of the GCR energy spectrum. An "anti-particle loop" – a massive scintillation detector surrounding the main "telescope" or strip or matrix detectors to determine the direction of particle arrival – can be used to screen out lateral particle passages. However, in many experiments to measure heliospheric particle fluxes, the registration of lateral particle passages in the instruments was absent – for example, on the geostationary GOES and Electro-L series satellites, in the proton channels on the ACE satellite, etc.; or did not work correctly, for example, on the IMP-8 satellite. Thus, the He GKL proton fluxes in a part of the GME channels of the IMP-8 spacecraft, despite the presence of its anti-collision loop, clearly have an account of lateral particle passages and, as a consequence, cannot be used for the development of models. Measurements of He GCR fluxes by the GME IMP-8 and SIS ACE instruments with energies of tens of MeV/nucleon have significant differences in the periods of solar activity minima.

These drawbacks of the currently available measurements of proton and nuclei fluxes of GCR with energies of 30–600 MeV/nucleon determine the relevance of the development of new techniques and new experimental measurements. On the basis of the analysis of the experiments performed and the data obtained by them, the requirements for the MODULATION experiment were formulated:

- Main measured parameter: fluxes of heliospheric protons and nuclei up to Fe, with separation by chemical and energy composition.
- Energy range: 30–600 MeV/nucleon.
- Determination of energy spectra of the fluxes with sufficient accuracy (achievable accuracy should be specified at the stage of instrument development, it is reasonable not worse than 10% for all types of particles).

- Determination of the trajectory of each incoming particle – to ensure accurate separation of particles by chemical composition and energies, and to screen out side passages of particles through the walls of the calorimeter.
- Separation of the lepton component (gamma quanta and electrons/positrons).
- Ensuring that the maximum particle flux is not less than 10^3 1/(cm^2·s·sr), preferably 10^4 1/(cm^2·s·sr).
- Conducting measurements in regions of space where there is no shielding effect of the Earth's magnetosphere on these fluxes: near-Earth orbit of the future manned station – near-polar regions; near-lunar orbit – all flight time.
- Carrying out measurements in the main stage – near-lunar orbit for a long time: 10 years.
- Measurement data in full volume, containing parameters of each particle registration, can be sent to Earth with a delay, including delivery on a solid-state drive by a transport ship. It is desirable to be able to receive reduced data (energy spectra with some step in particle energy and measurement time) with the least realisable delay through telemetry channels.

The MODULATION experiment should provide precision measurements of proton and nuclei fluxes up to Fe in solar proton events (excluding, possibly, extreme SPEs) in the energy range of at least 30–600 MeV/nucleon over a long period of time covering periods of low and high solar activity, which will be further used to improve numerical models of GCR and to study the physics of the heliosphere.

2 Methodology for Registration of Nuclei from Protons to Iron in the MODULATION Project

2.1 Instrument Design

One of the important advantages of the MODULATION project is that the main part of the scheme of the detecting part of the MODULATION instrument and the corresponding technical solutions are inherited from the NUCLEON-2 project developed earlier and tested in prototypes [15,16]. The MODULATION tool from the NUCLEON-2 project fully inherits the idea of particle energy measurement in the full energy absorption regime (Bragg regime), but in order to extend the energy range of the instrument it is proposed to supplement the method with a new method, which is related to the $dE/dX(E)$ method [17], but differs from it in a number of essential details (see below). It should also be noted that the NUCLON-2 instrument was designed primarily to measure the charge and isotopic composition of superheavy cosmic ray nuclei, while the MODULATION instrument is designed primarily to measure the energy spectra of nuclei from protons to iron, up to energies on the scale of 1 GeV/nucleon, although it will still be able to solve all the problems for which NUCLON-2 was designed.

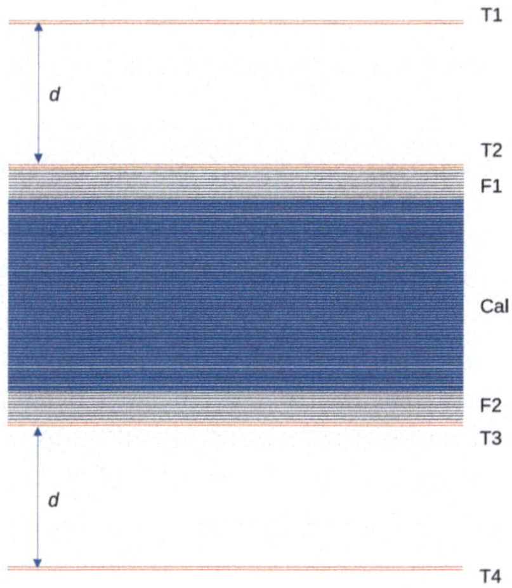

Fig. 1. Schematic of one tower of the MODULATION spectrometer. T1 and T2 – two double-layer silicon strip trackers at the top of the instrument, T3 and T4 – two of the same trackers at the bottom of the instrument (red lines). F1 and F2 – two thin-layer silicon calorimeters (300 mkm layer thickness). Cal – basic silicon calorimeter with 1 mm thick planes. The total thickness of the calorimeter F1+Cal+F2 is 60 mm.

The MODULATION instrument has a modular design and consists of individual "towers", of which there may be from one to several dozens depending on the operating conditions. A schematic diagram of one spectrometer tower is shown in Fig. 1. In the figure, T1 and T2 – two double-layer silicon strip trackers at the top of the instrument, T3 and T4 – two similar double-layer trackers at the bottom of the instrument (red lines). The trackers are designed to reconstruct the initial trajectory of the particle. The tracker should give a zenith angle accuracy of about 0.01 rad for a zenith angle of 45°, which means that the ratio h/d should be about 1/50, where h is the width of the tracker strip, d is the distance from the outer tracker to the calorimeter. If $d = 50$ mm, then $h = 1$ mm, if $d = 100$ mm, then $h = 2$ mm, etc. F1 and F2 – two thin-layer silicon calorimeters (300 mkm thick planes). Calorimeters F1 and F2 are needed specifically to extend the energy range of the instrument towards lower energies (tens of MeV/nucleon, the exact minimum energy depends on the nucleus, see below). Each of the F1 and F2 calorimeters should contain 10 to 20 silicon planes. Cal – a basic silicon calorimeter with each silicon plane 1 mm thick. It is assumed that the total thickness of calorimeters F1+Cal+F2 is 60 mm silicon. If, for example, calorimeters F1 and F2 each contain 10 planes of 300 mkm thickness, the main calorimeter Cal will contain 54 planes of 1 mm each. The instrument modelling (see below) was carried out under this assumption. The schematic of the instrument in Fig. 1 assumes that the spectrometer is exposed on both sides, top and bottom. It is assumed that the entrance of the instrument is exposed to outer space. It is for this case that the minimum energy thresholds are calculated below. In the case where the instrument will be placed in a container, it will of course be necessary to take into account the ionisation energy losses in the container walls to obtain the correct lower energy limits.

2.2 Mathematical Modelling of the Spectrometer

Below we will give an overview of the techniques for charge separation of nuclei and determination of their energy in the MODULATION spectrometer. These techniques are developed on the basis of mathematical modelling of the spectrometer. Note that the simulation of electromagnetic and nuclear interaction of charged particles with matter for moderate energies (tens of MeV – units of GeV per nucleon) is considered to be a relatively simple and well-studied task tested in numerous practical applications, so the results of the mathematical modelling of the device are sufficiently reliable for further development of the details of the device design. Nevertheless, it should be noted that before using the proposed solutions, it is mandatory to calibrate the corresponding techniques in beam experiments with the prototype instrument.

The FLUKA [18, 19] software package with the nuclear interaction generator dpmjet was used to simulate various aspects of the MODULATION spectrometer. Actually, there was no need to build a complete mathematical model of the spectrometer exactly corresponding to the device scheme in Fig. 1. For the search for minimum thresholds of energy determination it was enough to limit ourselves to the part of the spectrometer T1+T2+F1, and the trackers were effectively replaced by 300 mkm thick silicon planes, since the strip structure of the trackers did not play a role in the calculations. For the remaining problems, it was sufficient to replace the entire combination of F1+Cal+F2 calorimeters with a single effective calorimeter consisting of 60 silicon planes of 1 mm thickness each. In the following we will use a suitable model without further explanations.

2.3 Full Energy Absorption Mode. Minimum and Maximum Energy Thresholds of the Full Absorption Mode

In the regime of complete energy absorption the nucleus is completely braked in silicon due to ionisation losses. The braking process ends with a characteristic Bragg peak of ionisation losses, which takes place when the nucleus becomes strongly nonrelativistic. The characteristic curve of the nucleus deceleration is clearly visible in the cascade curve that is recorded by the spectrometer. In Fig. 2 (left panel), the simulated cascade curve for the full absorption mode of a carbon nucleus with vertical incidence and an initial energy of 180 MeV/nucleon is shown. For other nuclei and energies in the full absorption regime, the cascade curves have a similar appearance.

It is very important that, in the case of total energy absorption, the accuracy of the energy measurement is determined solely by the intrinsic accuracy and noise of the apparatus; the physical fluctuations of ionisation losses do not contribute to the measurement error. However, it is not always only ionisation losses caused by the electromagnetic interaction of the nucleus with matter that

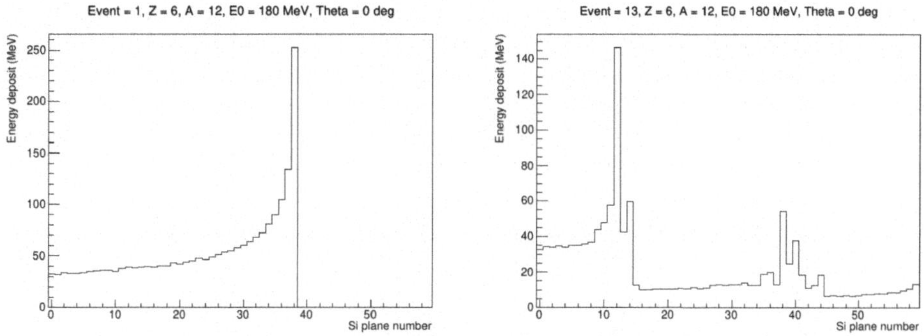

Fig. 2. Left: carbon, 180 MeV/nucleon, vertical drop from above, total energy absorption event (Bragg event). Right: the same, but there was a nuclear interaction.

slow down the nucleus. Nuclear interactions can also occur, resulting in fragmentation of nuclei, mass birth of π^0-mesons, due to which the energy release in the calorimeter becomes unpredictable. However, nuclear interaction events are easily filtered out of the total flux of events, and the efficiency loss due to such events is easy to estimate. In Fig. 2 (right panel), a typical nuclear interaction event (simulation) is shown. The effectiveness of the filtering algorithms for inelastic interaction events was practically demonstrated in beam tests of the NUCLEON-2 prototype [15,16].

The minimum energy threshold of the MODULATION instrument was determined from the following considerations. For nuclei coming from above, it was required that for all particles with zenith angles starting from 45° and less, the position of the Bragg peak of the cascade curve should be located no higher than in the seventh, counting from above, plane of the thin calorimeter F1 (plane number 6, since the planes are numbered starting from zero). In this way the lower threshold is determined with some redundancy, since for nuclei with smaller zenith angles the condition for the location of the Bragg peak in the seventh layer is realised for lower energies, and the condition of the seventh layer itself can be somewhat relaxed due to a stricter filtering of nuclear interaction events. The lower energy thresholds for all nuclei from protons to iron (for the main stable isotopes) were determined in this way. These thresholds are given in Table 1 (column E_{min}). In the same table, the registration efficiencies of nuclei at the minimum energy threshold relative to the nuclear interaction (column $Eff(E_{min})$) are given.

For each nucleus and for each zenith angle, there is a maximum energy at which the nucleus is completely retarded in the calorimeter only due to ionisation losses. In Table 1 (column E_{max} (fa)), the maximum energy thresholds at which the nucleus registers in the full energy absorption regime are given. The energies tabulated are determined from the following considerations. It is required that the full absorption mode be realised for a nucleus falling vertically (zero zenith angle). If the total absorption regime at some energy is realised for a vertically

Table 1. Energy thresholds and efficiency of the MODULATION spectrometer. E_{min} – minimum energy threshold (MeV/nucleon); E_{max} (fa) – maximum energy threshold for detection of nuclei in the full absorption mode; $Eff(E_{min})$ – efficiency of detection of nuclei at minimum energy (per cent, statistical error of calculation varies from 0. 13% for protons to 0.23% for iron); Eff (1 GeV) – detection efficiency at 1 GeV/nucleon (per cent, statistical error varies from 1.0% for protons to 2.2% for iron).

Nucleus	Z	A	E_{min}	E_{max} (fa)	$Eff(E_{min})$	Eff (1 GeV)
p	1	1	29	121	98.4	89.7
He	2	4	29	121	97.1	77.3
Li	3	7	33	140	96.6	76.0
Be	4	9	39	170	96.6	74.3
B	5	11	45	196	96.5	71.6
C	6	12	53	231	96.4	73.5
N	7	14	58	255	96.2	70.2
O	8	16	63	276	96.1	66.8
F	9	19	65	286	96.1	65.4
Ne	10	20	70	317	95.5	66.6
Na	11	23	73	326	95.9	64.5
Mg	12	24	78	355	95.7	63.3
Al	13	27	81	364	95.8	61.9
Si	14	28	85	391	95.8	59.8
P	15	31	88	400	95.3	58.7
S	16	32	92	426	95.4	57.1
Cl	17	35	94	435	95.4	57.9
Ar	18	40	93	430	94.7	55.4
K	19	39	99	467	95.4	54.0
Ca	20	40	104	491	95.3	56.2
Sc	21	46	103	477	94.7	56.4
Ti	22	49	104	487	94.5	52.2
V	23	51	106	505	94.5	51.3
Cr	24	52	109	526	94.5	52.4
Mn	25	55	112	532	95.0	52.8
Fe	26	56	115	558	94.9	48.8

falling nucleus, it is all the more realised for non-zero zenith angles and the same energy.

The column Eff (1 GeV) Table 1 shows the calculated efficiency of the device for the vertical fall of nuclei with energies of 1 GeV/nucleon, which is substantially above all thresholds of total absorption. At this energy, the particle completely passes the entire device if there were no nuclear interaction. Since the

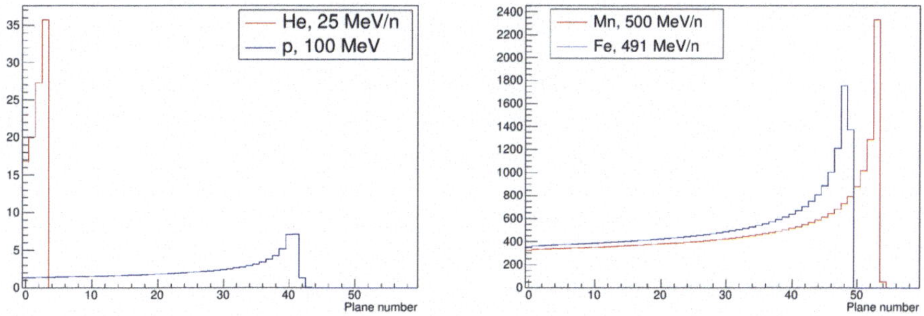

Fig. 3. Towards charge separation of nuclei in the full energy absorption regime. On the left is the separation of protons and helium for vertical incidence and total particle energy of 100 MeV, on the right is the separation of manganese and iron at total energy of 27500 MeV per particle, also for vertical incidence. The figures show the average cascade curves (averaged over several hundred events) for each nucleus.

nuclear interaction cross sections depend weakly on energy, approximately the same efficiencies (to within 1–2%) are also characteristic for energies E_{max} (np). All these efficiencies are noticeably less than 100%, including even the efficiency for protons, and fall to about 50% for nuclei near iron.

In the full absorption regime the charge separation of nuclei is not a difficult problem. As it was shown experimentally in beam experiments with the prototype spectrometer NUCLEON-2, in the full absorption regime not only nuclei are easily charge-separated, but for a given charge isotopic resolution up to $Z = 54$ [16] is possible. In the charge-isotope separation problem, it is important to realise that the energy and zenith angle of the particle are precisely known, so the charge separation problem is set at a fixed total energy of the particle and a fixed zenith angle. The Fig. 3 illustrates the neighbouring charge separation in proton-helium and manganese-iron pairs for vertical particle incidence. It can be seen that the cascade curves of protons and helium have nothing in common at all, while the cascade curves of manganese and iron differ very well.

2.4 Full Ionisation Loss Method in the Incomplete Energy Absorption Regime

The full energy absorption mode discussed in Sect. 2.3 provides a precise measurement of the energy of nuclei, but only in a very limited energy range. The upper energy limit of about 120 GeV/nucleon for protons and helium (see Table 1) is insufficient to fulfil the objectives of the MODULATION project. The same is true for all other nuclei.

To solve this problem, the MODULATION project proposes to use a new method, which is a close relative of the well-known method $dE/dX(E)$ [17]. The essence of the method $dE/dX(E)$ in the simplest variant consists in the following. Each nucleus and each substance is characterised by its own dependence of the specific ionisation energy in the substance dE/dX on the particle energy E.

Therefore, if the specific ionisation energy is measured for a known nucleus, the energy of the nucleus can be found from the curve $dE/dX(E)$. The idea of using this method in the MODULATION project should have been that for nuclei with energy greater than the limiting energy of total absorption in the calorimeter, which leave only a part of their energy in the calorimeter, by the measured cascade curve, calculating plane by plane the experimental specific ionisation energy dE/dX, by the known dependence $dE/dX(E)$ to restore the initial energy of the falling nucleus, i.e. to measure its energy. However, such a straightforwardly formulated idea faces some problems.

The first difficulty is that in the original formulation of the method the value dE/dX describes the specific energy loss for ionisation, which is purely local in nature, but the release of this lost energy in the detectors is nonlocal, so the value dE/dX is not directly measurable. The ionising particle creates an intense stream of delta electrons around it, which are absorbed by the detectors quite differently from where they were emitted. The difference in specific ionisation losses and specific density of absorbed ionisation energy in sufficiently precise experiments, to which the MODULATION project can be attributed, matter, so the $dE/dX(E)$ technique in its original simple formulation does not strictly make sense.

The second difficulty is that calculations of dE/dX values for specific *absorbed* ionisation energy, if we really wanted to consistently use only directly measurable differential quantities, present a difficult computational problem. Analytical estimates for precision experiments cannot be used because they give specific ionisation losses but not the specific density of the absorbed ionisation energy, and when calculating the specific density of the absorbed ionisation energy by the Monte Carlo method, even if the position of a small layer of matter relative to the rest of the detector is accurately taken into account (nonlocality effect), large statistical errors arise, especially for light nuclei and for energies on the scale of 1 GeV/nucleon and above. The large magnitude of the errors is due to the fact that, by implication, the matter layer must be thin to estimate dE/dX, and the relative magnitude of fluctuations in it will therefore be large.

In principle, a neat application of the dE/dX method for determining the energy in events with energies beyond the maximum total absorption energy, for the MODULATION project, would be as follows. Cascade curves of different nuclei should be simulated for all nuclei, all ranges of energies of interest, with sufficiently small energy steps, and for zenith angles corresponding to all cases of complete tower passage from the top to the bottom base in steps of about one degree. From these simulated cascade curves for each zenith angle, surfaces in the space "core energy – plane number – plane energy release" are plotted using interpolation. The experimental cascade curves are then approximated by reference curves from these surfaces, giving for each event both the charge of the nucleus and its initial energy. However, such an extreme programme involves huge amounts of Monte Carlo calculations and generally involves a rather complex and not very transparent methodology (multiple families of surfaces, complex interpolations, etc.).

Both mentioned problems can be solved in the following way. Instead of using directly detailed simulated cascade curves, one can take advantage of the simple fact that to each initial energy of a nucleus with a certain zenith angle there will correspond a quite definite average energy release $\Delta E(E)$ in the whole calorimeter. Then, if the nucleus is known, the initial energy E of the falling nucleus is immediately determined by the total energy release ΔE in the calorimeter and the measured zenith angle (determined by the trackers). For this purpose, only the $\Delta E(E)$ dependence needs to be calibrated for each angle, which can be done by simulating the experiment and then making corrections (if it really turns out to be necessary) based on the results of testing the technique in beam tests. In such a technique, firstly, one does not have to deal with sets of complex mathematical surfaces and the problems of interpolation along the cascading curves that form these surfaces; secondly, the problem of numerical errors of the Monte Carlo method disappears, since we now deal only with energy release in a very thick piece of silicon (6 cm in the MODULATION spectrometer) in which the total relative fluctuation of ionisation losses is small. At the same time, the method accurately accounts for all the subtle features of nonlocal energy release in the calorimeter substance.

The Fig. 4 shows the $\Delta E(E)$ curves for nuclei from protons to oxygen for vertical particle incidence on the calorimeter calculated by the Monte Carlo method with the FLUKA system (see Sect. 2.2).

The curves are shown only for energies above the maximum threshold of total absorption when the total energy absorption method (Sect. 2.3) fails. 10000 Monte Carlo trials were sufficient to calculate each point of each curve. The red dashed line in the figure illustrates the method for determining the initial total energy E of a nucleus from the total energy release in the calorimeter. Although Fig. 4 is plotted for vertical particle incidence, the methodology for non-zero zenith angles is no different, since the zenith angle will be known with sufficient accuracy to make no appreciable errors in the methodology. It makes sense to call the proposed method the total ionisation loss (TIL) method to contrast it with the differential loss method $dE/dX(E)$. In the following, we will consider all the peculiarities of the TIL method on the example of the vertical incidence of particles, implying that the methodology can be transferred to all non-zero zenith angles without significant changes. The Fig. 5 shows the $\Delta E(E)$ curves for heavy Cr, Mn, and Fe nuclei.

In order to discuss the TIL method, we note first of all that the events corresponding to purely ionisation energy losses by the nucleus in the calorimeter substance, without nuclear interactions, are distinguished as reliably as in the total energy absorption regime (Sect. 2.3) simply by the appearance of the cascade curve. In Fig. 6, the left column of the histograms shows examples of cascade curves for boron and carbon nuclei with ionisation losses only, and the right column shows events with nuclear interaction. It can be seen that events with nuclear interaction differ sharply from events with ionisation losses by their irregular shape, despite the presence of ionisation fluctuations in purely ionisation cascade curves.

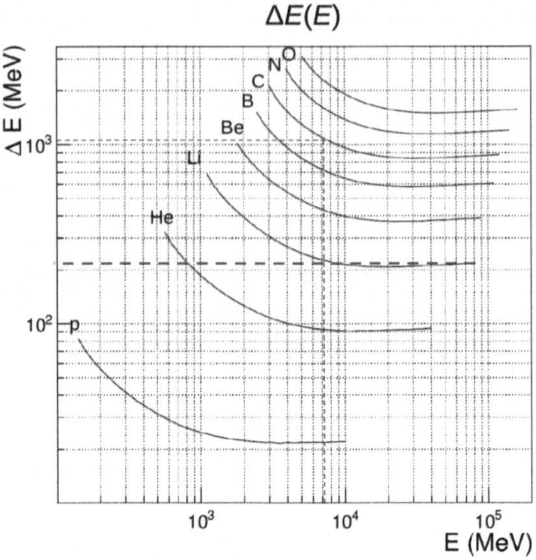

Fig. 4. Curves $\Delta E(E)$ for nuclei from protons to oxygen for a vertical fall of particles on the calorimeter. The red thin dashed line illustrates the method of determining the initial total energy E of a nucleus from the total energy release in the calorimeter. The blue thick dashed line illustrates the reason for the interference of the spectra of lithium and helium (see text). (Color figure online)

In Fig. 4 it can be seen that more than one type of nuclei with different initial energies can correspond to the same energy release ΔE in the calorimeter. From Fig. 5, which shows the $\Delta E(E)$ curves for Cr, Mn, and Fe, it can be seen that this situation is particularly characteristic of heavy nuclei. Consequently, the problem of determining the type of nucleus at a given energy release ΔE arises.

In Fig. 6, in the left column of the histograms, examples of the cascade curves of boron (upper histogram) and carbon (lower histogram) are shown for the same average energy release in the calorimeter $\Delta E = 1000$ MeV. One can see by eye the difference in the slopes of the cascade curves. Hence, for the same ΔE one can try to distinguish nuclei from each other by the slope of the cascade curves. To do this, the calorimeter can be divided into upper and lower halves (planes numbered 0 to 29 and numbered 30 to 59), calculate the total energy releases L for the upper half and R for the lower half, and consider the value $B = R/L - 1$. For an average horizontal curve there will be $B \approx 0$, for a curve with a positive slope there will be $B > 0$, with the larger B the steeper the curve.

If different initial energies E are considered for some nucleus, then on the plane $[\Delta E(E), B(E)]$ each nucleus will be characterised by its own trajectory of mean values $\Delta E(E), B(E)$. In addition, for each energy E there will be a characteristic spread (standard deviation) $\sigma\Delta E(E), \sigma B(E)$. To understand how well different nuclei will separate from each other in the TIL method, it is convenient to study the trajectories of different nuclei on the plane $[\Delta E, B]$.

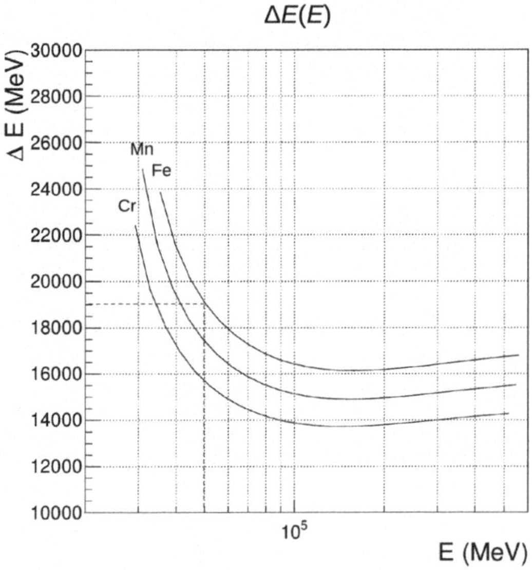

Fig. 5. Curves $\Delta E(E)$ for Cr, Mn, and Fe nuclei for the vertical fall of particles on the calorimeter. It can be seen that several different nuclei with different initial energies can correspond to the same energy release ΔE in the calorimeter.

The Fig. 7 shows the trajectories of nuclei from protons to oxygen on the $[DeltaE, B]$ plane. First, it is seen that protons cannot be confused with helium nuclei, since the trajectory of protons does not intersect with the trajectory of helium, even taking into account the statistical fluctuations of ΔE and B. Moreover protons do not intersect with any other trajectory, therefore measurements of the proton spectrum by the TIL method will be clean at any energies.

With helium nuclei, the situation is a little more complicated. Although the helium trajectory itself does not intersect with any other trajectory, the part of the helium trajectory within one standard deviation is overlapped by statistical fluctuations in the trajectory of the next element, lithium. However, the amount of lithium in cosmic ray spectra is negligible compared to helium, so the very small contamination of helium spectra by lithium can be completely neglected. Consequently, the helium spectrum can also be measured by the TIL technique without problems.

It is important to deal with the cause of the potential problem for the helium nucleus, since the same cause will cause a similar difficulty for other nuclei. The area of overlap between the errors in the determination of the B coefficient of lithium and the $[\Delta E, B]$ trajectory of helium occurs in the region where, on the $\Delta E(E)$ curves, the nearly horizontal high-energy section of the lithium curve is projected to the left onto the helium curve section (blue thick dashed line in Fig. 4). Ionisation fluctuations in the nearly horizontal high-energy section of the lithium curve are large, much larger than fluctuations in the steep low-energy section, so errors in the determination of the B coefficient are also

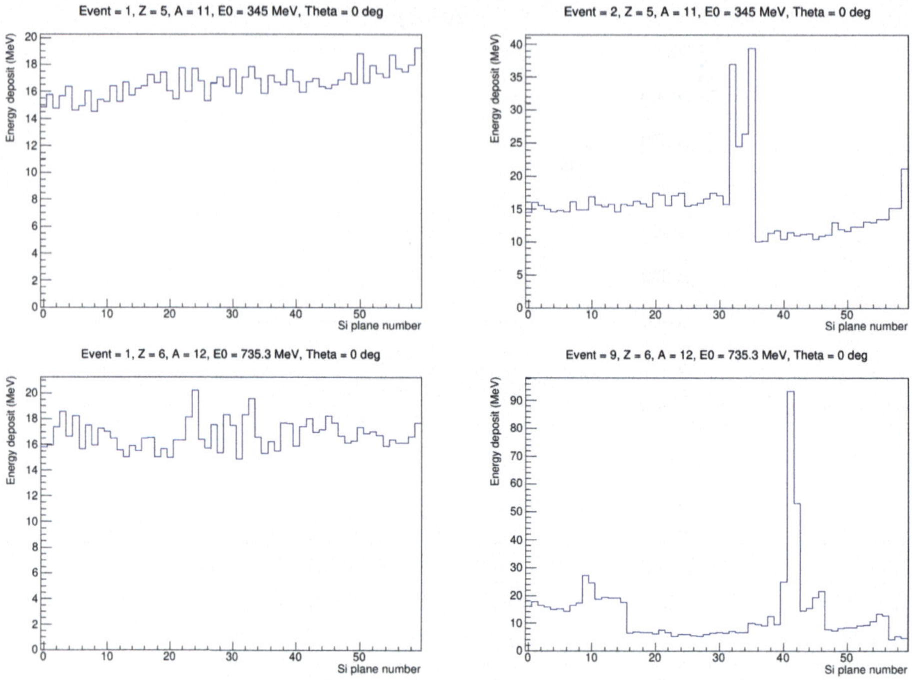

Fig. 6. Left column of histograms: Examples of cascade curves of boron (upper histogram) and carbon (lower histogram) for the same average energy release in the calorimeter $\Delta E = 1000$ MeV. The difference in the slopes of the cascade curves can be seen by eye. Right column: examples of boron (top) and carbon (bottom) events with nuclear interaction.

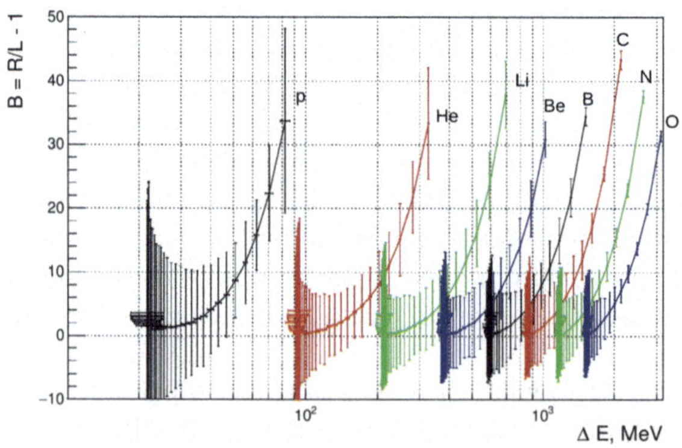

Fig. 7. Trajectories of nuclei from protons to oxygen on the plane $[\Delta E, B]$ (on the ordinate axis on this graph, as well as on other graphs, the value of B is expressed in per cent, i.e. the value $(R/L - 1) \times 100$ is shown).

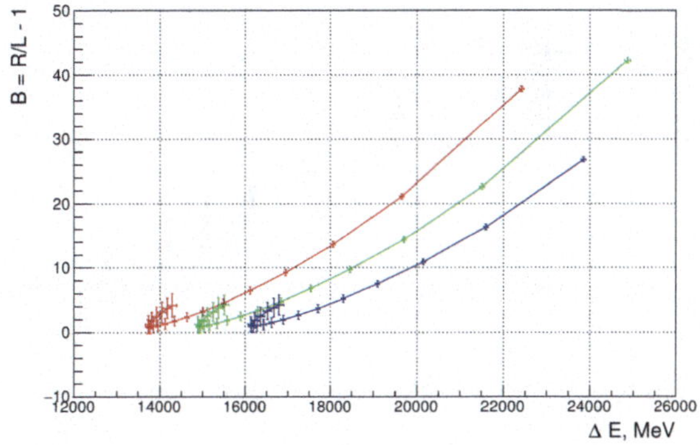

Fig. 8. Trajectories of Cr, Mn, Fe nuclei on the plane $[\Delta E, B]$.

large here, because of which the nearly horizontal spectrum sometimes gives a markedly positive B value, making the lithium cascade curve indistinguishable (in slope) from the steeper helium curve on average. The superposition of lithium on helium occurs over a rather narrow range of helium energies, from about 180 to 220 MeV/nucleon. In other words, the entire helium spectrum is measured without potential problems, except for the narrow "forbidden range" of energies (however, this does not matter for helium, because of the very low flux of lithium nuclei compared to the helium flux). The same situation will be reproduced for all other nuclei, as can be clearly seen in Fig. 7 and 8. A potentially problematic place is the neighbourhood of the intersection point of the leftward extrapolation of the horizontal high-energy portion of the $\Delta E(E)$ curve of some nucleus to the steep low-energy portion of the $\Delta E(E)$ curve of the previous charge nucleus.

The trajectories of Cr, Mn, and Fe nuclei in the plane $[\Delta E, B]$ are shown in Fig. 8. It can be seen that nothing fundamentally new arises in the region of heavy nuclei compared to lighter nuclei (Fig. 7).

The question arises whether, having in hand the experimental cascade curve, it is impossible to find any other measurable parameter in it, by which it would be possible to distinguish between high-energy events of one nucleus and low-energy events of the previous nucleus? If we compare the cascade curves of boron and carbon in the left column of the histograms in Fig. 6, which correspond to the same energy release ΔE in the calorimeter, it is easy to see that, in addition to the difference in the slopes of the cascade curves, the ionisation fluctuations are noticeably stronger in the carbon cascade curve corresponding to high energy than in the boron cascade curve corresponding to lower energy. Thus, the magnitude of the fluctuations is another parameter that is available directly in the measurements. The magnitude of the fluctuations can be accounted for in different ways, and the method we have chosen is as follows. First, to make the parameter structure uniform for all types of nuclei and all energies, the cascade curve is normalised by the energy ΔE, so that the area under the curve becomes equal to one. The curve is then multiplied by the number of calorimeter planes

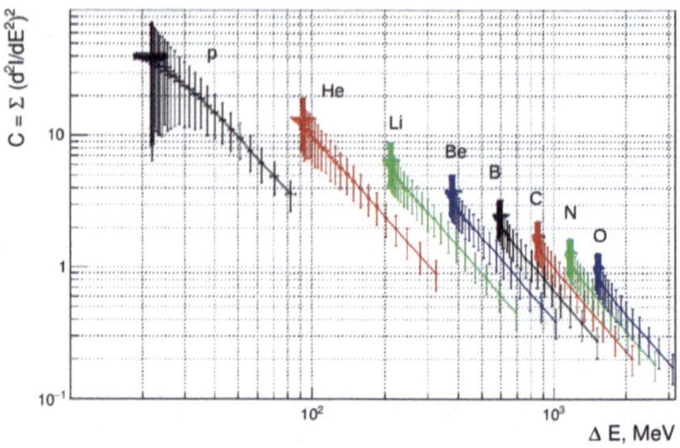

Fig. 9. Trajectories of nuclei from protons to oxygen on the plane $[\Delta E, C]$ (see text for explanation of the meaning of the parameter C).

(60 in the present study), so that the mean amplitude corresponding to one plane on the curve becomes equal to one. After this, the sum of squares C of the second derivatives of the cascading curve for the whole curve is calculated for all points from the second plane from the top to the penultimate one, where the i-second derivative is calculated using the standard numerical differentiation formula

$$D_i = S_{i+1} - 2S_i + S_{i-1}.$$

Here S_i is the i-th value of the cascade curve amplitude obtained after the normalisation described above. Obviously, the larger the fluctuations of the cascade curve, the larger will be the sum of squares of the second derivatives of C.

Using the sum of squares of the second derivatives of C, we can construct trajectories for each nucleus in the plane $[\Delta E, C]$ in the same way as we did earlier for the steepness parameter B. The Fig. 9 shows the trajectories of the nuclei from protons to oxygen on the $[\Delta E, C]$ plane. It can be seen that protons are easily separated from helium already by only one parameter C, without using the parametre B.

What is new in comparison with Fig. 7 is that the problem of the separation of He, Li, Be, and B nuclei is completely solved. There is a complete separation on the parameter C in the region where there is an overlap on the steepness parameter B, and, although in the low energy region there is no complete separation of trajectories on the parameter C alone, this is not necessary for the separation of nuclei, since in the low energy region there is a complete separation on the parameter B. However, for heavier nuclei the complete separation of trajectories does not occur in the three-dimensional space $[\Delta E, B, C]$, so for them the "forbidden intervals" of energies discussed above are preserved.

The most important question for the TIL method is the energy resolution that this technique gives for different nuclei at different energies of the incident nucleus. We determined the energy resolution by the Monte Carlo method. Particles with a certain energy are thrown at the instrument at a certain angle, for

Table 2. Energy resolutions (in per cent) for some nuclei, for the vertical incidence of particles on the instrument.

Energy, MeV/nucleon	protons	He	C	O	Si	Fe
200	2.7	1.4	0	0	0	0
400	7.2	3.7	1.0	0.6	0	0
600	12.0	6.8	2.0	1.4	0.6	0
800	19.8	11.1	3.4	2.4	1.2	0.5
1000	30.3	17.3	5.4	4.9	1.9	0.9
1200	–	25.5	7.8	5.9	3.0	1.4
1500	–	–	15.5	12.1	5.8	2.6
2000	–	–	–	15.1	11.1	7.6

each event number i the value of the ionisation energy release in the calorimeter ΔE_i is determined, then, by numerical solution of equation

$$\Delta E(E) = \Delta E_i$$

energy E_i is searched for, and the distribution of the found energies is used to determine their spread in terms of standard deviation, which is interpreted as the expected measurement error.

The calculated energy resolutions for p, He, C, O, Si, Fe nuclei for the vertical incidence of nuclei on the device for several initial energies are shown in Table 2 as an example. For non-zero zenith angles, the effective thickness of the instrument will be slightly larger, so the resolution will be slightly better. In other words, Table 2 gives the worst-case resolution constraints.

The table cells containing zeros indicate that this nucleus at a given energy is in the full absorption regime, when the energy is measured accurately, if one does not take into account the noise of electronics. The cells of the table containing a dash mean that at this energy for this nucleus the TIL method does not work. This occurs when the curve $\Delta E(E)$ fits the energy E too close to the flat section $d\Delta E(E)/dE \approx 0$ (see Fig. 4 and 5). It can be seen that the TIL method provides proton energies up to 1 GeV and iron energies up to 2 GeV/nucleon. According to Tables 1 and 2, the energy range of the MODULATION spectrometer fully satisfies the requirements for all nuclei from protons to iron.

The task of analysing the isotopic composition of heavy and superheavy nuclei for the NUCLEON-2 experiment, including beam tests with the prototype instrument, was investigated in [15,16]. The practical possibility of isotope separation was demonstrated on argon ($Z = 18$) and xenon ($Z = 54$) nuclei. All heavy isotope separation techniques developed for the NUCLON-2 project are transferred to the MODULATION project without any changes. We note here that isotopic analysis for lighter nuclei, including hydrogen isotopes, is a simpler task, so it will also be handled by the MODULATION project.

3 MODULATION Spectrometer Design Image

The registration of abundant, $Z = 1 - 26$, nuclei (born mainly in the nucleosynthesis of main-sequence stars) of cosmic rays in the energy region 30–1000 MeV has a number of peculiarities. First of all, it is connected with the difference of the range of nuclei of different charges, which creates certain difficulties in the creation of the apparatus. These difficulties have been overcome by the developed new approach of nuclei measurement in the TIL method, see Sect. 2.4. In modelling the presented approach, a spectrometer scheme was used, which almost completely repeats the STIKL spectrometer of the NUCLEON-2 project [15, 16]. Thus, the project is based on the ideology of the same general layout of both spectrometers, STIKL and the spectrometer tower of the MODULATION project, with the only significant difference in the amplitude range of the reading electronics.

Functionally, one tower of the MODULATION spectrometer is a silicon ionisation calorimeter. The main recording elements of the calorimeter are 60 layers of silicon calorimetric detectors each 1 mm thick and 20–40 layers of silicon detectors each 300 mkm thick (these figures may be slightly modified in the final implementation of the apparatus.). In addition to the calorimetric detectors, the tower has 8 strip silicon detectors each 300 mkm thick. These detectors provide information not only on the energy loss of the particle, but also on its coordinate (by the number of the triggered "strip"). Information on the coordinates of the particle from several layers of strip detectors allows to specify the angle of its entry into the calorimetric detectors.

Readout of pad and strip detectors is performed by charge sensitive amplifier (CSA) circuits. Then the signal is filtered from noise by RC-CR frequency filter circuit and digitised by ADC (at the moment of reaching the maximum amplitude). For strip detectors circuits and filters are implemented in specialised multichannel chips (one channel of the chip serves one strip detector), for pad detectors, the path "SNC-filter-ADC" is implemented on separate DUTs and serves a pair of detectors.

The fact of an "interesting" particle hitting the spectrometer tower is determined by the simultaneous operation of comparator circuits that monitor the signal level at the output of the four dedicated incident detectors.

The main structural element of the MODULATION spectrometer tower is the pad silicon detector, which has the shape of a regular hexagon inscribed in a circle with a diameter of 141 mm. Accordingly, the spectrometer tower has the shape of a regular hexagonal prism. The design appearance of one MODULATION spectrometer tower is shown in Fig. 10. In different variants of design the whole spectrometer can contain from one to several dozens of such towers arranged in a dense honeycomb structure. In the honeycomb structure it is possible to reconstruct events in which the trajectory of the incident particle passes through more than one tower. This allows us to increase the geometric factor of the whole device compared to the simple sum of geometric factors of single towers.

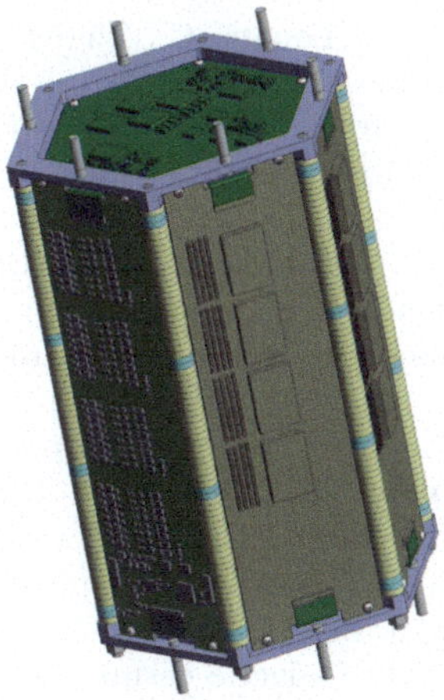

Fig. 10. Exterior view of one tower of the MODULATION spectrometer.

4 Conclusion

In conclusion, we note that the MODULATION project makes it possible to solve the main problems associated with the measurement of cosmic ray particle fluxes subjected to solar modulation and to provide measurements in a wide energy range for all cosmic ray nuclei, not only from protons to iron, which is the main objective of the mission, but, in fact, for superheavy nuclei up to actinoids, if circumstances allow including several dozens of spectrometer towers in the instrumentation and using electronics providing sufficient dynamic range. It can also be noted that in the minimum configuration (one spectrometer tower plus reading electronics) the mass of the equipment will be only about 20 kg, which will allow the first versions of the spectrometer to be sent into space in different ways, on different platforms, without significant costs.

References

1. Richardson, I., Cane, H., Von Rosenvinge, T., Mcguire, R.: IMP-8 GME energetic particle observations over three solar cycles. In: Proceedings of the 30th International Cosmic Ray Conference, vol. 1. Merida, Mexico, pp. 323–326 (2008)
2. George, J.S., Lave, K.A., Wiedenbeck, M.E., Binns, W.R., Cummings, A.C.: Elemental composition and energy spectra of galactic cosmic rays during solar cycle 23. Astrophys. J. **698**(2), 1666–1681 (2009)
3. Kühl, P., Gómez-Herrero, R., Heber, B.: Annual cosmic ray spectra from 250 MeV up to 1.6 GeV from 1995–2014 measured with the electron proton helium instrument onboard SOHO. Solar Phys. **291**, 965–974 (2016)

4. McKibben, R.B., Connell, J.J., Lopate, C., Zhang, M., Heber, B.: Intensities of galactic cosmic ray nuclei from the ecliptic to the south solar polar regions near solar maximum: observations from the ULYSSES COSPIN high energy telescope. In: Proceedings of the 27th International Cosmic Ray Conference, vol. 1. Hamburg, pp. 3893–3896 (2001)
5. Adriani, O., Barbarino, G.C., Bazilevskaya, G.A., Bellotti, R., Boezio, M., Bogomolov, E.A.: Time dependence of the proton flux measured by PAMELA during the 2006 July–2009 December solar minimum. Astrophys. J. **765**(2), 91–98 (2013)
6. Shikaze, Y., Haino, S., Abe, K., Fuke, H., Hams, T.: Measurements of 0.2 to 20 GeV/n cosmic-ray proton and helium spectra from 1997 through 2002 with the BESS spectrometer. Astropart. Phys. **28**(1), 154–167 (2007)
7. Abe, K., Fuke, H., Haino, S., Hams, T., Hasegawa, M.: Measurements of cosmic-ray proton and helium spectra from the BESS-polar long-duration balloon flights over Antarctica. Astrophys. J. **822**, 65 (2016)
8. Badhwar, G.D., O'Neill, P.M.: Galactic cosmic radiation model and its applications. Adv. Space Res. **17**(2), 7–17 (1996)
9. O'Neill, P.M., Golge, S., Slaba, T.C., Badhwar, G.D.: 2014 Galactic Cosmic Ray Flux Model Description. NASA/TP-2015-218569 (2015)
10. ISO 15390, Space environment (natural and artificial) ?- Galactic cosmic ray model. ISO, 2004
11. OST 134-1044 2007 Modification 1. Apparatus, instruments, devices and equipment of spacecrafts. Calculation methods radiation conditions on board of spacecrafts and establishing requirements for resistance of radio-electronic equipment of spacecrafts to charged particles. Spacecraft radio-electronic equipment to the effects of charged particles in outer space. of natural origin. FGUP TsNIIMash, Moscow (2016). (In Russian)
12. Kuznetsov, N.V., Popova, E.P., Panasyuk, M.I.: Empirical model of long-time variations of galactic cosmic ray particle fluxes. J. Geophys. Res. **122**(2), 1463–1472 (2017)
13. Tylka, A.J., Adams, J.H.J., Boberg, P.R.: A revision of the cosmic ray effects on micro-electronics code. IEEE Trans. Nucl. Sci. **44**(6), 2150–2160 (1997)
14. Matthia, D., Berger, T., Mrigakshi, A.I., Reitz, G.: A ready-to-use galactic cosmic ray model. Adv. Space Res. **51**(3), 329–338 (2013)
15. Bulatov, V., Filippov, S., Karmanov, D.: NUCLEON-2 mission for the investigation of isotope and charge composition of cosmic ray ions. Adv. Space Res. **64**(12), 2610–2618 (2019)
16. Vasiliev, O., et al.: Current status of the NUCLEON-2 mission. Phys. Part. Nucl. Lett. **18**(3), 36–51 (2021)
17. Kurashov, A.A.: Identification of Ionising Radiation of Medium and Low Energies. Atomizdat, Moscow (1979)
18. Ferrari, A., Sala, P.R., Fassò, A., Ranft, J.: FLUKA: a multi-particle transport code. CERN-2005-10, INFN/TC_05/11, SLAC-R-773 (2005)
19. Bohlen, T.T., Cerutti, F., Chin, M.P.W., et al.: The FLUKA code: developments and challenges for high energy and medical applications. Nucl. Data Sheets **120**, 211–214 (2014)

Geophysical Fields and Their Interaction

Geomagnetic Measurements at the Pleshchenitsy Geophysical Observatory (Minsk, Republic of Belarus)

Genady A. Aronov, Arkady G. Aronov, Ihar A. Saladukhin(✉), Aliaksei A. Kadymau, and Valiantsina M. Laziuk

Center of Geophysical Monitoring of the National Academy of Sciences of Belarus, 1/3 Kuprevich str., Minsk 220141, Republic of Belarus
saladukhin@cgm.by

Abstract. Continuous observations of the geomagnetic field elements have been carried out at the Pleshchenitsy Geophysical Observatory in Belarus since 1960 in order to study the annual mean values of the magnetic field components. The paper presents the secular variations of the geomagnetic field elements *D*, *H*, *Z*, *F* within the period of 1960–2020 in comparison with the results of the earlier investigations (starting from 1875). A significant long-term increase of some geomagnetic field components was revealed suggesting a continuous movement of the Earth's magnetic pole. So, during the last 63 years the magnetic declination has been increased on the average with a gradient about 4 arcmin/year, along with this during the last 14 years its increase became faster and exceeded 8 arcmin/year. The total geomagnetic field vector modulus has been also tending to increase with time, namely, from 49621 to 51485 nT for the period under consideration. In addition, the periodic change of the degree of the geomagnetic field disturbance was determined, which may be due to the solar activity variations within the 11-year cycles.

Keywords: geomagnetic field elements · geomagnetic field disturbance · solar activity cycle

1 Introduction

Data obtained by instrumental measurements suggest that the annual mean values of the geomagnetic field elements are continuously changed [1–4]. Such long-term variations are called secular variations. The secular variations of the geomagnetic field components are essentially influenced by the position of the Earth's magnetic pole in the Northern Hemisphere under the effect of three independent sources: two major magnetic anomalies (the Canadian and Siberian ones) and the main (dipole) field of the Earth [5]. As it was estimated for the years of 2015–2018, a speed of the magnetic pole movement was not uniform and varied within 37–72 km/year [6]. However, the magnetic field sources are located not only inside the Earth, but also in the circumterrestrial space. These sources are responsible for regular (diurnal, seasonal, 11-year) variations as well as irregular

A. Dmitriev et al. (Eds.): STRPEP 2023, SPEES, pp. 93–103, 2023.
https://doi.org/10.1007/978-3-031-50248-4_10

variations due to the solar activity influence upon the magnetosphere and ionosphere of the Earth [3, 7]. Just these irregular variations cause the geomagnetic activity.

The geomagnetic activity is formed as a response of the processes occurring in the magnetosphere to the changes of the character of the solar wind streams, i.e., in fact, reflects the solar corona structure transformation during the solar activity development. The most pronounced manifestations of the solar activity (magnetic storms) are associated with the influence of coronal mass ejections and high-speed solar wind streams [8]. So, the next peak of the solar activity is expected to occur at the beginning of 2026 [9]. The geomagnetic disturbances in many cycles reach their maximum frequency not near the maximum of the sunspot numbers, but some years later [10, 11]. This is due to the fact that the maximum number of storms with a gradual beginning, that are caused by high-speed solar wind streams flowing from the coronal holes, takes place 2–3 years after the solar activity maximum. While the storms with a sudden beginning caused by the coronal mass ejections are well correlated with the curve of the sunspot numbers. Numerically, a degree of the geomagnetic disturbance is usually described by K-indexes [1, 12, 13].

The effects of the impact of the space environment which are mostly pronounced during magnetic storms can cause the problems with the positioning of navigation systems, satellite electronic circuit failures, interference to radio communications, appearance of extra currents in power lines, induced currents in pipelines, damage of power systems like transformers [9, 14]. The storms are caused by an intensification of the ring current (magnetospheric electrons and ions with an energy of 10–300 keV) and its movement closer to the Earth, which results in the depression in the geomagnetic field H-component [15]. Therefore, the study of the magnetic field variation pattern still remains a topical problem as it serves as a basis of the surrounding outer space investigations.

2 Historical Investigations of the Earth's Magnetic Field in Belarus

The characteristics of the magnetic field of the Earth was measured in Minsk for the first time in 1875 by I.N.Smirnov, an Associate Professor of the Kazan University (Russia), who was among the pioneers of the magnetic survey in Belarus [16]. Later, the determinations of the values of geomagnetic elements in Minsk were repeated by the staff of the Main Geophysical Observatory (St. Petersburg, Russia) in 1904, 1924 and 1927. Subsequent measurements of the geomagnetism elements were carried out at the base observation station of the secular variations in Minsk, which was created by the Institute of Terrestrial Magnetism (now Institute of Terrestrial Magnetism, Ionosphere and Radio Wave Propagation or IZMIRAN) [16]. The Earth's magnetic field measurements were made there in 1938, 1945, 1947, 1948, 1949, 1954, 1959. Since 1960 the absolute and variational measurements of the geomagnetic field elements are performed at the Pleshchenitsy Geophysical Observatory located in Pleshchenitsy settlement near Minsk [17], which will be described in the next section.

All the above-mentioned measurements make possible an analysis of the changes, that were experienced by individual Earth's magnetic field elements during the period from 1875 to 1960, i.e., for 85 years. For this purpose all the data of observations made at the Minsk station were reduced to the values obtained at the Pleshchenitsy Geophysical

Observatory on the basis of simultaneously determined values of the geomagnetic field elements at both stations in 1959 and 1960 [16].

The curves of the annual mean values of the field elements *D, H, Z, F* (Fig. 1) show that the magnetic field of the Earth in the region of Minsk changed considerably since 1875 till 1960. So, during this period the magnetic declination *D* changed from the western one equal to 3°2′ in 1875 to the eastern one about 1913 and reached 4°55′ till 1960. Hence, the overall change of the magnetic declination made up more than 8°. The vertical component *Z* and an absolute value of the full vector of Earth's magnetic field strength *F* have increased for the above-mentioned period by 2520 and 1980 nT, respectively (Fig. 1).

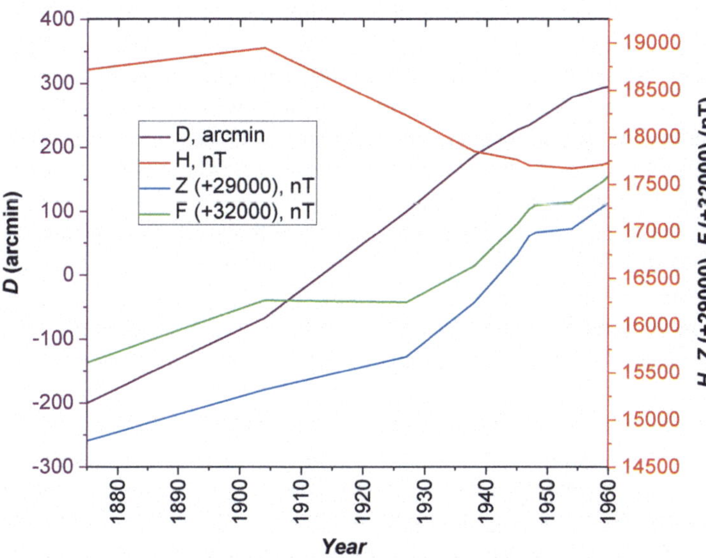

Fig. 1. Changes in the annual mean values of the magnetic declination *D*, horizontal *H* and vertical *Z* components, as well as of the full geomagnetic field vector modulus *F* during the period of 1875–1960

The established increase in annual mean values of the geomagnetic field elements *D, Z, F* is mostly stipulated by the physical and chemical processes in the Earth's interior that result in the continuous migration of the Earth's magnetic poles [4–6]. In addition, the field secular variations are influenced by even faster processes due to the solar activity. The solar activity may either increase or decrease, to some extent, the secular variations of the geomagnetic field elements [2]. A combination of these two major impacts is responsible for the secular variation behaviour in certain region of the Earth's surface.

3 Description of the Pleshchenitsy Geophysical Observatory

The complex Pleshchenitsy Geophysical Observatory for investigation of geomagnetic field and seismic activity was created in 1958 within the framework of the participation of the Academy of Sciences of the Belarusian SSR in investigations for the International

Geophysical Year program and in the further international and national geophysical projects and programs. This was the first similar scientific institution in the territory of Belarus, which until 1963 was a part of the Institute of Geological Sciences of the Academy of Sciences of the Belarusian SSR. Scientific investigations in physics of the Earth as new field of research in Belarus have started at the Pleshchenitsy Observatory.

The research activity of the observatory contributed to scientific communications with the leading geophysical institutes of the former USSR. At present, the Pleshchenitsy Geophysical Observatory is a part of the Center of Geophysical Monitoring of the National Academy of Sciences of Belarus.

The Pleshchenitsy Geophysical Observatory (MNK code) is situated 65 km away from Minsk, on the northwestern outskirts of the Pleshchenitsy settlement, Logoisk district, Minsk region, Republic of Belarus. Its geographical coordinates are $\varphi = 54.4186°$N, $\lambda = 27.7958°$E.

The observations of the geomagnetic field variations since 2002 till 2009 were carried out using an automated digital magneto-variation station (manufactured by the Institute of Terrestrial Magnetism, Ionosphere and Radio Wave Propagation, Russia) which continuously recorded the time changes of the horizontal *H,* vertical *Z* components and of the angular component, namely, declination *D*.

Since 2010 the main measuring instrument is a three-component fluxgate magnetometer LEMI-022 (manufactured by the Lvov Center of the Space Research Institute of the National Academy of Sciences of Ukraine and the National Space Agency of Ukraine), which serves to measure the northern *X*, eastern *Y*, vertical *Z* components of the Earth's magnetic field strength and their variations in the frequency range from 0 to 0.3 Hz.

To determine the base-line values of the geomagnetic field elements D_0, H_0, Z_0 and F_0, the absolute observations are carried out with a fluxgate fDI magnetometer LEMI-204 based on a nonmagnetic theodolite 3T2KP-NM, as well as with a scalar Overhauser magnetometer MINIMAG (manufactured by Scientific Production Enterprise "Geologorazvedka", St. Petersburg, Russia). The mean squared error of angle measurement is $2''$, and the resolution of the magnetometer MINIMAG is 0.03 nT. The absolute values of the *D, H, Z* components and full geomagnetic field vector modulus *F* are regularly measured. These are used to monitor the operation of the magnetic variometer and to determine the base-line values of the field elements.

At present, a three-component digital magneto-variation station "Quartz-7" (manufactured by the Institute of Terrestrial Magnetism, Ionosphere and Radio Wave Propagation, Russia) to perform variational measurements, as well as fDI-magnetometer MinGeo based on a nonmagnetic theodolite Theo-020B (MinGeo Ltd, Hungary) and a scalar Overhauser magnetometer POS-1 (Institute of Physics and Technology of the Ural Federal University, Russia) for absolute observations are adjusted with an aim to put them into operation.

4 Observations of the Geomagnetic Field Elements at the Pleshchenitsy Geophysical Observatory

Continuous variational observations of the geomagnetic field elements *D, H, Z* are carried out at the Pleshchenitsy Geophysical Observatory since 1960 in order to study the secular variations of the geomagnetic field in Belarus.

For processing data obtained with a fluxgate magnetometer LEMI-022, a software has been developed at the Center for Geophysical Monitoring of the National Academy of Sciences of Belarus and permits the following procedures to be performed:

- visualization of obtained data on display;
- formation of a daily table of the geomagnetic field variations with minute data;
- formation of a monthly table of the geomagnetic field variations with hourly data;
- determination of daily three-hour *K*-index values;
- determination of characteristics of the magnetic storms;
- input of new corrections to base-line values;
- export of the data processing results into the database.

Data are processed according to the standard procedure. The average values of the field elements for a day, month, year are determined from their average hourly values. So, according to the observation data, the annual mean values of the magnetic field components in 2022 were as follow: $D = 9.271°$, $H = 17749$ nT, $Z = 48329$ nT, $F = 51485$ nT.

During the whole year of 2022 an increase of the *D, Z, F* element values was observed, like as in the previous years [2]. The secular variations of the geomagnetic field elements are calculated as a difference between their annual mean values for two selected years (Table 1).

Table 1. Secular variations (last column) of the annual mean values of the geomagnetic field elements between the years of 2022 and 2021

Elements	Year: 2022	Year: 2021	Difference
D (in ang. Values)	$9^\circ 16' 12''$	$9^\circ 08' 09''$	$0^\circ 08' 03''$
H (nT)	17749	17766	–17
Z (nT)	48329	48258	71
F (nT)	51485	51423	62

The annual mean values of the *D, H, Z, F* elements of the Earth's magnetic field according to data obtained at the Pleshchenitsy Geophysical Observatory within 1960–2022 are plotted in Fig. 2.

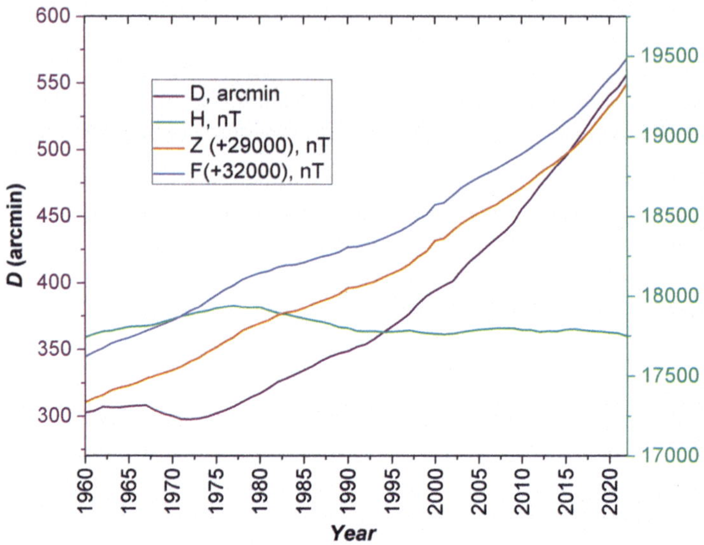

Fig. 2. Changes in the annual mean values of the magnetic declination *D*, horizontal *H* and vertical *Z* components, as well as of the full geomagnetic field vector *F* during the period of 1960–2022

Secular variations of the annual mean values of the geomagnetic field elements between the years of 2022 and 1960 are as follow:

- 253.56 arcmin ($4^\circ 13'34''$) for the geomagnetic field declination *D*; the average annual variation is 4.09 arcmin ($4'5''$), though since 2009 the secular variation averages $8'34''$ per year, which is almost 2.1 times the average long-term variation for 63 years of observations;
- 7.4 nT for the horizontal component *H* of the geomagnetic field; the average annual variation is 0.1 nT;
- 1991.2 nT for the vertical component *Z* of the geomagnetic field; the average annual variation is 32.1 nT;
- 1864.5 nT for the full geomagnetic field vector *F*; the average annual variation is 30.1 nT.

An increase of the annual mean values of the observed geomagnetic field elements *D, Z, F* is indicative of the continuing displacement of the Earth's magnetic poles.

5 Analysis of the Geomagnetic Field Disturbance Carried Out at the Pleshchenitsy Geophysical Observatory

Monthly reviews of the magnetic field state composed at the Pleshchenitsy Observatory describe the magnetic field disturbance using three-hourly values of the *K*-indexes (9-scores scale) and the magnetic storm characteristics. The following scale of *K*-indexes reflecting the deviation of the most disturbed component of the geomagnetic field was adopted at the Pleshchenitsy Observatory (Table 2).

Table 2. Scale of *K*-indexes adopted at the Pleshchenitsy Geophysical Observatory

K-index value	0	1	2	3	4	5	6	7	8	9
Upper limit (nT)	5	10	20	40	70	120	200	330	550	>550

Magnetic storms are described by their duration (beginning and end), as well as by an amplitude of the *D, H, Z* element variations in accordance with the scale presented in the Table 3.

Table 3. Scale of magnetic storms adopted at the Pleshchenitsy Geophysical Observatory

Storm type	*D* (arcmin)	*H* (nT)	*Z* (nT)
minor storm	19–26	80–125	40–90
moderate storm	27–38	126–200	91–140
major storm	39–55	201–270	141–250
severe storm	>55	>270	>250

As it is known, the impact of the solar plasma stream (solar wind) on the Earth's magnetosphere, the internal magnetospheric changes and the magnetosphere and ionosphere interaction influence the geomagnetic field disturbance which manifests itself as irregular magnetic field variations [3, 7, 8]. In this context, the dynamics of the geomagnetic field disturbance determined from the data obtained at the Pleshchenitsy Observatory were analyzed in comparison with the solar activity within four adjacent 11-year cycles: 21st (1976–1986), 22nd (1986–1996), 23rd (1996–2008), 24th (2008–2019) (Fig. 3). The solar activity is usually described by the Wolf numbers (*W*) that are sunspot relative numbers. The annual mean Wolf numbers are taken in accordance with the Solar Bulletin published by the American Association of Variable Star Observers (AAVSO) [18].

Long-term variations of the magnetic storm number suggest that phases of storm reduction are followed by phases of increasing number of storms (Fig. 3). The maximum value (62) of the magnetic storm number was observed in 1989 and the minimum one (2) was recorded in 2009. The period between the minimum values is 11–12 years that coincides with the current period of solar activity. At the same time, the maxima of

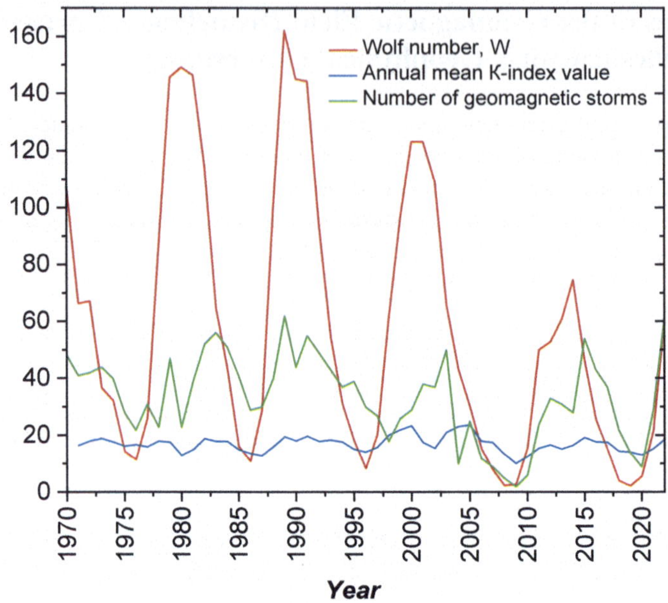

Fig. 3. Dynamics of the geomagnetic field disturbance

the magnetic storm number do not coincide with the maxima of Wolf numbers, but are shifted toward the phase of decreasing solar activity that is reported by other authors too [10].

It should be also noted that an increase of the geomagnetic field disturbance is observed in 2021–2022 simultaneously with and an increase of the solar activity. So, 61 magnetic storms were recorded and processed in 2022, which exceed by 32 the number of storms recorded in 2021 (29) and is 1.86 times the average value (32.8) of long-term observations over 52 years (1970–2021). During the 2023 first half-year 24 storms were recorded, which is indicative of a high degree of the geomagnetic field disturbance in the current year.

How does the frequency of solar storms correlate with the solar activity? The number of solar events in the years near the solar cycle maximum is really larger than that at the solar cycle minimum. So, up to several solar flares per day can be observed sometimes at the cycle maximum. However, it was noted that the solar flare activity and mass ejections are not rare events [19] and they are sometimes characterized by extreme parameters in the years near the solar cycle minimum, especially, in the phase of the solar activity decrease [10, 11]. Therefore, a question about a connection between the cycles of the solar activity and its manifestation in the interplanetary environment like such high-energy processes as solar flares and coronal mass ejections requires further study.

So, in the 21st cycle of the solar activity the maximum of annual mean relative number of sunspots was observed within 1979–1980. In the subsequent years the intensity of the solar activity was decreasing and reached its minimum in 1986. Nevertheless, the geomagnetic activity during the solar activity decrease phase remained rather high and even slightly increased in 1983 as it follows from the number of recorded magnetic disturbances. The similar phenomenon was observed in the next cycles, namely, in 1995 (22nd cycle), 2003 (23rd cycle), and 2015 (24th cycle) in the solar activity reduction phase.

This phenomenon was discussed both in the earlier [20], and more recent [9, 13] works, where it was indicated that the maxima of the geomagnetic field disturbance appeared in the *W* decrease phase are mainly associated with recurrent perturbations which have gradual beginning, but the maxima of the geomagnetic field disturbance observed near the solar activity maxima should be associated with flare magnetic storms. This assumption is confirmed by long-term observations carried out at the Pleshchenitsy Geophysical Observatory.

Long-term variations of the annual mean values of total *K*-indices show the more complicated pattern, these are also subject to fluctuations, but their period is less than the 11-year solar activity period (Fig. 4). Nevertheless, in the years when the Wolf numbers *W* have the minimum values (1976, 1986, 1996, 2008, 2019), the annual mean values of *K*-index also tends to the minimum values. In addition, with the solar activity increasing the annual mean values of *K*-index also rises, but, in most cases, with one year lag (Fig. 4). Such a delay is also noted in some other works [8, 13]. However, after the early growth stage a decrease of the annual mean values of *K*-index is observed. This drop could be sufficiently great in some cases, like in 1980 and 2001, though the solar activity in these years still remained high.

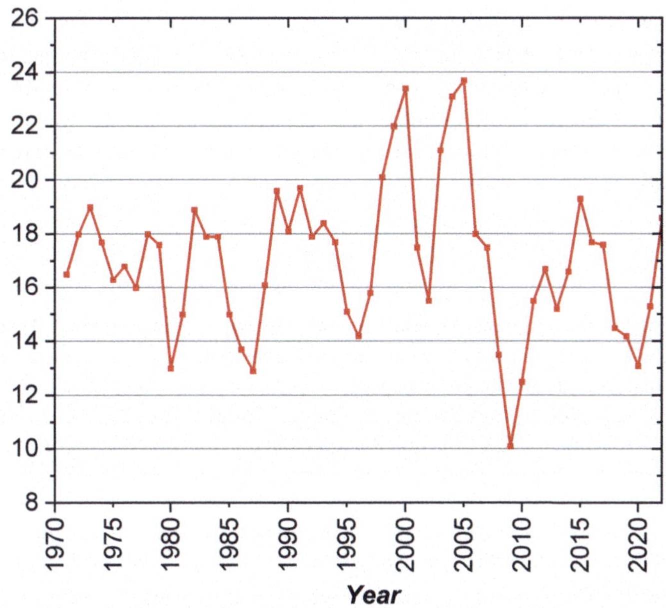

Fig. 4. Time series of the annual mean values of the total *K*-indices during 1971–2022

The maximum values of the annual mean total K-indices (>23) were recorded in 2000 and 2005, the minimum ones (<13) were observed within 2009–2010. In 2021–2022 a growth of the annual mean values of the total K-index was outlined which testifies to the onset of the next activation period within the 25th solar cycle.

Therefore, it may be concluded that a definite correlation between such a parameter of the geomagnetic field disturbance as K-index and the activity of the Sun takes place. However, like as in the case with the quantity of the magnetic storms, there is one or two years lag between a stage of growth and a stage of decrease with respect to the analogous phases of the11-year solar cycles, as well as more than one maximum of the K-index annual mean values within one solar cycle (see Figs. 3 and 4). These facts point once again to a contribution of the recurrent magnetic disturbances, which have gradual beginning, into geomagnetic activity along with the flare magnetic storms [8].

6 Conclusions

The main result of the work of the Pleshchenitsy Geophysical Observatory (Minsk, Republic of Belarus) is the continuous determination of the secular variations of the Earth's magnetic field elements *D, H, Z, F* on the basis of data from variational and absolute observations carried out since 1960. Annual reviews of the geomagnetic field state are composed and the annual mean values of the geomagnetic field elements *D, H, Z, F* are calculated. An increase in the geomagnetic field elements *D, Z, F* values observed is indicative of the continuing displacement of the Earth's magnetic pole in the Northern Hemisphere.

A periodic character of the change of the geomagnetic field disturbance which corresponds to the solar activity cycles with a shift towards the solar activity decrease phase is observed. In the average, the geomagnetic activity level in 2022 was determined to be higher than that in the previous year and the geomagnetic field was considered to be rather disturbed.

References

1. Parkinson, W.D.: Introduction to Geomagnetism. Scottish Academic Press, Edinburg (1983)
2. Karatayev, G.I., Karagodina, O.I.: Space-time characteristics of the magnetic declination in the territory of Belarus and practical aspects of its monitoring. Litasfera **29**, 127–135 (2008). (in Russian), https://lithosphere.by/wp-content/uploads/2018/01/127_135_Karat_Karag.pdf
3. Adushkin, V.V., Riabova, S.A., Spivak, A.A.: Geomagnetic Effects of Natural and Industrial Processes. GEOS, Moscow (2021). (in Russian), https://doi.org/10.34756/GEOS.2021.16.37855
4. Dobrica, V., Stefan, C., Demetrescu, C.: Planetary scale geomagnetic secular variation foci in the last 400 years. Glob. Planet. Change **199**, 103430 (2021). https://doi.org/10.1016/j.gloplacha.2021.103430
5. Kuznetsov, V.V.: Physics of the earth: textbook-monograph. Novosibirsk (2011). (in Russian), https://www.geokniga.org/bookfiles/geokniga-fizika-zemli_5.pdf
6. Regi, M., Di Mauro, D., Lepidi, S.: The location of the earth's magnetic poles from circum-terrestrial observations. J. Geophys. Res. Space **126**, e2020JA028513 (2020). https://doi.org/10.1029/2020JA028513

7. Kilifarska, N.A., Bakhmutov, V.G., Melnyk, G.V.: The hidden link between earth's magnetic field and climate. Elsevier Inc., Amsterdam, Oxford, Cambridge (2020). https://doi.org/10.1016/C2018-0-01667-9
8. Bezrodnykh, I.P., Morozova, E.I., Petrukovich, A.A., Kozhukhov, M.V.: Dynamics of solar and geomagnetic activity. II. Periodic variations in solar and geomagnetic activity. In: Matters of Electromechanics. VNIIEM Proceedings, vol. 171, No. 4, pp. 24–38. JC "VNIIEM Corporation", Moscow (2019). (in Russian), https://jurnal.vniiem.ru/text/171/24-38.pdf
9. Owens, M.J., Lockwood, M., Barnard, L.A., Scott, C.J., Haines, C., Macneil, A.: Extreme space-weather events and the solar cycle. Sol. Phys. **296**, 82 (2021). https://doi.org/10.1007/s11207-021-01831-3
10. Obridko, V.N., Kanonidi, H.D., Mitrofanova, T.A., Shelting, B.D.: Solar activity and geomagnetic disturbances. Geomagn. Aeron+. **53**(2), 157–166 (2013). (in Russian), https://doi.org/10.7868/S0016794013010148
11. Reyes, P.I., Pinto, V.A., Moya, P.S.: Geomagnetic storm occurrence and their relation with solar cycle phases. Space Weather **19**(9), e2021SW002766 (2021). https://doi.org/10.1029/2021SW002766
12. Love, J.J., Chulliat, A.: An international network of magnetic observatories. Eos **94**(42), 373–374 (2013). https://doi.org/10.1002/2013EO420001
13. Matzka, J., Stolle, C., Yamazaki, Y., Bronkalla, O., Morschhauser, A.: The geomagnetic *Kp* index and derived indices of geomagnetic activity. Space Weather **19**(5), e2020SW002641 (2021). https://doi.org/10.1029/2020SW002641
14. Amiantov, A.S., Zaitzev, A.N., Odintzov, V.I., Petrov, V.G.: Variations of the Earth's magnetic field – a database of magnetic observatories of Russia for the period 1984–2000. CD-ROM. IZMIRAN, Moscow (2001). (in Russian). http://wdcb.ru/stp/data/geo_min.val/Variational_Measurements/Database_Earth_Magnetic_Field_Variations/Variations_of_the%20Earth_Magnetic_Field_Database.pdf
15. Lakhina, G.S., Tsurutani, B.T.: Geomagnetic storms: historical perspective to modern view. Geosci. Lett. **3**, 5 (2016). https://doi.org/10.1186/s40562-016-0037-4
16. Babushnikov, M.S.: On the secular variations of the geomagnetic field in the Minsk region. In: Khotko, Z.P. (ed.) Study of the Earth's electromagnetic field at the Pleshchenitsy (Minsk) geophysical station, pp.37–44. BSSR Academy of Sciences Publ., Minsk (1966). (in Russian)
17. Khotko, Z.P., Kuznetsov, Y.N.: Geophysical Observatory Pleshchenitsy (Minsk). Navuka i Tekhnika Publ., Minsk (1993). (in Russian)
18. Yearly American Sunspot Numbers. Solar Bulletin. American Association of Variable Star Observers – Solar Division. http://www.wdcb.ru/stp/data/solar.act/sunspot/AAVSO/table_yearly.txt
19. Grodji, O.D.F., et al.: A study of solar flare effects on the geomagnetic field components during solar cycles 23 and 24. Atmosphere – Basel. **13**(1), 69 (2022). https://doi.org/10.3390/atmos13010069
20. Vitinsky, Y.I., Ol, A.I., Sazonov, B.I.: The Sun and the Atmosphere of the Earth. Gidrometeoizdat Publ., Leningrad (1976). (in Russian)

Advanced Instruments for Geo and Helio Environment Monitoring on the Cubesat Format Spacecraft

Vitaly Bogomolov[1,2](✉), Sergey Svertilov[1,2], Vladislav Osedlo[1], Victor Bengin[1], Ivan Zolotarev[1], Anatoly Iyudin[1], Oleg Nechaev[1], Ivan Yashin[1], Georgy Antonyuk[1], Ivan Kucherenko[1,2], and Mikhail Korzhik[3]

[1] D.V. Skobeltsyn Institute of Nuclear Physics, M.V. Lomonosov Moscow State University, Leninskie Gory 1(2), GSP-1, 119991 Moscow, Russia
bogovit@rambler.ru

[2] Physics Department, M.V. Lomonosov Moscow State University, Leninskie Gory, 119991 Moscow, Russia

[3] Institute for Nuclear Problems of Belarusian State University, Bobruiskaya 11, 220030 Minsk, Belarus

Abstract. Within the framework of the Moscow University project "Constellation-270", scientific instruments for the cubesat format satellites were developed, including the universal detectors of cosmic radiation DeCoR, DeCoR-2, and the KODIZ instrument, which is a combined detector of cosmic rays and other space particles. The DeCoR and DeCoR-2 instruments are designed to study fast variations in electron fluxes, as well as gamma-ray bursts of various nature. The DeCoR instrument is a scintillation spectrometer of gamma rays and electrons, its energy range is 0.05–2.0 meV, effective area is about 18 cm^2. It successfully operated on satellites of the cubesat format SiriusSat-1,2, AmurSat, VDNKh-80, Norbi, DEKART, etc. The DeCoR-2 instrument is a modified version of the DeCoR instrument, characterized by an increased sensitive area up to ~60 cm^2 and a wider energy range of 0.03–3.0 meV. The KODIZ instrument is designed to test equipment designed to detect radiation-dangerous fluxes of solar cosmic rays. The instrument includes a Cherenkov detector for detecting relativistic protons, semiconductor detectors and neutron detector. All mentioned devices are now operating in space. Modified instrument with pixelated detector based on GAGG:Ce scintillators is planned for launch in the next year

1 Introduction

The study of ionizing radiation fluxes in outer space is one of the most important areas of space research. These studies are extremely relevant both from the point of view of studying the physical patterns of formation and spatio-temporal variations of energetic particle and gamma radiation fluxes, and from the point of view of determining their influence on the functioning of spacecraft. According to available expert estimates, more than half of the failures and malfunctions in the operation of onboard spacecraft systems occur due to adverse effects on materials and equipment elements of spacecraft of the space environment, the main role among which is played by radiation effects [1].

A. Dmitriev et al. (Eds.): STRPEP 2023, SPEES, pp. 104–112, 2023.
https://doi.org/10.1007/978-3-031-50248-4_11

Physical processes associated with changes in the radiation situation in near-Earth space are determined primarily by energetically active processes in the Sun and, for the most part, are the trigger for the development of geophysical phenomena in the Earth's magnetosphere. The question of the role of acceleration in the pulsed phase of flares in the formation of the longest and most intense events in solar cosmic rays (SCR), in particular, solar proton (SPE) and electron events, remains open. The appearance of energetic particles on the Sun in the post-flare phase can be explained within the framework of various models, and there are no problems with the acceleration of electrons, although the conditions for acceleration, motion and loss of energy by electrons in loops after the flare differ from the case of protons and nuclei. The indisputable evidence of electron acceleration, which is the detection of electromagnetic radiation in the source, is not yet a guarantee of the effective release of particles into interplanetary space.

The interrelation of physical processes on the Sun and near the Earth is not unambiguous due to the peculiarities of the processes of transfer, acceleration and death of charged particles under various conditions characteristic of electromagnetic processes inside the Earth's magnetosphere and beyond. For example, active processes in the Sun can lead to the generation of powerful coronal emissions of solar plasma and energetic particles, but not all such phenomena necessarily lead to dangerous disturbances of radiation fields in near-Earth space. On the other hand, "invisible" manifestations of the Sun's activity on its reverse side can lead to "unexpected" radiation threats. In this regard, monitoring the flows of charged particles in near-Earth space, as well as patrolling solar flare activity in the hard range of the electromagnetic spectrum are extremely necessary for early warning of radiation hazards.

Further progress on this issue requires correlated observations of hard X-ray bursts and energetic solar particles, which can be implemented within the same space experiment. As follows from the above, the space weather forecasting experiment should provide constant monitoring of solar flare activity through observations in the hard range of the electromagnetic spectrum, as well as the detection of energetic charged particles (primarily electrons and protons) of solar cosmic rays. The scientific goals of such an experiment are determined by the relevance of studying the solar flare activity, since it is this activity that ultimately determines the properties of interplanetary and near-Earth space. In particular, measurements of energetic particle fluxes in the polar regions of the Earth, where particles can freely penetrate without being deflected by the Earth's magnetic field, will allow us to estimate particle fluxes in interplanetary space outside the magnetosphere.

From the point of view of radiation exposure, the flows of energetic charged particles in near-Earth space, capable of penetrating into the spacecraft body and destroying its electronic components, play a decisive role. Basically, these are streams of energetic electrons and protons from the Earth's radiation belt, captured, quasi-captured and spilled out, as well as short-term intense streams of energetic particles from powerful solar flares. It is also necessary to take into account the possibility of the appearance of relativistic and subrelativistic energy electrons in low orbits, accelerating at high-altitude electromagnetic discharges.

It should be noted that real flows of charged particles near the Earth, even in geomagnetically calm conditions, are subject to significant medium- and long-term fluctuations associated with solar and geomagnetic activity, including changes in the Earth's magnetic field and changes in the density of the upper atmosphere. Even more significant changes in the flow can occur briefly during disturbances of the magnetosphere. For example, variations in the fluxes of relativistic electrons of the outer belt according to measurements on board the GOES and CORONAS series satellites from 1994 to 2011 [2] and variations in radiation doses according to measurements on the GLONASS spacecraft, whose orbit is also located in the area of the outer radiation belt, from 2006 to 2010 [3] showed that that the variations in the values of electron fluxes and doses measured on these satellites are more than 2 orders of magnitude.

In addition, significant streams of energetic particles emitted during powerful solar flares can penetrate into the region of the Earth's orbit. During such events, proton fluxes with energies of tens and hundreds of MeV, capable of penetrating beyond the protective shield of the spacecraft, can exceed the background values of proton fluxes of galactic cosmic rays by 3–4 orders of magnitude or more for up to several days. Such events are even less predictable, so the existing models of energetic particle fluxes SCR [4, 5] are probabilistic.

Thus, the actual flows of energetic charged particles in the orbits of satellites may differ significantly from the model values. For this reason, new experimental measurements are needed that can provide information about the radiation situation in the orbits of operated satellites at the current time. To predict the radiation situation, it is necessary to observe the activity of solar flares in the hard range of the electromagnetic spectrum simultaneously with monitoring measurements of charged particle fluxes in near-Earth space. The corresponding instruments can also be used to observe transients in X-ray and gamma radiation, both of astrophysical and atmospheric origin.

When planning measurements, it should be taken into account that particles of atmospheric, ionospheric, magnetospheric, solar and galactic origin are observed in a wide range of energies and intensities; spectra of electrons, protons, neutrons and gamma rays in various energy ranges - from low-energy ionospheric plasma particles to particles of cosmic rays of solar and galactic origin, as well as electromagnetic radiation (in the gamma range) of solar, atmospheric and astrophysical origin. To implement these tasks, it is necessary to develop and manufacture a complex of new generation devices with high temporal and spectral characteristics, as well as sensitivity.

Monitoring observations of rapid changes in hard X-ray radiation fluxes, including during solar flares, as well as measurements of charged particle fluxes in near-Earth space can also be carried out on small cubesat spacecraft. In this case, the DeCoR device (cosmic radiation detector) and its improved analog DeCoR-2 were designed to study short-term variations of various components of cosmic radiation, primarily high-energy electrons and gamma quanta caused by dynamic processes in the magnetosphere (acceleration and deposition of electrons) in the atmosphere (high-altitude electrical discharges leading to generation so-called electromagnetic transients, including in the gamma-ray range), hard X-ray and gamma-ray radiation from solar flares, as well as gamma-ray bursts of astrophysical nature.

Within the framework of the Moscow University space project "Constellation-270" in 2023, on June 27, several nanosatellites were launched into a sun-synchronous orbit. Among them is the Avion satellite, made in the cubesat-6U standard, and a number of satellites in the cubesat-3U standard: Monitor-2, 3, 4, Sirius-SINP-3U, UTMN2. All of these satellites were equipped with one or more instruments developed at the Moscow State University designed to study the temporal and spectral parameters of electrons and gamma rays. The subject of the study will be cosmic gamma-ray bursts, solar flares, electron precipitation and variations of particle fluxes associated with solar activity.

2 Instruments for Radiation Monitoring on Cubesats

2.1 The DeCoR Instrument

The devices of the DeCoR family (Detectors of Cosmic Radiation) are designed for spectral and temporal measurements of gamma radiation and electrons in the energy range from tens of keV to several MeV. They are scintillation spectrometers using a combination of plastic and crystal scintillators for separate registration of charged particles and gamma radiation. Devices of this type, developed at the Nuclear Research Institute of Moscow State University, were launched in 2018 as a payload of scientific and educational nanosatellites SiriusSat-1 and SiriusSat-2 in the Cubesat-1U format, which successfully operated in orbit until December 2020 [6]. The equipment of these small satellites combined monitoring measurements, during which the counting rates in several channels were transmitted to Earth with time resolution of 1 s, with the possibility of an event-by-event recording, during which the amplitude characteristics and the exact time of all interactions in the detector are recorded. Taking into account the significant limitations on the data amount that could be transmitted from orbit, the option of conducting a space experiment was chosen, when the recording of event-by-event data in the non-volatile memory of the device was activated at a pre-selected point of the orbit. The amount of memory allowed to store the results of detailed measurements for approximately 5 orbits. The most interesting data sections selected after viewing the results of monitoring conducted at the same time, were transmitted to Earth during a number of subsequent communication sessions. This mode of operation made it possible, in particular, to study the characteristics of fast variations of subrelativistic electron fluxes in the region of the gap between the belts [7].

The device, called DeCoR (Cosmic Radiation Detector), was modified for a number of cubesats launched in subsequent years. Its detector element is a combination of a plastic scintillator ~3 mm thick and a CsI (Tl) crystal ~10 mm thick with a sensitive area of 18 cm^2, viewed by two miniature PMT. Figure 1 shows a block diagram explaining the operation of the DeCoR device. The electronics of the device analyzes the shape of the pre-amplified pulse from the output of the photodetectors, determining the type of scintillator in which the interaction occurred. To do this, separate integration of the initial part of the signal (for the first ~0.5 μs from the moment of interaction) and its subsequent part is done, and signals of the so-called fast and slow components are formed, separately for each PMT. These generated signals are digitized using the ADC of the microcontroller, which conducts their further processing.

Similarly to SiriusSat satellite instruments, the output data are generated both in the monitoring format (the counting rate in several channels corresponding to a certain type of particles and energy) and in the form of a detailed recording in an event-based format, when a set of amplitudes and an exact timer time with a resolution of ~1 μs are recorded

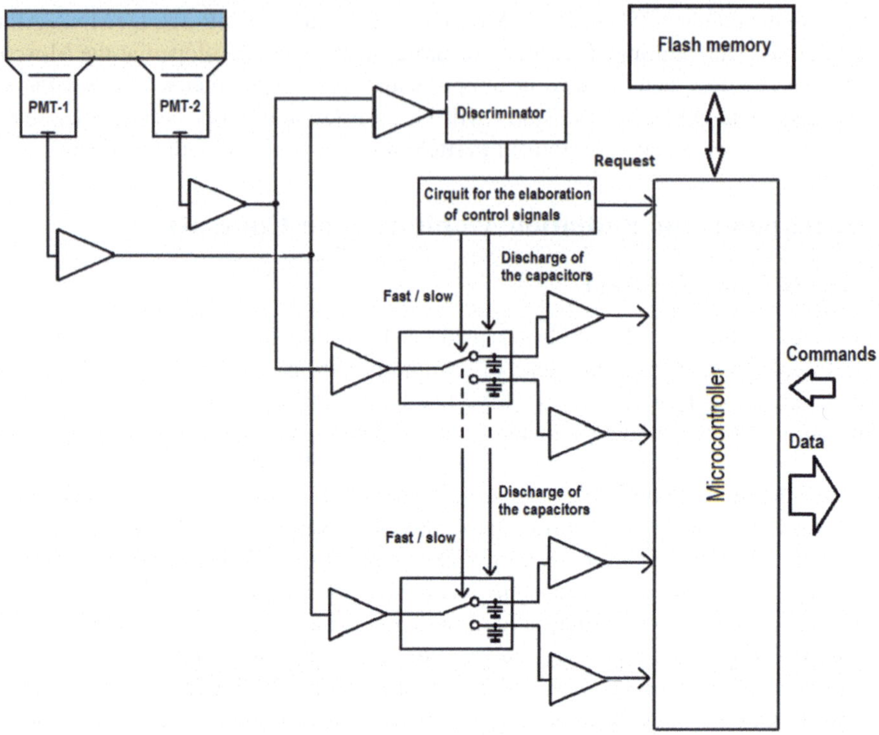

Fig. 1. Block diagram of the DeCoR device

for each interaction in the detector. The data are stored in the non-volatile payload memory, then they can be transmitted either directly to the satellite radio transmitter or to the on-board computer memory. Thus, during a space experiment, it is possible to select the most important sections of data to be transmitted to Earth in primary form. The particle detectors in the DeCoR devices were supplemented with a triaxial magnetometer, which allows taking into account the orientation of the device relative to the Earth's magnetic field when analyzing electron flux variations. With the help of DeCoR devices, near-Earth electron fluxes have been studied for several years on the VDNKH-80, Norbi satellites and several other Cubesat format nanosatellites [8, 9]. Since these satellites were launched into polar orbit, the DeCoR instruments observed cases of solar cosmic rays coming to Earth and electron precipitations in polar region.

In June 2023, a number of Cubsat-3U (Monitor-2, Monitor-3, Monitor-4, UTMN-2, Sirius-SINP-3U) and Cubsat-6U (Avion) spacecraft equipped with DeCoR-2 instruments optimized for detection and study of the gamma-ray bursts of various nature were launched into a solar synchronous orbit. This device has an effective area increased to ~64 cm^2, which is necessary both to increase sensitivity when searching for weak bursts, and to improve the time resolution, which is determined primarily by the statistics of the recorded gamma quanta. A composite scintillation detector consisting of a 3 mm thick plastic scintillator and a 9 mm thick CsI (Tl) in the shape of a square with a side of 8 cm is viewed by an assembly of silicon photomultipliers (SiPM). The use of a combination of a plastic scintillator with which gamma quanta practically do not interact, and a crystal with a high efficiency of gamma quanta registration provides separate detection of gamma radiation and electrons in the energy release range from 20 keV to 1 meV. This makes it possible to distinguish between cases of increased readings caused by cosmic gamma-ray bursts and electron precipitation, which is very important when conducting an experiment to study gamma-ray bursts in polar orbit.

DeCoR-2 devices use two identical electronics boards that register events in half of the scintillation detector. Each of these boards contains its own chips of analog electronics and power converters, as well as its own microcontroller that processes signals, accumulates data and transmits it to on-board systems via the CAN interface. This architecture of the device not only increases reliability, but also increases the dynamic range of the device, and also allows one to compare the flow variations observed by independent parts of the detector.

2.2 The KODIZ Instrument

The complex radiation detector KODIZ is designed for use on small spacecraft, including cubesat format satellites. In the device, due to the use of semiconductor detectors and a Cherenkov detector, both the registration of the cosmic radiation dose rate and the registration of proton fluxes with an energy greater than 330 meV are provided. A distinctive feature of the device is the small dimensions and weight of the device compared to traditional Cherenkov detectors. The device can be used to detect the appearance of a stream of particles of solar cosmic rays capable of creating an additional radiation load on board high-altitude aircraft.

A distinctive feature of this device is the registration of light that occurs in the Cherenkov detector simultaneously by both the PMT and the SiPM detector. This is done in order to be able to compare the results of Cherenkov light registration by various photodetectors and evaluate the possibility of switching, in the future, to SiPM detectors. The device provides registration with a time resolution of 10 s of the dose rate of cosmic radiation (electrons, protons and heavier nuclei) in the range from 10^{-8} to 10^{-5} Gy/s., flows of protons and nuclei with $Z > 1$ with energy greater than 30–50 meV/nucleon in the range from 10^1 to 10^4 particles/cm^2, flows protons and nuclei with $Z > 1$ with energy greater than 330 meV / nucleon in the range from 10^1 to 10^3 particles/cm^{2}s, as well as fluxes of thermal and epithermal neutrons in the range from 10^1 to 10^3 neutrons/cm^{2}s.

The detector system of the device includes:

- two silicon ion-implanted semiconductor detectors with a thickness of 0.3 mm and an area of 1 cm^2;
- Cherenkov detector - plexiglass with a diameter of 38 mm and a thickness of 20 mm (an area of 11.34 cm^2) viewed, as well as a SiPM detector;
- two neutron detectors based on lithium glass (BC-720) with a diameter of 38 mm, a thickness of 20 mm and 5 mm, viewed by the HAMAMATSU R5611A.
- Neutron detectors can also be used to register Cherenkov radiation, actually tripling the effective area of registration of high-energy protons.

3 Examples of Radiation Measurements with DeCoR Instruments

The first experience of using the DeCoR device in the configuration described above took place within the framework of the "Universat-SOCRAT" project of Lomonosov Moscow State University, during which the AmurSat and VDNKh-80 nanosatellites were launched in 2019 [8]. With the help of these satellites and the Norby and DEKART satellites launched a year later, also equipped with DeCoR devices, space weather was monitored, in some cases solar cosmic rays were observed from DeCoR [9].

In June 2023, several new nanosatellites equipped with one or more DeCoR devices of various configurations were launched. The Monitor-3, Monitor-4, Sirius-SINP-3U and UTMN2 satellites are equipped with the DeCoR-2 device installed in the end of the cubesat 3U as the only payload. Control and data acquisition from these spacecraft is carried out in the amateur VHF frequency range (~435 MHz). The average volume of scientific information from these satellites is ~0.5 MB per day per receiving station. The Avion and Monitor-2 satellites, made on platforms developed in Kaluga, have a radio module that transmits data in the S-band, which allows increasing the volume of transmitted scientific information to ~100 MB per day. Their payload includes a complex of several DeCoR detector nodes of various configurations complementing each other. In particular, a spectrometric module based on a large-sized CsI(Tl) crystal is installed on these satellites, which allows expanding the energy range of measurements of gamma-ray spectra, which is very important when studying various gamma transients, such as solar flares and cosmic gamma-ray bursts.

In Fig. 2 an example of the readings of the DeCoR device installed on the Avion satellite is presented. One can see variations in the detector counting rate corresponding to both the satellite's passage through the inner radiation belt (South Atlantic Anomaly)

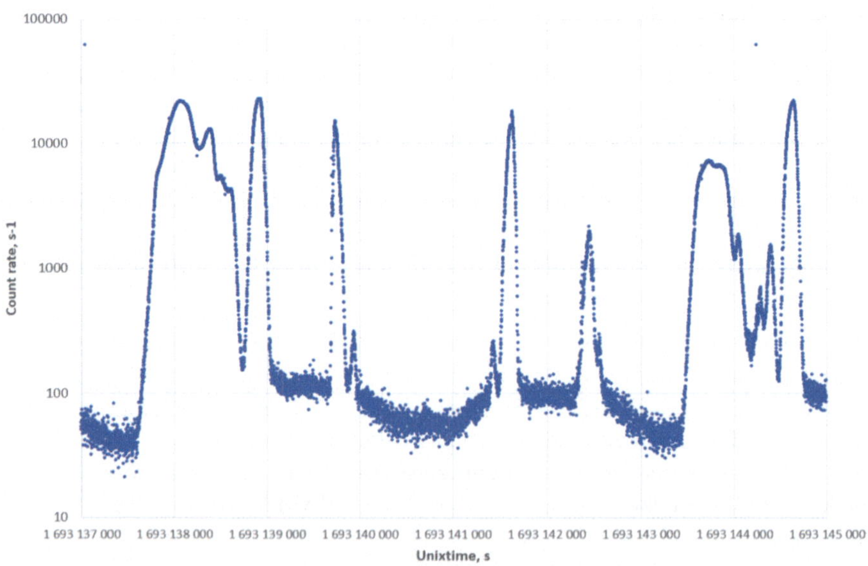

Fig. 2. Readings of the DeCoR device installed on the Avion satellite

and the outer radiation belt, as well as the zones of existence of quasi-captured particles at L~2.2.

4 Next-generation Space Radiation Detectors for Nanosatellites

The improvement of the cosmic radiation detectors developed at the Moscow State University, Skobel'tsyn Institute for Nuclear Research should, while maintaining the mass and dimensional characteristics of the device, improve its capabilities for spectral and temporal analysis, as well as for determining the nature of the observed particle flux or gamma radiation. One of the areas of development is the creation of position–sensitive detectors based on an assembly of 50–100 small scintillation crystals viewed by semiconductor photodetectors. A sketch of such a detector is shown in Fig. 3. The sensitive area is about 60 cm^2, similar with the DeCoR-2 device. Scintillation crystals GAGG:Ce are supposed to be used as detecting elements having a high density of 6.6 g/cm^3, a relatively short illumination time of ~90 ns and energy resolution of 5%–6% for 662 keV [10]. As in the DeCoR devices of previous generation, a combination of a thin layer of plastic scintillator with crystals will be used, while the place of interaction will be determined by the shape of the pulse at the output of the photodetectors.

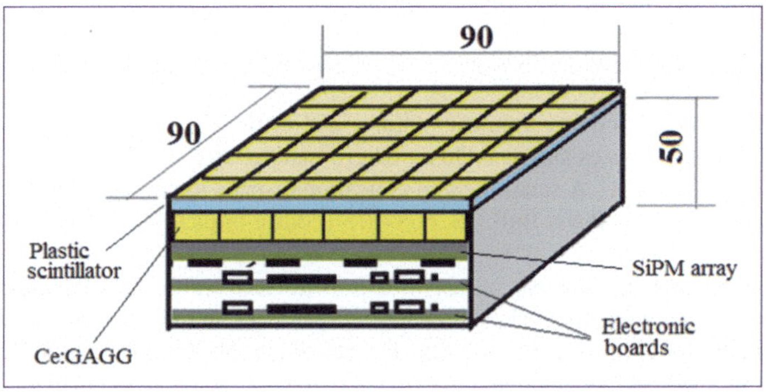

Fig. 3. Sketch of a position sensitive gamma-ray spectrometer.

5 Conclusion

A number of universal instruments have been developed at the M.V. Lomonosov Moscow State University, D.V. Skobeltsyn Institute of Nuclear Physics for the study of the temporal and spectral characteristics of cosmic radiation. The DeCoR, DeCoR-2 and KODIZ devices were successfully launched into orbit in 2018–2023 as payload of several cubesat format satellites. The measurements carried out with their help are important both for fundamental studies of the dynamics of radiation belts, solar-terrestrial physics, etc., and for developing a methodology for monitoring the flux of solar cosmic rays and other energetic particles on small satellites, which is important from the point of view of radiation safety in orbit. In the near future, an advanced device of the DeCoR family is expected to be launched into space, having a pixelated detector based on promising GAGG:Ce scintillators. With its help, as well as by comparing the results of simultaneous measurements on several satellites, new important scientific results will be obtained.

Acknowledgements. Authors from Moscow State University and Belarus State University are grateful for the support of the Russian Science Foundation Project No. 23–42-10005 and the Belarus Foundation for Fundamental Research F23-RSF-074.

References

1. Novikov, L.S., Panasyuk, M.I.: Study of space radiation and its influence on spacecraft materials and equipment. Voprosy atomnoy nauki i tehniki. Seria: Physika radiatcionnogo vozdeystvija na radioelectronnuy apparaturu, no. 4, pp. 3–13 (2002). (In Russian)
2. Mjagkova, I.N.: Solar wind influence on the outer radiation belt dynamics and structure – experiments on satellites of KORONAs series. In: Sixth every year conference Plasma physics in Solar System, 14–18 February 2011, Moscow. Space Research Institute RAS. Abstracts, p. 68 (2011). (in Russian)
3. Tverskaya, L.V., Balashov, S.V., Vedenkin, N.N., et al.: Outer radiation belt of relativistic electrons in the minimum of 23 cycle of solar activity. Geomag. Aeron. **52**(6), 740–745 (2012)
4. Nymmik, R.A.: Model of particle fluxes and averaged energy spectra of solar cosmic rays. Cosm. Res. **31**(6), 51–59 (1993)
5. ISO/DTR 18147. Space environment (natural and artificial) — The method of the solar energetic protons fluences and peak fluxes determination. 2013
6. Bogomolov, V.V., et al.: A scientific and educational space experiment on the siriussat-1,2 satellites. Mosc. Univ. Phys. Bull. **75**, 665–675 (2020). https://doi.org/10.3103/S0027134920060077
7. Prokhorov, M.I., et al.: Analysis of fast variations in electron fluxes in the gap region using the normalized range method based on measurement data on the siriussat-1 satellite. Cosm. Res. (English translation of Kosimicheskie Issledovaniya) **60**(4), 241–253 (2022). https://doi.org/10.1134/S0010952522040062
8. Bogomolov, V.V., et al.: A first experience of space radiation monitoring in the multi-satelliteexperiment of moscow university in the frameworkof the universat-socrat project. Mosc. Univ. Phys. Bull. **75**, 676–683 (2020). https://doi.org/10.3103/S0027134920060089
9. Bogomolov, A.V., et al.: Space weather effects from observations by Moscow university cubesat constellation. Universe **8**, 282 (2022). https://doi.org/10.3390/universe8050282
10. Bogomolov, V.V., et al.: The timing and spectral characteristics of detectors based on a Ce:GAGG inorganic scintillator using photomultiplier tubes and silicon photodetectors. Instrum. Exp. Tech.Exp. Tech. **63**(5), 633–640 (2020). https://doi.org/10.1134/S0020441220050097

Oscillator Associated with One Two-Mode Dynamo Model with Memory

Gleb Vodinchar(✉)

Institute of Cosmophysical Research and Radio Wave Propagation FEB RAS, Mirnaya Str., Paratunka 684034, Kamchatka, Russia
gvodinchar@ikir.ru

Abstract. Real cosmic dynamo-systems are complex oscillatory systems. They demonstrate complex dynamic regimes, including regular and chaotic inversions, bursts, and vacillations. Dynamic memory effects play an important role in these systems. In this paper, we study the hereditary oscillator, which can be considered as a simple model of a two-mode dynamo with memory. The model implements hereditary feedback in the quenching of the alpha-effect by the field energy. Numerical simulation of dynamic regimes for two types of kernels was carried out. These kernels correspond to two fundamentally different quenching options. In the first case, quenching is determined by the current and past values of the phase variable and its derivative. This means that quenching is activated instantly. In the second case, quenching depends only on past values. This corresponds to a delay in quenching. The dynamic regimes observed in the simulation are interpreted from the point of view of the theory of oscillators.

Keywords: dynamo-systems · oscillators · system with memory · heredity

1 Introduction

The magnetic fields of space objects: planets, stars and galaxies are created by the action of hydromagnetic dynamo mechanism [1–3]. For the real magnetic fields regular and chaotic oscillations, vacillations, bursts, excursions (short changes in polarity after which the polarity is restored) are observed [4–6].

The physical idea of $\alpha\omega$-dynamo for space objects was proposed by [7]. The quenching of field generation (feedback) is important in this type of dynamo – a large-scale magnetic field affects the turbulent generator (α-effect), providing a self-consistent nonlinear mechanism for generating a finite field. An important property of this feedback is spatial non-locality and memory [8–11].

Thus, the real dynamo systems are complex oscillatory systems with memory. Therefore, we can try to describe the dynamo systems as the oscillators with memory (hereditary oscillators). One of these oscillators was studied in [12], where α-quenching was ensured by the helicity of the field. An oscillator with the quenching by the field energy is studing in this paper.

A. Dmitriev et al. (Eds.): STRPEP 2023, SPEES, pp. 113–120, 2023.
https://doi.org/10.1007/978-3-031-50248-4_12

2 Two-Modes $\alpha\omega$-Dynamo with Memory as Hereditary Oscillator

For the two-modes approximation of a large-scale magnetic field, the $\alpha\omega$-dynamo with memory can be described by the equations [12]:

$$\begin{aligned}\frac{dB^T}{dt} &= \omega B^P - \eta^T B^T,\\ \frac{dB^P}{dt} &= \alpha_0(1-w)B^T - \eta^P B^P,\\ w(t) &= \frac{1}{T_K}\int_0^t K\left(\frac{t-\tau}{T_K}\right)\frac{Q\left(B^T(\tau), B^P(\tau)\right)}{B_0^2}\,d\tau,\end{aligned} \tag{1}$$

where $B^T(t)$ and $B^P(t)$ are the amplitudes of the toroidal and poloidal modes; $1/\eta^T$ and $1/\eta^P$ are the decay timescale of the modes; α_0 – measure of the helicity in the absence of a strong magnetic field; ω – differential rotation measure; B_0 – typical value of the field; $K(\cdot)$ – some dimensionless memory kernel; T_K – kernel timescale; $Q(\cdot,\cdot)$ – is some quadratic form. In this paper, it is considered that $Q\left(B^T, B^P\right) = |B^T|^2 + |B^P|^2$, i.e. the quenching is provided by field energy.

Now let's make the system (1) dimensionless. Let's decay time of the poloidal field $1/\eta^P$ is the time unit. The new dimensionless variables

$$B^T = B_0\eta^P\sqrt{\frac{T_K}{\alpha_0}}x(t), \quad B^P = B_0\frac{\eta^P\eta^T}{\omega}\sqrt{\frac{T_K}{\alpha_0}}y(t), \quad w = \frac{\eta^T\eta^P}{\alpha_0\omega}z(t),$$

and to the new dimensionless parameters:

$$D = \frac{\alpha_0\omega}{\eta^T\eta^P}, \quad \sigma = \frac{\eta^T}{\eta^P}, \quad p = T_K\eta^P, \quad s = \frac{\omega}{\eta^T}.$$

These parameters: D – is the dynamo-number; σ^{-1} – is the dimensionless time of toroidal field decay; p – is the dimensionless timescale of the kernel; s – ratio of toroidal and poloidal modes scales.

Then the model (1) takes the form:

$$\begin{aligned}\frac{dx}{dt} &= \sigma(y-x),\\ \frac{dy}{dt} &= (D-z)x - y,\\ z(t) &= \int_0^t K\left(\frac{t-\tau}{p}\right)\left(sx^2(\tau) + \frac{1}{s}y^2(\tau)\right)d\tau.\end{aligned} \tag{2}$$

The system (2) has only one stationary point: $x = y = z = 0$. It is easy to show that it will be stable if and only if $D < 1$. It is clear that field generation will occur in case $D > 1$. Exactly in this case small initial values of the field will increase. In what follows, only this case of a "working" dynamo will be considered.

The system (2) can be reduced to the following equation:

$$\frac{d^2x}{dt^2} + (1+\sigma)\frac{dx}{dt} + \frac{1}{s}K(0)x^3 - x\left[\sigma(D-1) - u(t) - w(t) + \frac{1}{s}K(t/p)x_0^2\right], \tag{3}$$

where

$$\begin{aligned} u(t) &= \frac{1}{\sigma s}\int_0^t K\left(\frac{t-\tau}{p}\right)\left(\frac{dx}{d\tau}\right)^2 d\tau, \\ w(t) &= \int_0^t J\left(\frac{t-\tau}{p}\right)x^2(\tau)\,d\tau, \\ J(\cdot) &= \frac{\sigma(1+s^2)}{s}K(\cdot) + \frac{1}{sp}K'(\cdot), \\ x_0 &= x(0). \end{aligned} \tag{4}$$

So, the two-mode dynamo model (2) with the initial conditions $x(0) = x_0, y(0) = y_0$ are equivalent to the Eqs. (3, 4) with the initial conditions $x(0) = x_0, x'(0) = \sigma(y_0 - x_0)$. For planetary and stellar dynamo systems it is reasonable to assume that $x_0 = 0$. It is related to the fact that a small external field, which is poloidal, is required to start the dynamo system at the initial moment [1]. Therefore, we always assume in the future that $x_0 = 0$.

The Eqs. (3) can be considered as a oscillator. The potential

$$\begin{aligned} U(x) &= Ax^4 - Bx^2, \\ 4A &= \frac{K(0)}{s}, \quad 2B = \sigma(D-1) - u(t) - w(t). \end{aligned} \tag{5}$$

Representing the dynamo model in the form of an oscillator allows us to study various dynamic modes from the general physical point of view of the theory of oscillations. It seems that such a view will allow us to better understand the physical processes in dynamo systems. Next, we will consider some results of numerical simulation and their interpretation from the point of view of oscillator theory. To solve the integro-differential equation, we used the difference scheme described in paper [12]. It has been modified for the oscillator studied in this paper.

3 Simulation Results

In the numerical simulation we shall use the value $\sigma = 3.37$. This value of the σ is determined by the ratio of the eigenvalues of the largest-scale toroidal and poloidal modes of free decay of the field. Also, the value of $s = 1$ was always used in the simulation.

The form of the potential (5) graph is determined by the signs of the coefficients A and B. The coefficient A is fixed, but the coefficient B is determined by the current and previous values of the phase variable x. Therefore, it is the $B(x)$ that determines the memory in the system.

It is important to note that for the E energy of the oscillator

$$\frac{dE}{dt} = -(1+\sigma)\left(\frac{dx}{dt}\right)^2 - 2x^2\frac{dB}{dt}. \tag{6}$$

The potential $U(x)$ degree is 2 or 4 depending on the value of $K(0)$. If $K(0) \neq 0$, then potential $U(x)$ degree is 4. This corresponds to the case of instantaneous response in α-quenching. If $K(0) = 0$, then potential $U(x)$ is quadratic. This corresponds to the case of response delay in α-quenching.

In this paper we consider two types of kernels with exponential asymptotics. In the first case, the kernel is $K(r) = \exp(-r)$ and the α-quenching is switched on instantly. In the second case, the kernel is $K(r) = r\exp(-r)$ and there is a delay in the α-quenching.

3.1 Instant α-Quenching: $K(r) = \exp(-r)$

In this case $A = 1/(4s) > 0$. When $B > 0$ the potential has two potential wells. These wells correspond to stable generation of the field of two polarities. If $B > 0$, the potential has one well $x = 0$. However, such a stable regime is impossible for $D > 1$, because $x = y = 0$, will be the unstable stationary point of the system (2). Therefore, one can expect either a change in polarity or bursts. In this case, a change in polarity or a bursts should be accompanied by B going into the negative area.

The some simulation results are shown in Figs. 1 and 2. For small values of the dynamo-number D, the regime of a field generating of fixed polarity is set. In this case, the coefficient B is set to a positive value. The system is in one of two potential wells. Such a regime is shown in Fig. 1. It can be noted that the only change of polarity in the transition mode ($t \sim 4$) was accompanied by a short excursion of B into the negative area. In this case, two potential wells were connected. Then they parted again, but x managed to change sign.

For large values of the dynamo-number D, the oscillator is in a chaotic regime of dynamics, similar to the dynamics of the Lorenz system. Such a case is shown in Fig. 2. There are growing vacillation around a fixed value of one sign, then a reversal and fluctuations around a fixed value of another sign. It is important to note, however, that these are not vacillations near stationary points. The oscillator has no non-zero stationary points. It is clearly seen that each inversion is accompanied by a short-term excursion of B into the negative area. In this case, the merging of potential wells occurs – the possibility of field reversal.

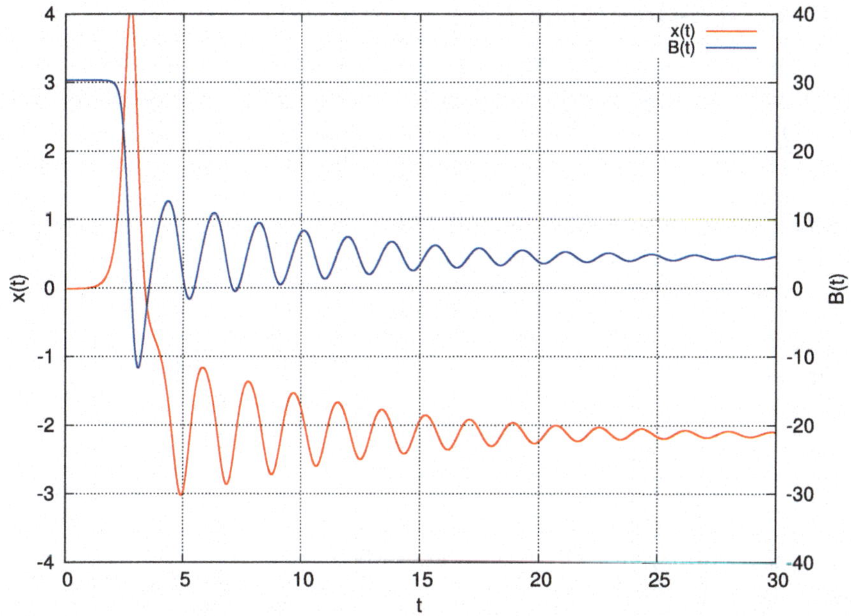

Fig. 1. Results of simulation: $K(r) = \exp(-r)$, $D = 10$ – asymptotically stationary regime.

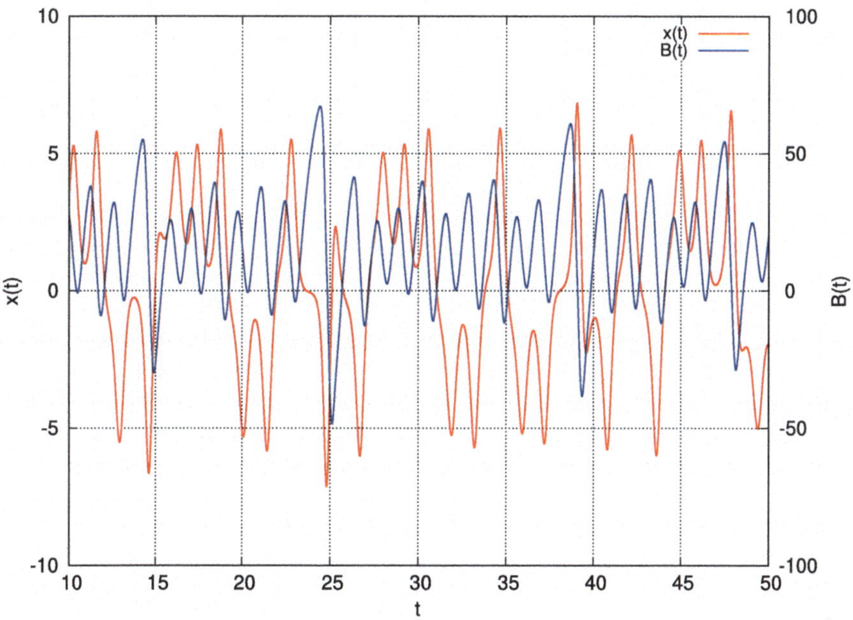

Fig. 2. Results of simulation: $K(r) = \exp(-r)$, $D = 30$ – chaotic vacillations and chaotic reversals.

3.2 Delay of α-Quenching: $K(r) = r \exp(-r)$

In this case $U(x)$ is quadratic. When $B < 0$ the potential has one potential wells $x = 0$. However, it was already noted earlier that for $D > 1$ the oscillator cannot be stable in zero position. When $B > 0$ the potential has one peak $x = 0$. Then x can roll in any direction, but the feedback should stop this drift. Therefore, oscillations around the zero level or bursts (regular or chaotic) can be expected. These oscillations must be accompanied by a sign change of B. Such dynamics in the form of bursts is shown in Fig. 3.

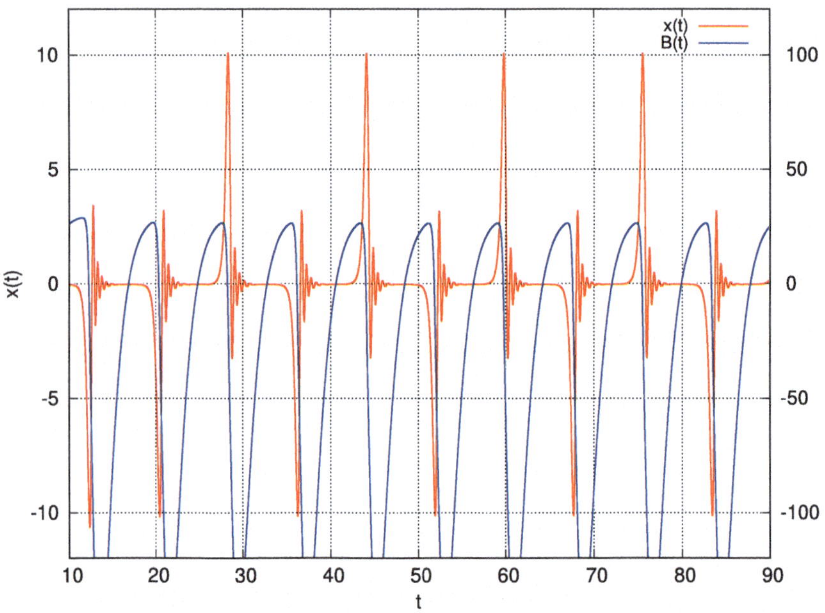

Fig. 3. Results of simulation: $K(r) = r \exp(-r)$, $D = 10$ – regular dynamo-bursts.

At small values of the dynamo number, another unexpected regime was discovered. It is shown in Fig. 4. These are the approximation of x to a stationary nonzero value and the approximation of the potential to a rectilinear form. We can interpret this mode in this way. The variable x rolls over one parabola branch. As the branch rises, x approaches zero. When it goes down, x goes away from zero. It turns out balancing, which leads to asymptotic stabilization.

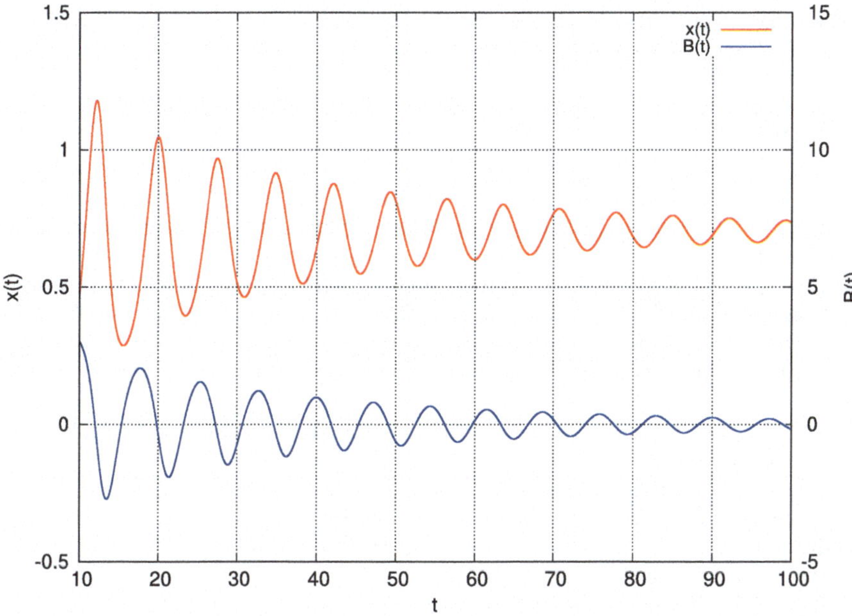

Fig. 4. Results of simulation: $K(r) = r\exp(-r)$, $D = 2$ – asymptotically stationary regime.

4 Discussion

In this article:

- A two-mode $\alpha\omega$-dynamo model with hereditary α-quenching by field energy has been obtained. The general idea of building such a model was previously described in the work of the author [11].
- The model is convert to the form of an oscillator with a memory. The potential of the oscillator is determined, by the current and previous values of the squares of the phase variable and phase velocity.
- Numerical simulation of oscillators dynamic for two type kernels is carried out. These kernels correspond to the instantaneous quenching and quenching delay in the dynamo system. It is shown that the simulations results are well interpreted from the oscillators point of view.

It should be noted that the modes of oscillator dynamics that we observed in this paper are very similar to those in [12]. In this work, feedback was provided by helicity, not energy. It can be assumed that they are determined by the fact that the α-quenching is provided by the quadratic expression of the field components. The specific form of this expression (helicity or energy) does not qualitatively change the situation.

In the author's opinion, the description of the dynamo system as an oscillator will make it possible to better comprehend the processes in this system from a general physical point of view.

Acknowledgments. The work was carried out within the framework of realization of the State Task №AAAA-A21-121011290003-0.

References

1. Zeldovich, Y.B., Rusmaikin, A.A., Sokolo, D.D.: Magnetic Fields in Astro-physics. The Fluid Mechanics of Astrophysics and Geophysics. Gordon and Breach, New York (1983)
2. Merril, R.T., McElhinny, M.W., McFadden, P.L.: The Magnetic Field of the Earth: Paleomagnetism, the Core, and the Deep Mantle. Academic Press, London (1996)
3. Stix, M.: The Sun. An Introduction. Astronomy and Astrophysics Library. Springer, Berlin, Heidelberg, New York (1989). https://doi.org/10.1007/978-3-642-97129-7
4. Ruzmaikin, A.: The solar cycle as a strange attractor. Comments Astrophys. **9**(2), 85–96 (1981)
5. Ermushev, A.V., Rusmaikin, A.A., Sokoloff, D.D.: Fractal nature of the sequence of reverscals of the geomagnetic field. Magnetyhydrodynamics **4**, 326–330 (1992)
6. Gaffin, S.: Analysis of scaling in the geomagnetic polarity reversal record. Phys. Earth Planet. Inter. **57**, 284–29 (1989)
7. Parker, E.N.: Hydromagnetic dynamo models. Astrophys. J. **122**, 293–314 (1955)
8. Krause, F., Radler, K.-H.: Mean-field Magnetohydrodynamics and Dynamo Theory. Berlin, New York (1988)
9. Hori, K., Yoshida, S.: Non-local memory e ects of the electromotive force by uid motion with helicity and two-dimensional periodicity. Geophys. Astro. Fluid **102**, 601–632 (2008). https://doi.org/10.1080/03091920802260466
10. Brandenburg, A.: Memory e ects in turbulent transport. Astrophys. J. **706**, 712–726 (2009). https://doi.org/10.1088/0004-637X/706/1/712
11. Vodinchar, G., Feshchenko, L.: Fractal properties of the magnetic polarity scale in the stochastic hereditary $\alpha\omega$-dynamo model. Fractal Fract. **6**(6), 328 (2022). https://doi.org/10.3390/fractalfract6060328
12. Vodinchar, G.: Hereditary oscillator associated with the model of a large-scale $\alpha\omega$-dynamo. Mathematics **8**(11), 2065 (2020). https://doi.org/10.3390/math8112065

Atmosphere and Lithosphere Interaction Could Triggered the 2023 Mw 7.8 Turkey Earthquake

Victor Volkov[1], Mstislav Dubrov[2(✉)], Igor Larionov[3], Jan Mrlina[4], Vaclav Polak[4], Dmitriy Aleksandrov[2], and Sergey Golovachev[2]

[1] Shmidt Institute of Physics of the Earth RAS, Bolshaya Gruzinskaya 10, Moscow 123995, Russia

[2] Fryazino Branch of Kotel'nikov Institute of Radioengineering and Electronics RAS, Vvedensky 1, Fryazino, Moscow Region 141190, Russia
mnd139@ire216.msk.su

[3] Institute of Cosmophysical Research and Radio Wave Propagation FEB RAS, Mirnaya 7, Paratunka, Kamchatka Region 684034, Russia

[4] Institute of Geophysics CAS, Bocni II/1401, 14131 Praha, Czech Republic

Abstract. Comprehensive data analysis of atmospheric and cyclonic activity based on worldwide meteorological and oceanology data as well as the comparison with tilts and strains precise measurements by far distanced instruments have been performed. The earlier proposed model of earthquakes triggering due to atmosphere, ocean and lithosphere interaction was confirmed. The interaction develops as successively arising hurricanes (typhoons) activity in form of spatial-temporal swings of the lower pressure areas over the tectonic plates. The process started 1–2 months before the 2023 Mw 7.8 Turkey earthquake and after some cyclones reduction, it resumed. It was at this time that a major seismic shock occurred. This study considers the cyclones interaction in the Indian Ocean, North Atlantic Ocean and Mediterranean Sea during December 2022–January 2023. Excitations of Indo-Australian, African, Eurasian and Arabian tectonic plates progressed as NW-SE spatial and temporal swings over seismogenic area and were accompanied by tilt-baric and strain-baric disturbances detected by instruments installed in Central and East Europe and Far East regions. Tilt-baric effects of 1.2 mas/hPa and strain-baric events were observed for the most intensive cyclones 2–7 weeks before the earthquake.

Keywords: Tiltmeter · Strainmeter · Tropical Cyclone · Earthquake Triggering

1 Introduction

Temporal and spatial correlations between the strongest tropical cyclones (typhoons, hurricanes) and earthquakes were detected and discussed during recent decades. Different physical mechanisms of relationship between these two main natural disasters, which are similar in order of realized energy in the environment, has been considered in a number of publications [1–4].

A. Dmitriev et al. (Eds.): STRPEP 2023, SPEES, pp. 121–130, 2023.
https://doi.org/10.1007/978-3-031-50248-4_13

The connection of major earthquakes and typhoons in the northwestern part of the Pacific Ocean was demonstrated by means of satellite images of cloud covers accompanying the earthquakes occurring [1]. The intra-annual analysis of cyclogenesis in this most active basin allowed the seasonal dynamics of seismicity to be proposed [2].

The triggering of slow earthquakes as well as common earthquakes due to typhoons passage similarly to the passage of teleseismic waves from large regional events has been shown in [3] and [4].

Another model of a physical link between the two hazard types suggested the rocks erosion induced by very wet tropical cyclones [5]. Although this approach required the long time delays (from a few months up to few years) between tropical cyclone passage and earthquake happening [5].

Most of the studies mentioned above [1, 2, 5] are based entirely on statistical comparison of seismic catalogues data and typhoons (hurricanes) services information. The applying instrumental methods, such as the borehole strain-meters [3], laser interferometers [4] and tiltmeters [6] to detect accompanying earth deformations, promoted the observed phenomena investigation sufficiently.

The results of many years of experimental searching [7–9], and [10] allowed uncovering a physical entity of the interrelated processes. The proposed model assumes an atmosphere, ocean and lithosphere interaction to be a possible drive of triggering mechanism of major earthquakes [11]. We have described this process as successive development of cyclonic activity and arising hurricanes (typhoons) in form of a specific spatial-and-temporal motions of lower pressure areas over the Earth's tectonic plates.

The process starts 4–7 weeks before an earthquake and after some cyclonic activity descending it resumes wherein it was at this time that occurring a power seismic shock becomes highly probable. Investigation of the decade 1997–2007 major earthquake series showed that duration of the above period of cyclones system swinging within earthquake preparation continuance can increase up to 1–2 months for the strongest M 8–9 seismic events.

This study considers the initiation of powerful Turkey earthquake (6 February 2023) that could be a result of tropical cyclones interaction in the Indian Ocean and extra tropical windstorms in North Atlantic Ocean and Mediterranean Sea.

2 Earthquakes and Tropical Cyclones Interrelation Background

There are distinct physical justifications for coupling these two most terrible disasters on the globe.

Solar irradiation amounts the substantial part of atmosphere-ocean interactive energy releasing in tropical cyclones activity. We have shown [10] the daily energy outcome of a powerful tropical cyclone (Category 5 SSHWS) can reach the value about $1.5 \cdot 10^{19}$ J, which is equivalent to the energy released by a strong earthquake of $M_W > 8.3$.

An active zone of this rotating geophysical "engine" can spread over 10^3 km that is an order of size of a preparation zone of large regional earthquake. Lower pressure disturbs carth crust and being at vicinity of fault zone close to failure can trigger an earthquake.

A common period of tropical cyclone development including its highest activity phase alternates from a few days up to a few weeks. It is just the period, which characterizes the time scale of short-term earthquake precursors. Disturbances in atmosphere and lithosphere caused by powerful tropical cyclone (typhoon, hurricane) in the World Ocean are spread over wide areas.

In this way, typhoons and hurricanes are usually accompanied by a variety of earthquake precursory phenomena, including abnormal behavior of ultra-wideband (0.002 MHz–3 Hz) Earth's oscillations, which are available to be recorded at far distances up to 1,000–10,000 km.

Although not all of these phenomena may be due to earthquake preparation processes in the Earth's crust or mantle, at the same time, they are detected and studied for a long time included classic authors [12] and our earlier publications [8, 9].

3 Analysis of the 2022–2023 Tropical Cyclogenesis Preceding the 2023 Mw 7.8 Turkey Earthquake

Tropical cyclones transition from the Northern Hemisphere to the southern one in the end of November and the beginning of December 2022 was characterized by the calm condition period in one and a half weeks duration that was observed in atmosphere and World Ocean [13–15] and [16]. This was about two months before 06.02.2023 Turkey earthquake.

Meanwhile, tropical cyclones activity moved from NW Pacific to the Indian Ocean and sometime later, to the Southern Pacific. Last system of the 2022 Eastern and Central Pacific hurricanes dissipated on October 23 and last system of the 2022 Atlantic hurricane season dissipated in the northeastern Caribbean Sea on November 11. A series of extra tropical cyclones of the 2022–2023 European windstorm seasons succeeded the cyclonic activity in North Atlantic Ocean (Table 1).

Severe tropical storm MANDOUS, which developed in the Bay of Bengal on 6 December 2022, was the first notable system (after tropical cyclone ASANI, 7–12 May 2022) for this round of cyclonic activity in the North Indian Ocean [14]. Arising of the succeeded low pressure events have progressed as NW-SE spatial and temporal swings of cyclonic disturbances between Atlantic and Indian Ocean basins during December 2022–January 2023 (see Fig. 1).

Extra tropical cyclone EFRAIN from North Atlantic hitting the Iberian Peninsula and France was the most intense, strongest storm of the 2022–2023 European windstorm season (955 hPa ~ C2, SSHWS). Cyclone EFRAIN came after tropical storm MANDOUS in the Indian Ocean. Tropical storm PAKHAR (the last system of the 2022 NW Pacific

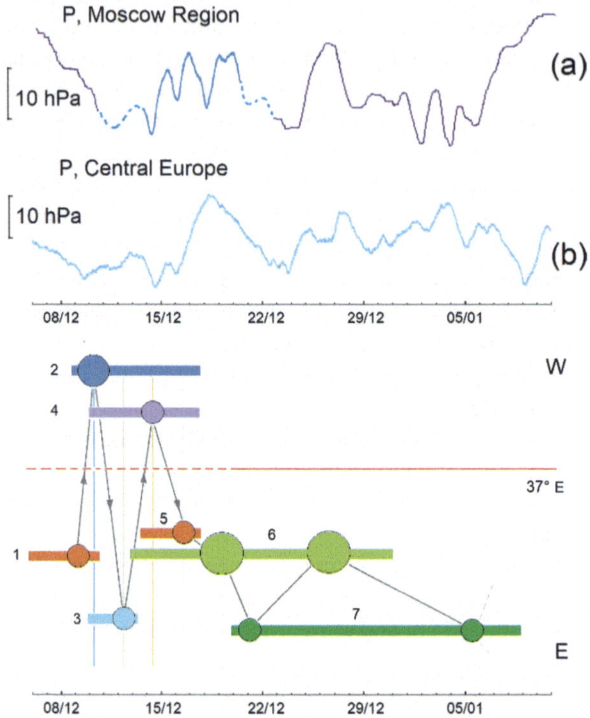

Fig. 1. Atmospheric pressure variations in Moscow Region (a) and Central Europe (b) during December 2022–January 2023; 1…7 – tropical cyclones in the Indian Ocean and European windstorms in North Atlantic (see Table 1), circles denote the time intervals in the lowest pressure of cyclones

typhoon season) succeeded the cyclone EFRAIN, GAIA came after PAKHAR and so on, as shown in Fig. 1 and Table 1.

These spatially and temporary swinging the lower pressure systems excited the Indo-Australian, African and Eurasian tectonic plates bordering around the Arabian tectonic plate (Fig. 2). It is important that spatial distribution of cyclones active zones formed triangular patterns having the epicenter of impending earthquake near their medians (marked by star in Fig. 2).

This period of cyclonic activity has ended by swinging between tropical cyclone DARIAN (C4 SSHWS) and tropical storm ELLIE excited the eastern part of the Indo-Australian tectonic plate [15]. The atmospheric condition was featured by stochastic air pressure variations without any sign of synchronizing or correlation of signals recorded at spatially distanced observational points (Fig. 1).

Some cyclonic activity weakening happened in 9–13 January 2023. This pause was fulfilled by very rare subtropical cyclone in the Southern Atlantic that moved from the Brazilian coast toward the African tectonic plate.

The next round of cyclones development started 14.01.2023 (Table 1). Extra tropical cyclone FIEN occurred in Northern Atlantic and tropical storm IRENE in Southern

Table 1. Tropical cyclones in the World Ocean and European windstorms in North Atlantic and Mediterranean Sea during December 2022–February 2023 [13–15] and [16]

No	Name	Category (SSHWS)	Duration	Basin
1	MANDOUS	TS[1]	06–10 December	North Indian Ocean
2	EFRAIN	ETC2[1]	10–17 December	Northern Atlantic
3	PAKHAR	TS	10–12 December	NW Pacific
4	GAIA	ETC	10–17 December	Northern Atlantic
5	ARB 03	TS	14–17 December	North Indian Ocean
6	DARIAN	C4	13–21–30 December	SW Indian Ocean
7	ELLIE	TS	20 December–8 January	Aust. Indian Ocean[2]
8	HALE	TS	07–08 January	Southern Pacific
9	Subtropical ccl	STC	07–10 January	Southern Atlantic
10	FIEN	ETC1	14–20 January	Northern Atlantic
11	IRENE	TS	14–19 January	Southern Pacific
12	GE'RARD	ETC2	15–17 January	Northern Atlantic
13	CHENESO	C2	16–29 January	SW Indian Ocean
14	HANNELORE	ETC1	19–28 January	Mediterranean
15	DINGANI	TL(C2)[1,3]	27 January–9 February	Aust. Indian Ocean
16	FREDDY	C1	04–14 February…[4]	Aust. Indian Ocean
17	BARBARA	ETS[1]	05–08 February	Mediterranean
18	GABRIELLE	C2	05–10 February…[5]	Aust. Indian Ocean

[1] TS – Tropical storm; ETC(S)–Extra tropical cyclone (storm); TL – Tropical low
[2] Australian region cyclone
[3] DINGANI – Tropical Low; 9 – 15 February– C2 SSHWS, SW Indian Ocean
[4] FREDDY –C1; 14 February–15 March – C5 SSHWS, SW Indian Ocean
[5] GABRIELLE – 10–11 February, Southern Pacific

Pacific basin. Another cyclone – European windstorm GE'RARD developed next day and tropical cyclone CHENESO (C2 SSHWS) formed 16.01.2023 in SW Indian Ocean.

The above swinging of three tectonic plates (I, II, III) surrounding the Arabian plate (IV) resumed. This process became more severe when Mediterranean "hurricane" [16] – Medicane HANNELORE was involved on 20.01.2023 (see left Fig. 3). The HANNELORE's competition with tropical cyclone CHENESO in SW Indian Ocean was their alternate impacts up to the end of January 2023. The less intensity of cyclone HANNELORE (C1 SSHWS equivalent) was more than compensated by its proximity (Mediterranean and Balkans) to the region of upcoming disaster – epicenter location was near 37.2° N and 37.0° E. The next tropical cyclones DINGANI, FREDDY and GABRIELLE arising in the Indian Ocean went to the eastern part of the Indo-Australian

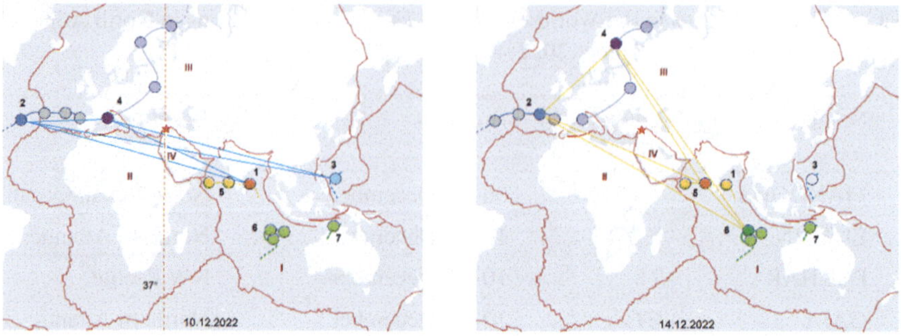

Fig. 2. Tracks and spatial distribution of active zones (1…7, see Table 1) for tropical cyclones in the Indian Ocean and European windstorms in North Atlantic. The lower pressure areas (bright circles) excite tectonic plates: I – Indo-Australian, II – African, III – Eurasian and IV – Arabian on 10 and 14 December 2022

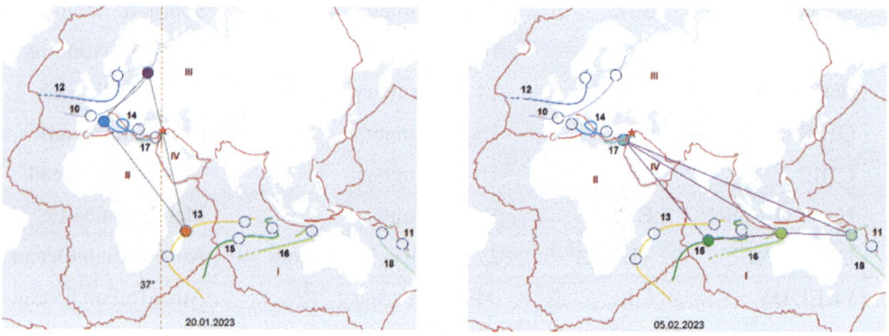

Fig. 3. The same as Fig. 2 on 20 January and 5 February 2023

tectonic plate in the very beginning of February (Fig. 3, right).Meanwhile, the Mediterranean storm BARBARA affected directly the area of the 6 February 2023 Turkey earthquake, preceded for tens hours the main Mw 7.8 shock ($t_0 = 01$ h 17 m 35 s UTC) and accompanied all the strong aftershocks.

4 Tilt-Baric and Strain-Baric Processes Preceding and Accompanying the 2023 Mw 7.8 Turkey Earthquake

The Earth's deformations (tilts and strains) and atmospheric pressure variations were monitored at three observational points: in Central Europe (Geophysical Observatory Pribram), Moscow Region (testing site Fryazino) and Kamchatka peninsula (observational site Karymshina). Instruments installations, measuring techniques and precision characteristics were described in detail in our previous publications [6–8, 10] and [11].

As noted above, atmospheric pressure variations were random during December 2022 without any sign of coherency between the signals recorded at the observational points Pribram and Fryazino spatially distanced by 1600 km. That was the period when the windstorms and tropical cyclones started the NW-SE swinging between Northern

Atlantic (European areas) and Indian Ocean basin. That was 1.5–2 months before the 2023 Mw 7.8 Turkey earthquake.

A different situation developed 2–3 weeks before this earthquake. The high degree of correlation and even signs of synchronizing could be distinguished between the wideband tilt-baric and strain-baric processes recorded at observational points being 1600–8100 km apart (Fig. 4, Fig. 5, and Fig. 6).

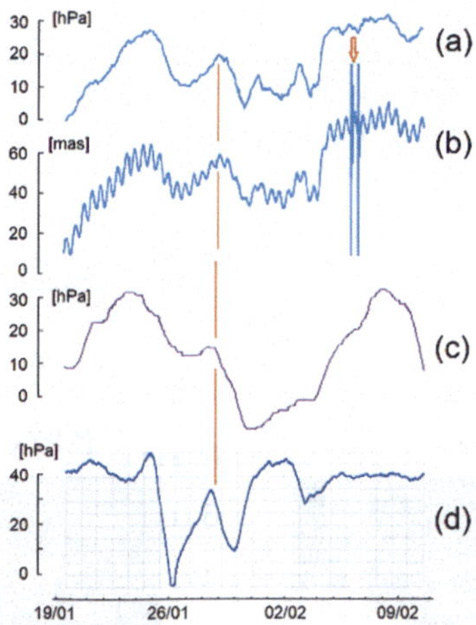

Fig. 4. Ultra-long period tilt-baric variations in Pribram (a, b), Fryazino (c) and Karymshina (d) 1–2 weeks before the 2023 Mw 7.8 Turkey earthquake (denoted by arrow)

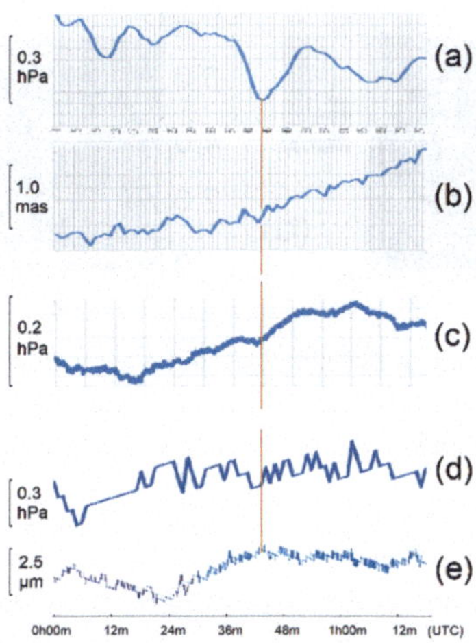

Fig. 5. Long-period tilt-baric and strain-baric variations in Pribram (a, b), Fryazino (c) and Karymshina (d, e) 1 h before the 2023 Mw 7.8 Turkey earthquake

Strong correlation (R_P ~ 0.8–0.9) was observed between NS component of tilts and atmospheric pressure variations with characteristic periods about 20–100 h that recorded in Pribram (Fig. 4 a, b). The value of tilt-baric variations reached the level of a few tidal amplitudes wherein tilt-baric coefficient was near 1.2 mas/hPa. Good conformity of baric variations at ultra-long periods up to 400 h was seen under 1600 km distance (observational points Pribram and Fryazino) and kept in part up to 8100 km in Karymshina point (Fig. 4 c, d). The direct estimations of correlation coefficient are hugely intricate due to wave shapes blur under their propagation at far distances.

The long period 5–20 min tilt-baric and strain-baric variations appeared by less clear correlation features 1–1.5 h before earthquake though some consistency of recorded baric strokes were noted at distanced points (Fig. 5 a-e). The strain-baric coefficient for strainmeter in Karymshina point was evaluated to be about (2–3) $\cdot$ 10^{-8} hPa^{-1} (Fig. 5 d, e; the linear thermal trend 2.4 K/h was removed).

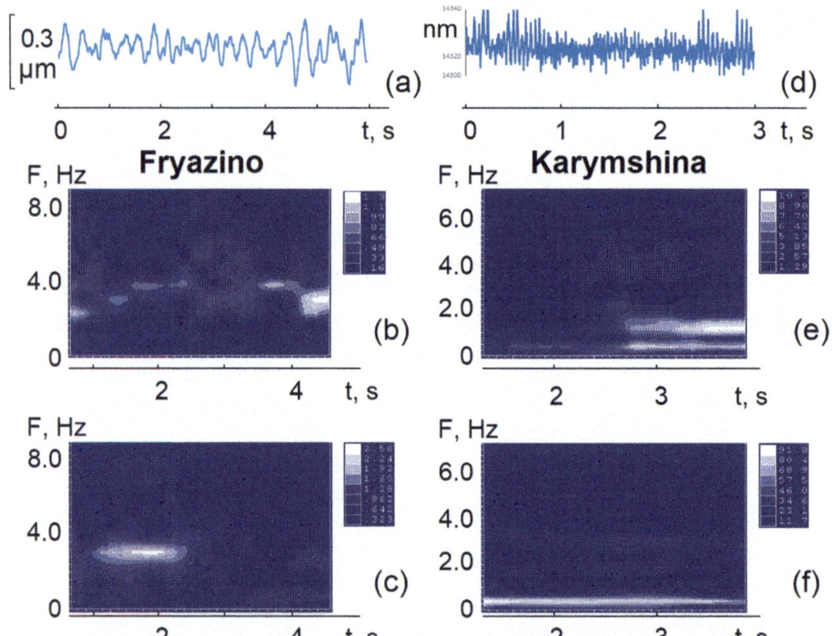

Fig. 6. Micro-seismic oscillations in Fryazino (a) and Karymshina (d) and their time-frequency diagrams 1 min (b, e) and few seconds (c, f) before the main Mw 7.8 shock (t_0 = 01 h 17 m 35 s UTC) of the 2023 Turkey earthquake

A special behavior of micro-seismic oscillations in 1–10 Hz band were recorded by the two distanced laser strainmeters (Fryazino and Karymshina) just before the 2023 Mw 7.8 Turkey earthquake (Fig. 6 a-f). The synchronization phenomenon of high frequency microseisms, which we observed before strong seismic events earlier [7, 10], appeared this time clearly. The speckle structure of time-frequency diagram (2–4 Hz, Fryazino) and spectral pick splitting (0.4–1.2 Hz, Karymshina), which were seen 1 min before earthquake (Fig. 6 b, e), collapsed into narrow bands in 3.1 Hz (Fryazino) and 0.4 Hz (Karymshina) respectively a few seconds preceding the main shock (Fig. 6 c, f).

5 Conclusion

The comparative analysis of atmosphere condition, cyclonic activity in the World Ocean and wideband Earth's deformations (tilts and strains) reveal the three geospheres interaction could be a trigger of the 2023 Mw 7.8 Turkey earthquake. The newly developed methods and original techniques have been applied in this study. The spatially distributed at distances 1600–8100 km tilt meters and laser strainmeters allowed the ultra-long periods up to 400 h variations and micro-seismic oscillations in 1–10 Hz band to be analyzed. Their behavior has been connected with time intervals preceding earthquake for one minute and up to the hundreds hours. The obtained result will be useful in earthquake prediction application.

Acknowledgements. The work was carried out within the framework of the state tasks FMWU-2022-0012 (IPE RAS), 075-01110-23-01 (IRE RAS) and AAAA-A17-117080110043-4 (IKIR FEB RAS). Some work on recording and processing data in Czech observatory was supported by the CzechGeo/EPOS projects (No.LM-2010008 and LM-2015079).

References

1. Morozova, L.I.: Typhoons and seismicity. Dokl. Earth Sci. **410**, 1132–1135 (2006)
2. Yaroshevich, M.I.: Intra-annual dynamics of seismic activity in the cyclonic zone of the north-western part of the Pacific Ocean. Doklady Earth Sci. **431**(1), 409–412. Springer International Publishing AG. (2010)
3. Liu, C., Linde, A., Sacks, I.: Slow earthquakes triggered by typhoons. Nature **459**, 833–836 (2009). https://doi.org/10.1038/nature08042
4. Dubrov, M.N., Golovachev, S.P.: Earthquake and Hurricane Remote Monitoring by Ground-Based Interferometry. In: ISPRS Technical Commission VIII Symposium, 9–12 August 2010, Kyoto, Japan, WG-VIII/1 TS-23, Abstract Paper N064015, ISPRS Archives, vol. 38, no. 8, pp. 254–255 (2010). http://www.isprs.org/proceedings/XXXVIII/part8/
5. Wdowinski, I., Tsukanov, S.: Disaster triggers disaster: Earthquake triggering by tropical cyclones. AGU Fall Meeting Abstracts, American Geophysical Union, Fall Meeting, San Francisco, California, 5–9 December 2011, abstract id. U53E-06 (2011). http://adsabs.harvard.edu/abs/2011agufm.u53e..06w
6. Volkov, V.A., Dubrov, M.N.: Geodynamical observations using spatially distributed gravimeters, tiltmeters, and laser strainmeters. Bulletin d'Information des Marees Terrestres **148**, 11971–11986 (2014). ISSN 0542–6766
7. Dubrov, M.N., Alyoshin, V.A.: Laser strainmeters: new developments and earthquake prediction applications. Tectonophys. **202**, 209–213 (1992)
8. Petrova, L., Volkov, V.: Dynamic features of seismo-gravitational oscillations of the Earth. Trans. (Doklady) Russ. Acad. Sci./Earth Sci. Sect. **351A**, 1477–1480. Pleiades Publishing, Inc. (1996)
9. Aleksandrov, D.V., Dubrov, M.N., Larionov, I.A., Marapulets, Y.V. Shevtsov B. M.: Earth strain and acoustic monitoring of geodynamic processes in seismic and aseismic zones using highly sensitive, spatially separated instruments. J. Volcanol. Seismol. **13**(3), 193–200 (2019)
10. Volkov, V., Mrlina, J., Dubrov, M., Smirnov, V., Golovachev, S., Polak, V.: Atmosphere, ocean and lithosphere interaction as a possible drive of earthquake triggering. Geodesy Geodyn. **11**(6), 442–454 (2020). https://doi.org/10.1016/j.geog.2020.07.001

11. Kost'ak, B., Mrlina, J., Stemberk, J., Chan, B.: Tectonic movements monitored in the bohemian massif. J. Geodyn. **52**, 34–44 (2011). https://doi.org/10.1016/j.jog.2010.11.007
12. Rikitake, T.: Earthquake Prediction. Elsevier Scientific Pub. Co., Amsterdam, New York (1976)
13. Knapp, K.R., et al.: International Best Track Archive for Climate Stewardship (IBTrACS). NOAA Centers for Environmental Information, NOAA (2023) https://www.ncei.noaa.gov/products/international-best-track-archive
14. Tropical Cyclones - Tropical Weather Outlook: Archive. Regional Specialized Meteorological Centre (RSMC), New Delhi, India (2023). https://rsmcnewdelhi.imd.gov.in/archive.php
15. Bath, M.: Tropical Cyclone Track Maps for Australia and the Southern Hemisphere. Australian Severe Weather (2023). https://australiasevereweather.com/cyclones/index.html
16. Gencic, D., et al.: The first medicane of the 2023 season formed on 21 January over the Adriatic. European Organization for the Exploitation of Meteorological Satellites (EUMETSAT), 30 January 2023. https://www.eumetsat.int/medicane-northern-adriatic

Analysis of Cosmic Ray Variations During Geomagnetic Storms in 2019–2022

O. Mandrikova and B. Mandrikova(✉)

Institute of Cosmophysical Research and Radio Wave Propagation FEB RAS, Mirnaya Str., 7, Paratunka, 684034 Elizovskiy District, Kamchatka Region, Russia
555bs5@mail.ru

Abstract. The dynamics of cosmic ray (CR) variations during magnetic storms of different strength was under investigation. Neutron monitor data from high-latitudinal and polar stations were used. Events of 2019–2022 were analyzed. The analysis was based on the hybrid method developed by the authors. It includes application of the SSA (Singular Spectrum Analysis) of nonlinear schemes in a wavelet bases. The investigation showed the presence of anomalies (Forbush effects) of complicated form in CR variations before magnetic storms. The anomalies were generated during CME arrivals and increase in IMF southern component fluctuations. During low-strength events, local increases in CR intensity were observed. During moderate magnetic storms, long anomalous decreases (Forbush decreases) were observed. CR variation components having strong correlation with Dst index were also detected during the study. The results confirm the importance of taking into account the CR dynamics in magnetic storm forecast methods.

Keywords: space weather · cosmic rays · geomagnetic storms · data analysis

1 Introduction

Scientific and technical development extends the possibilities of investigation of the environment and the processes in the near-Earth space (NES) making them an important research object. One of the problems of the NES investigation is the problem of its inhomogeneity. It is known that the less a subject is investigated, the more homogeneous it is from the point of view of a researcher. In the course of investigation of a subject, its inhomogeneities are revealed that is reflected in the complexity of continuously improving models and methods for investigation of natural processes. The problem of inhomogeneity is associated with the problems of information obtaining and preliminary processing. This problem becomes more complicated by a wide spectrum of its application in different spheres of science and activities (in Astrophysics, Geophysics, space weather forecasting, Medicine, industries etc.) and during solution of completely different tasks (positioning, tele- and radio communications, global and local monitoring, ecology and so on). Thus, the problems of investigation of NES processes are solved by applying different methods, such as physical, statistical, spectral, and the means of artificial intelligence and hybrid approaches.

A. Dmitriev et al. (Eds.): STRPEP 2023, SPEES, pp. 131–140, 2023.
https://doi.org/10.1007/978-3-031-50248-4_14

The object of this investigation is cosmic ray (CR) intensity variations recorded by neutron monitors. Anomalous phenomena and processes on the Sun are reflected in non-periodic variations of cosmic rays (Forbush effects and GLE events). Anomalous changes in CR flux, preceding and accompanying magnetic storm development, were described in the papers by the researchers from different scientific teams [1–3]. For example, in the paper [1], the impact of a space weather extreme event (on 6 September 2017) on the ionosphere and primary cosmic rays was under the study. The authors [1] estimated cross-correlation between hourly time series of CR secondary flux from several sites, geomagnetic activity indexes and between the parameters of the interplanetary magnetic field and the solar wind (SW speed, Dst Index, Proton Channel 16–20 meV). The investigations [1] confirmed the correlation between geospheric and geomagnetic parameters. The paper [2] analyzes radiation background time series (measured inside a multilayer structure), geomagnetic activity and CR intensity. The investigations were carried out on the basis of linear correlation analysis and a new correlation measure based on a one-dimensional cumulant component of the fourth order. The results [2] showed that there is direct (as well as indirect) relation between geomagnetic activity data and CR data. Analysis, carried out by other authors [3], showed the correlation between CR changeability and Dst with Dst delay by 3–4 h. The obtained delay of several hours confirms the importance of taking into account the CR in space weather.

In the paper, we investigated cosmic ray (CR) flux dynamics during magnetic storms of different strength. Events of 2019–2022 were under analysis. The analysis was carried out on the bases of the hybrid approach developed by the authors. It implies application of nonlinear wavelet schemes and singular spectrum analysis («Caterpillar» method). It is known that wavelets make it possible to study the structure of nonstationary data in detail, to detect informative components and to suppress noise. The papers [4–6] present the examples of effective application of wavelet analysis for investigation of natural data, including space environment parameters. Singular spectrum analysis (SSA) method, which was first suggested by the Russian researchers [7], allows us to analyze nonstationary time series. The SSA makes it possible to detect different components in an initial time series such as trend, periodic variations and noise. In this case, one does not need to define a priori the parametric form. The basis is formed from the inner structure of the series under investigation. In view of the mentioned advantages, the SSA method is successfully used to investigate geophysical monitoring data. For example, using the SSA, it was shown in the paper that variations of the global pressure at the sea level contain many quasi-periodic or periodic components besides the trend. The periods, obtained by the authors [8], are typical for the space-time evolution of the Earth rotation.

This work continues the investigations of the paper [4]. High sensitivity and adaptability of the developed method allows us to detect anomalies in CR variations of different intensity and duration. The authors of the paper considered weak and moderate magnetic storms and investigated CR variations dynamics during disturbed periods. The analysis showed formation of anomalous structures of complicated form in CR variations before magnetic storm development. Application of the method also made it possible to detect cosmic ray variation components having strong correlation with Dst index. The observed delay of several hours confirms the investigation results of the paper [3].

2 Description of the Methods

2.1 Singular Spectrum Analysis

On the basis of the *Singular spectrum analysis* we transform the recorded F data of a ground neutron monitor [9] into a matrix that is followed by singular decomposition, grouping and transition to its detected components [10]. The algorithm of the method [10] realization is described below:

1. Transformation of the recorded F into a trajectory matrix

$$X = [X_1, \ldots, X_K] = \begin{bmatrix} f_1 & \cdots & f_K \\ \cdots & \cdots & \cdots \\ f_L & \cdots & f_N \end{bmatrix},$$

   where f_i is the initial series element, L is the window length, N is the initial series length.
2. Singular decomposition of the trajectory matrix X.
   Assume that $S = XX^T$, $\lambda_1, \ldots, \lambda_L$ are *eigen values of* S, taken in nonincreasing order ($\lambda_1 \geq \cdots \geq \lambda_L \geq 0$), and $U_1, \ldots, U_L$ is the orthonormalized system of *eigen vectors* of the matrix S.
   We assume that $d = \text{rank}X = max\{i : \lambda_i > 0\}(d = L)$ and $V_i = X^T U_i/\lambda_i (i = 1, \ldots, d)$. Then the singular decomposition of the matrix X is represented as

$$X = X_1 + \cdots + X_d,$$

   where $X_i = \sqrt{\lambda_i} U_i V_i^T$ are *elementary matrixes,* $\sqrt{\lambda_i}$ is the singular number, which serves as data dispersion measure; U_i and V_i are the left and right singular vectors of the matrix X.
   Thus, the trajectory matrix X can be represented as

$$X = \sum_i \sqrt{\lambda_i} U_i V_i^T.$$

3. Grouping of the set d of elementary matrixes from item 2 on m of non-overlapping subsets X_{I_i}, $I_i \in \{I_1, \ldots, I_m\}$:

$$X = X_{I_1} + \ldots + X_{I_m}.$$

4. The matrixes X_{I_i} of the grouped decomposition are hankelized (averaged over anti-diagonals). Using the correspondence between hankel matrixes and time series, recovered series $\tilde{F}^{(k)} = (\tilde{f}_1^{(k)}, \ldots, \tilde{f}_N^{(k)})$ are obtained. The initial series $F = (f_1, \ldots, f_N)$ is decomposed into the sum m of recovered series where each value of initial series equals

$$f_i = \sum_{k=1}^{m} \tilde{f}_i^{(k)}, i = 1, 2, \ldots, N.$$

This decomposition is the main result of SSA algorithm for the analysis of cosmic ray variations (of any analyzed time series on the whole). Taking into account the presence of diurnal variations, we used the window length $L = 1440$ counts (corresponds to a day, neutron monitor data with minute sampling were used). Figure 1 a and i show the NM data at Oulu station [9] for two different time intervals containing both calm days and periods of geomagnetic disturbances. Figure 1 b-h, j-p shows the corresponding seven components obtained on the basis of the SSA (item 2 of the SSA algorithm). Analysis of Fig. 1 a-p shows that NM initial data have nonsationary structure and contain high level of noise. The detected components included the trend, periodic components, local features and noise variations (Fig. 1 b-h, j-p).

Matrix grouping in the SSA algorithm (item 3 of the algorithm) was based on the estimate of confined dispersion fraction [11]. The graph of the first 30 eigen values is illustrated in Fig. 1 q. The dashed line in Fig. 1 q separates eigen values corresponding to the components used in the analysis. The confined dispersion fraction was estimated by the formula

$$\frac{\sigma_1^2 + \sigma_2^2 \ldots + \sigma_p^2}{\sigma_1^2 + \sigma_2^2 \ldots + \sigma_p^2 + \ldots + \sigma_d^2}.$$

Results of the estimates of confined dispersion fraction for the time intervals under consideration, 8–28 March 2023, 29 October–4 November 2022, are presented in Tables 1–2. The component with the 1st eigen value determines the trend and the components with the 2nd-4th eigen values include diurnal periodicities of CR data. The results obtained after the addition of these components were used in the investigation.

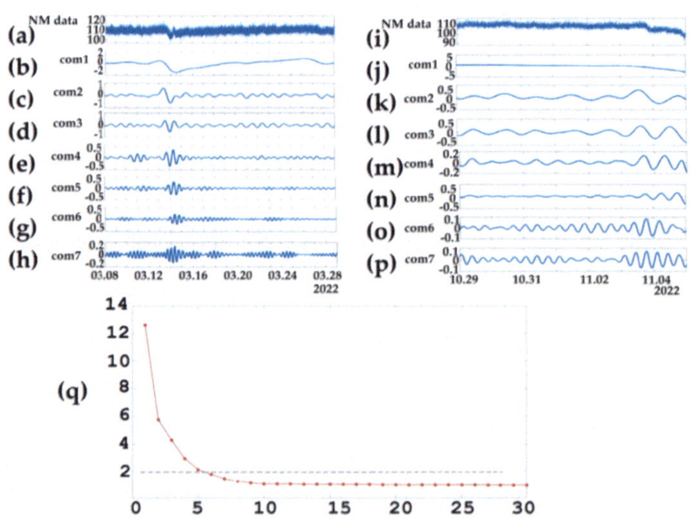

Fig. 1. Investigation of NM data by the SSA method

Results of the estimates of confined dispersion fraction:

Table 1. March 2023

Component	Confined dispersion fraction
com1	0.8942
com1 + com2	0.9288
com1 + com2 + com3	0.9501
com1 + com2 + com3 + com4	0.9607
com1 + com2 + com3 + com4 + com5	0.9655
com1 + com2 + com3 + com4 + com5 + com6	0.9681
com1 + com2 + com3 + com4 + com5 + com6 + com7	0.9703

Table 2. 27 October–4 November

Component	Confined dispersion fraction
com1	0.7508
com1 + com2	0.8490
com1 + com2 + com3	0.9074
com1 + com2 + com3 + com4	0.9291
com1 + com2 + com3 + com4 + com5	0.9395
com1 + com2 + com3 + com4 + com5 + com6	0.9470
com1 + com2 + com3 + com4 + com5 + com6 + com7	0.9526

2.2 Adaptive Algorithm for Anomaly Detection

The adaptive algorithm for anomaly detection (AAAD) was proposed for the first time by the authors in the paper [12]. The algorithm includes the following operations:

1. We represent a discrete time series $F[n]$ in the form of a series [13, 14]:

$$F[n] = \sum_{j=0}^{J} \sum_{k=1}^{K} WF\left(\frac{1}{2^j}, \frac{k}{2^j}\right) \Psi_{jk}[n],$$

   where $\Psi_{jk} = 2^{\frac{j}{2}} \Psi\left(2^j n - k\right)$ are basic wavelets, $j, k \in N$, $WF\left(\frac{1}{2^j}, \frac{k}{2^j}\right) = \left\langle F, \Psi_{jk} \right\rangle$ are the coefficients of function F decomposition into a series, J is the largest scale of decomposition into a wavelet series, K is the series length.
2. We apply a threshold function

$$P_{T_j^l}\left[WF\left(\frac{1}{2^j}, \frac{k}{2^j}\right)\right] = \begin{cases} WF\left(\frac{1}{2^j}, \frac{k}{2^j}\right), \ if \left|WF\left(\frac{1}{2^j}, \frac{k}{2^j}\right)\right| \geq T_j^l, \\ 0, \ if \left|WF\left(\frac{1}{2^j}, \frac{k}{2^j}\right)\right| < T_j^l, \end{cases}$$

where $T_j^l = t_{1-\frac{\alpha}{2},l-1}\hat{\sigma}_j^l$, $t_{\alpha,N}$ are the α-quantiles of Student's distribution [15], $\hat{\sigma}_j^l$ is the root-mean-square deviation of the coefficients, it is estimated in a moving time window of the length l, $\hat{\sigma}_j^l = \sqrt{\frac{1}{l-1}\sum_{m=1}^{l}\left(WF\left(\frac{1}{2^j},\frac{k}{2^j}\right)-\overline{WF\left(\frac{1}{2^j},\frac{k}{2^j}\right)}\right)^2}$.

We obtain the series representation as

$$\hat{F}[n] = \sum_{j=0}^{J}\sum_{k=1}^{K} P_{T_j^l}\left[WF\left(\frac{1}{2^j},\frac{k}{2^j}\right)\right]\Psi_{jk}[n].$$

3. For the detected anomalies, their intensity at the time $t = k$ can be estimated as

$$E_k = \sum_{j=0}^{J} P_{T_j^l}\left[WF\left(\frac{1}{2^j},\frac{k}{2^j}\right)\right],$$

which is positive in case of function values *anomalous increase* and negative in case of function values *anomalous decrease.*

3 Results of Data Processing

Neutron monitor (NM) minute data from the high-latitudinal Oulu, Inuvik and Tule stations and the polar Sopo station (www.nmdb.eu) were applied in the experiment.

Figure 2 shows the processing results of NM data from Oulu station (Fig. 2 e) for the period 8–24 March 2022 [9]. Red color in Fig. 2 indicates the SSA component of CR intensity, blue color is the Dst index data [16]. The times of geomagnetic activity increase are marked by the orange vertical line. The time of the moderate magnetic storm commencement is illustrated by the red vertical line. According to space weather data [17], the NES state on 8 and 9 March was characterized as very calm. Southern component fluctuations were about Bz = ±3 nT. Based on the processing results, CR intensity on 8 and 9 March was within the background values (Fig. 2 d, f, g). By the end of the day on 10 March an inhomogeneous accelerated flux from coronal mass ejection (CME on March 6) arrived. Southern component fluctuations grew up to Bz = −9 nT. The results of processing show anomalous changes (Forbush effects) in CR intensity data during the CME arrival (Fig. 2 d, f, g, blue color shows CR intensity decrease, yellow color shows the increase). We should note that formation of Forbush effects on 10–12 March correlates with sudden increase in solar wind (SW) density (Fig. 2 b) and intensification of IMF southern component fluctuations (Fig. 2 c). At the end of the day on 11 March, at 23:00 UT, weak magnetic storm was observed. Dst index decreased to –51 nT (Fig. 2 d). According to the processing data (Fig. 2 d, f, g), sudden anomalous increase in CR intensity occurred at that period. CME (on 10 March) arrived the next day on 13 March. IMF southern component fluctuations were intensified to Bz = −24 nT, SWW increased to 600 km/s. A moderate magnetic storm (minimum Dst = −85) was recorded at 10:48 UT [16]. Anomalous decrease in CR intensity (Forbush decrease) was observed at that period (Fig. 2 d, f, g). It occurred at the time of the magnetic storm observation. During the period, preceding and accompanying the moderate magnetic storm (on 8–23 March 2022), strong correlation of the CR component, detected on the basis of the SSA, (red color in Fig. 2 d) with geomagnetic activity Dst index (blue color in Fig. 2 d) was observed. This result also confirms the importance of taking into account the CR parameters for space weather forecasting.

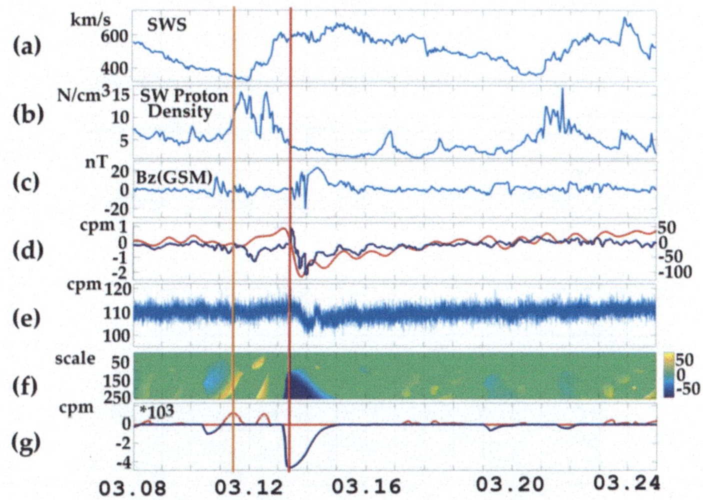

Fig. 2. a) SW velocity data; b) SW proton density; c) Bz data (GSM); d) blue color is the Dst index, red color is the result of SSA method application; e) NM data at Oulu station; f) and g) results of AAAD application to the NM signal approximated by the SSA. (Color figure online)

Figure 3 illustrates the results of processing of the NM data at Inuvik, Tule and the South pole stations for the period 24–29 May 2019 [9]. Two weak magnetic storms occurred on 27 May 2019. The times of magnetic storm commencements are indicated by orange lines in Fig. 3. Based on space weather data [17], NES was calm on 24 and 25 May. Based on the processing data, CR intensity was within the background level (Fig. 3 h-q). At the end of the day on 26 May, an inhomogeneous accelerated flux from a coronal mass ejection (CME on 21 May) arrived. SE velocity increased to 400 km/s (Fig. 3 a), Bz component fluctuations grew to Bz = –9,5 nT (Fig. 3 c). Sudden increase in SW density was also observed (Fig. 3 b). Results of the processing show Forbush effect occurrences at all the stations under analysis during that period (Fig. 3 h-q). General clearly expressed dynamics in CR was observed (Fig. 3 h-q). It confirms the results validity and the method sensitivity. According to IZMIRAN data [18], at the end of the day on 26 May, Forbush effect of amplitude 1.1 was recorded. Then, on 28 May, CME arrived [17], SW velocity and density increased (Fig. 3 a, b), Bz component fluctuations increased to Bz = -8 nT. Based on the results of the processing, anomalous decrease in CR intensity (Forbush decrease) was observed at Inuvik and Tule stations (Fig. 3 h, i, k, l, o, p). According to the data [17], Forbuch effect of the amplitude 0.5 was recorded at 15:00 on 28 May, the magnetic storm occurred at 19:00 [17].

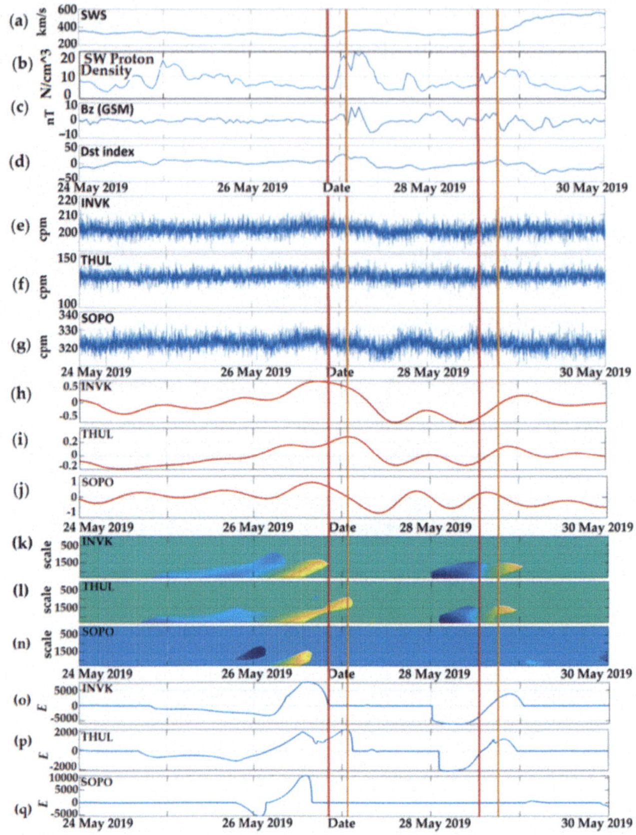

Fig. 3. a) solar wind velocity; b) SW proton density; c) data of interplanetary magnetic field Bz component; d) Dst index data; e), f), g) are the initial signals of INVK, THUL, SOPO stations, respectively; h), i), j) signal, detected by the singular spectrum analysis method at INVK, THUL, SOPO stations, respectively; k)-q) results of AAAD application to NM signals approximated by the SSA.

4 Conclusions

The work results confirmed the complicated dynamics of CR during disturbed periods and proved the efficiency of the method developed for the investigation. Event analysis showed anomalies of complicated form in CR variations before magnetic storms. Anomaly occurrences were observed during CME arrivals to the Earth region and they correlated with the increases in solar wind density and intensification of the IMF southern component fluctuations. During the events of low strength, local increases were detected in CR dynamics. During the moderate magnetic storm, long anomalous decrease in CR intensity (Forbuch decrease) was recorded. Application of the method also made it possible to detect CR variation components having strong correlation with the Dst index. Similar to the paper [3], Dst delay by several hours was observed. The obtained results agree with the investigations [3, 19, 20] and confirm the importance of taking into account the CR dynamics in the tasks of space weather.

Aknowledgements. The work was carried out in the framework of the State Task on the subject (2021—2023) "Physical processes in the system of near space and geospheres during solar and lithospheric impacts», registration number is AAAA-A21-121011290003-0.

The authors are grateful to the institutes making space weather forecast and supporting neutron monitor stations (www.nmdb.eu, http://spaceweather.izmiran.ru), the data of which were used in the paper.

References

1. Kolarski, A., et al.: Impacts of ex-treme space weather events on september 6th, 2017 on ionosphere and primary cosmic rays. Remote Sens. **15**(5), 1403 (2023)
2. Iglesias-Martínez, M.E., et al.: Correlations between background radiation inside a multilayer interleaving structure, geomagnetic activity, and cosmic radiation: a fourth-order cumulant-based correlation analysis. Mathematics **8**(3), 344 (2020)
3. Badruddin, B., Aslam, O.P.M., Derouich, M., Asiri, H., Kudela, K.: Forbush decreases and geomagnetic storms during a highly disturbed solar and interplanetary period, 4–10 September 2017. Space Weather. **17**, 487 (2019)
4. Mandrikova, O., Mandrikova, B.: Hybrid method for detecting anomalies in cosmic ray variations using neural networks autoencoder. Symmetry **14**(4), 744 (2022). https://doi.org/10.3390/sym14040744
5. Wang, X., Huang, L., Wang, Y., Yuan, H.: Influence of Alfvén ion-cyclotron waves on the anisotropy of solar wind turbulence at ion kinetic scales. Universe **9**(9), 399 (2023). https://doi.org/10.3390/universe9090399
6. Balasis, G., et al.: Investigation of dynamical complexity in swarm-derived geomagnetic activity indices using information theory. Atmosphere **14**(5), 890 (2023). https://doi.org/10.3390/atmos14050890
7. Belonin, M.D., et al.: Factor Analysis in Petroleum Geology. VIEMS, Moscow (1971)
8. Lopes, F., Courtillot, V., Le, Mouël, J.L., Gibert, D.: Triskeles and symmetries of mean global sea-level pressure. Atmosphere **13**, 1354 (2022)
9. Real Time Data Base for the Measurements of High-Resolution Neutron Monitor. https://www.nmdb.eu
10. Danilov, D.L., Zhiglyavsky, A.A.: Principal Components of Time Series: The Caterpillar Method. Presskom, St. Petersburg, Russia (1997)
11. Pattanayak, S.: Pro Deep Learning with TensorFlow: A Mathematical Approach to Advanced Artificial Intelligence in Python. Apress, Bangalore, India (2017)
12. Goodfellow, I., Bengio, Y., Courville, A.: Deep Learning. The MIT Press, Cambridge, MA, USA (2016)
13. Chui, C.K.: An Introduction to Wavelets; Wavelet Analysis and Its Applications. Academic Press, Boston, MA, USA (1992)
14. Daubechies, I.: Ten lectures on wavelets. In: CBMS-NSF Regional Conference Series in Applied Mathematics, Society for Industrial and Applied Mathematics, Philadelphia, PA, USA (1992)
15. Witte, R.S., Witte, J.S.: Statistics, 11th edn. Wiley, New York, NY, USA (2017)
16. Geomagnetic Equatorial Dst Index. https://wdc.kugi.kyoto-u.ac.jp/dstdir/
17. Institute of Applied Geophysics. http://ipg.geospace.ru/

18. IZMIRAN Space Weather Forecast Center. Catalog of Forbush Effects and Interplanetary Disturbances. http://spaceweather.izmiran.ru/rus/fds2019.html
19. Belov, A.V., Dorman, L.I., Gushchina, R.T., Yanke, V.G.: Temporal and latitudinal dependence of the temperature effect for neutron component of cosmic ray. Proc. Int. Cosm. Ray Conf. **4**, 1141–1144 (1995)
20. Papailiou, M., Mavromichalaki, H., Belov, A., Eroshenko, E., Yanke, V.: Precursor effects in different cases of forbush decreases. Sol. Phys. **276**, 337–350 (2011)

Analysis of Ionospheric Parameters Based on Threshold Wavelet Filtering (from foF2 Data of Paratunka (Russia) and Wakkanai (Japan) Stations)

Oksana Mandrikova and Yurii Polozov(✉)

Institute of Cosmophysical Research and Radio Wave Propagation FEB RAS, Kamchatsky Krai, Paratunka, Mirnaya Str., 7, 684034 Paratunka, Russian Federation
oksanam1@mail.ru, up_agent@mail.ru

Abstract. In this paper we analyze the critical frequency of the ionospheric layer F2 (foF2) during magnetic storms from 2018 to 2022. The structure of the ionosphere is dynamically changing. It contains local inhomogeneities of different character and intensity. Anomalous changes in the regular course of ionospheric parameters have negative impact on various aspects of life, functioning of spacecrafts and the stable operation of radio communications. The foF2 data were analyzed using the method of adaptive threshold wavelet filtering developed by the authors. The results showed the presence of anomalies of different intensity and duration in foF2 variations during magnetic storms. The comparison of the results from different stations revealed significant dependence of ionospheric dynamics on the parameters of interplanetary medium and the magnetosphere, as well as on local factors. The effects of anomalous increase and anomalous decrease before magnetic storms of different intensities were detected.

Keywords: Ionosphere · Wavelets · Magnetic storm · Data analysis

1 Introduction

Many studies of ionospheric parameters are related to its global influence on human life. A number of important tasks related to theoretical calculations and practice of radio wave propagation can be solved only with the help of knowledge of current ionospheric processes. Moreover, significant volume of information on the ionosphere state is presented space weather forecasting and description. At the same time, the ionosphere is a complex combination of physical processes that are not always can be theoretical calculated or approximated. The quality and efficiency of natural data processing can be improved by applying modern mathematical tools with elements of artificial intelligence.

The parameter of the ionospheric layer critical frequency F2 considered in this paper is used in various fields of geophysics or taken into account in the operation of technical systems. Observations of foF2 allow us to estimate the state of the atmosphere. For example, the review [1] shows that analysis of foF2 made it possible to conclude that

A. Dmitriev et al. (Eds.): STRPEP 2023, SPEES, pp. 141–149, 2023.
https://doi.org/10.1007/978-3-031-50248-4_15

the amount of atomic oxygen in the thermosphere decreased and turbulent diffusion increased in the turbopause region. Studying the dynamics of foF2 is also of great interest in solving space weather problems [1–6]. The behavior of foF2 during magnetic storms is described in many papers, for example, the reviews [4, 5]. Ionospheric perturbations associated with electron concentration increase and decrease usually occur after the magnetic storm onset. However, a number of papers show that ionospheric anomalies can occur before magnetic storms [4–7]. The nature of such events is currently under investigation [5, 8]. Detection and interpretation of such anomalies is a difficult task, requiring analysis of large statistics of events [5, 6]. Moreover, complex internal structure of ionospheric data and low efficiency of traditional methods for studying the time series make the procedure of their analysis difficult [1].

The foF2 parameter includes regular daily and seasonal variations, long-term trend changes [1], fluctuations related to solar activity, etc. Analysis of foF2 time series faces the problem of significant nonstationarity of the data. For example, in [2], the median database [3], the linear regression method, and the EMMD method [2] were used to construct trend changes of foF2 and hmF2. The modeling results showed the difference between the actual and predicted trends for different ionospheric stations. According to the authors [1], this is due to daily and diurnal variations of ionospheric parameters, which were not considered in [2].

In this work, the behavior of foF2 during disturbed periods is studied in detail using the method developed by the authors. The method is based on the combination of threshold functions with wavelet processing. Wavelets are a set of filters that allows one to extract structures of different complexity and duration from the data [9]. Wavelet processing has a wide range of applications in different spheres associated with noisy data processing and information retrieval [9–11]. The advantages of wavelets are also a set of flexible constructs that make it possible to adapt the tool to a particular task. They also allow one to decompose a time series into informative components.

In this work, we analyzed data from Paratunka (53.0° N and 158.7° W, Russia, Kamchatka region) and Wakkanai (45.16° N and 141.75° W, Japan, the resource [12]) stations, which make foF2 observations. The data under analysis were obtained during storms in 2018–2022. The developed method allowed us to study in detail the behavior of foF2 time series for different stations, and identify anomalous changes in the ionospheric variation before and during the events. Special attention was paid to ionospheric parameters preceding magnetic storms.

2 Method Description

We represent the original foF2 time series in the form $y(t_n)$. Application of the discrete wavelet transform [9] leads to

$$y(t_n) = \sum\nolimits_{j,k} Wy(j, t_k)\Psi_{jk}(t_n) \tag{1}$$

where $\Psi_{jk}(t_n) = 2^{\frac{j}{2}}\Psi\left(2^j t_n - t_k\right)$ are basic wavelets, $j, k \in \mathbb{N}$, $Wy(j, t_k) = \langle y, \Psi_{jk}\rangle$ are wavelet coefficients of y decomposition.

The obtained sets of wavelet coefficients $Wy(j, t_k)$ contain the information about the anomalous state of the ionosphere. Based on adaptive thresholds, they can be extracted from the representation (1):

$$\Pi_{T_j^u}\left[Wy(j, t_k)\right] = \begin{cases} Wy(j, t_k), & if\,|Wy(j, t_k)| \geq T_j^u, \\ 0, & if\,|Wy(j, t_k)| < T_j^u. \end{cases} \quad (2)$$

where $T_j^u = Y\hat{\sigma}_j^u$, $\hat{\sigma}_j^u$ is the standard deviation of the coefficients estimated in a moving window of length u, $\hat{\sigma}_j^u = \sqrt{\frac{1}{u-1}\sum\limits_{k=1}^{u}(Wy(j, t_k) - \overline{Wy(j, t_k)})^2}$, where Y is the threshold coefficient.

The threshold coefficient is determined based on the local ionospheric state and is empirically calculated for different observation stations based on the minimization of the posterior risk [13].

As a result of (2), the original expression (1) is transformed into the following form

$$y(t_n) = \sum\nolimits_{j,k} \Pi_{T_j^u}\left[Wy(j, t_k)\right]\Psi_{jk}(t_n).$$

The intensity of anomalies in the ionosphere at time $t = t_k$ is described by the absolute wavelet coefficients $Wy(j, t_k)$ remaining after applying the adaptive thresholds. The intensity of an anomaly at a time point can be estimated by the function

$$I_k = \sum\nolimits_{j=0}^{J} \Pi_{T_j^u}\left[Wy(j, t_k)\right]. \quad (3)$$

The derived expression (3) allows us to represent the anomalous ionospheric increases (I_k positive) and decreases (I_k negative) in the foF2 data.

3 Results of the Method Application

The foF2 data for 2018–2022 of the Paratunka (IKIR FEB RAS, Kamchatka Krai, Russia) and Wakkanai (WDC NICT, Japan) stations were used for the analysis. The foF2 data were obtained by the vertical radiosonde method. The time series have hourly resolution and contain manual corrections from the data of the initial ionograms. Operations (1)–(3) were implemented based on a Daubechies wavelet of order 3 [9], which provides the minimum approximation error in foF2. The time window duration in expression (2) was $u = 168$ hours, threshold coefficient $Y = 2.5$. The interplanetary magnetic field data (IMF Bz), solar wind parameters (SWS and SWD, obtained from the ACE satellite in GSM coordinates [14]), and DST index data [15] were used as space weather factors in the analysis.

During the first analyzed period (Fig. 1) a moderate magnetic storm occurred. Its sudden onset was registered at 0.22 UT on April 20, 2018 at Novosibirsk station [16]. On April 20, the anomalous disturbance of the geomagnetic field was observed at all latitudes. Before the event, an inhomogeneous accelerated flux from the coronal hole (CIR) arrived at the end of the day on April 19 [16]. The results of the foF2 data processing show a large anomalous increase of electron concentration at the analyzed stations during the CIR arrival (Fig. 1b, c, e, f). The exceedance of the anomalous threshold was recorded at Paratunka station several hours earlier than at Wakkanai station. The formation of a high-intensity positive ionospheric anomaly (in the vicinity of 200 conventional units, values are indicated on the vertical axis of Fig. 1c, f) was observed in the early hours of April 20. It occurred on the background of a significant increase in the solar wind (SW) density (Fig. 1j) and a significant southward turn of the IMF component (Fig. 1h). According to space weather data [16], on April 20 the SW velocity (Fig. 1i) increased to 650 km/s and the fluctuations of the IMF southern component (Fig. 1h) increased to Bz $= \pm 19$ nT at 08.12 UT. The observed general anomalous dynamics of the ionospheric process at the analyzed stations (Fig. 1b, c, e, f) indicates the large scale of the disturbance. The positive ionospheric anomaly reached its maximum intensity during the main phase of the magnetic storm. During the recovery phase, the electron concentration decreased significantly and a negative ionospheric anomaly of high intensity (more than 200 conventional units) was formed. The recovery of the foF2 time variation lasted for more than 2 days. The results of the moving median (Fig. 1a, d) agree with the results of the method, confirming their reliability.

During the next period under analysis (Fig. 2), a strong magnetic storm with a gradual onset occurred on August 25, 2018. Due to the arrival of a heterogeneous accelerated flux from the coronal mass ejection (CME) occurred on August 20, the disturbance was registered on August 25 during some hours of the day at mid-latitudes and at some high-latitude stations [16]. The SW velocity during that period increased to 450 km/s (Fig. 2i), the fluctuations of the southern component of the IMF increased to Bz $=$ ± 11 nT (Fig. 2h). At the Paratunka station, the magnetic storm onset was recorded at 21:40 UT. The results of the foF2 data processing show a smooth anomalous increase of electron concentration at Paratunka station before the storm (Fig. 2b, c). A positive ionospheric anomaly of moderate intensity (not exceeding 100 conventional units) was formed by the end of the day on August 23, the duration of the anomaly was about 2 days (Fig. 2c). The anomaly reached its maximum intensity on August 24 during the local increase of the SW density (Fig. 2j) and intensification of IMF southern component fluctuations (Fig. 2h). At the time of the peak of the positive anomaly intensity, according to space weather data [16], a magnetic storm occurred at high latitudes (at 12:00 UT on August 24 at Barentsburg station, the storm onset was gradual). At mid-latitudes, the geomagnetic field was very quiet during that period. It should be noted that the anomaly in the ionosphere was observed at Paratunka station (53.0° N and 158.7° W) and was absent at Wakkanai station (45.16° N and 141.75° W).

During the magnetic storm, a general anomalous dynamic of the ionospheric process was observed at the analyzed stations (Fig. 2). During the initial phase of the storm, electron concentration of the ionosphere was anomalously decreased. At the beginning of the day on 26 August, an inhomogeneous accelerated flux from two coronal holes arrived [16]. The SW velocity started to increase and reached more than 600 km/s on 27 August (Fig. 2i), the IMF southern component fluctuations increased to Bz $= \pm 17$ nT (Fig. 2h). During that period, the ionosphere showed a sharp increase in electron concentration and a moderate positive ionospheric anomaly (about 200 conventional units, Fig. 2c, f) occurred. During the recovery phase, similar to the event considered above, a long (about 2 days) negative ionospheric anomaly was formed at the analyzed stations (Fig. 2b, c, e, f). At Paratunka station the negative ionospheric anomaly had higher intensity (more than 200 conventional units) than that at the Wakkanai station (close to 200 conventional units). The moving median data (Fig. 2a, d) confirm the reliability of the results of the developed method.

Figure 3 shows the results of foF2 data processing during a moderate magnetic storm that occurred on October 22, 2022. Before the storm on October 20, the fluctuations of the IMF southern component increased to $Bz = -5$ nT, the SW velocity in the first half of the day increased to 470 km/s due to IMF fluctuations, and decreased to 400 km/s by the end of the day [16]. The SW density increased smoothly during that period (Fig. 3j). The results of the foF2 data processing show general anomalous dynamics of the ionospheric parameters at the analyzed stations (Fig. 3b, c, e, f). In the second half of the day on October 20, the electron concentration increased and a positive ionospheric anomaly appeared, which reached its maximum intensity at the end of the day on October 20 at both stations.

Against the background of the positive anomaly, a brief decrease in the concentration was observed in the middle of the day on October 21. It occurred simultaneously at Paratunka (Fig. 3b, c) and Wakkanai (Fig. 3e, f) stations.

At the beginning of the day on 22 October, an inhomogeneous flux from the CIR arrived, the fluctuations of the IMF southern component were intensified to $Bz = -11$ nT, and the SW velocity increased to 415 km/s [16]. The geomagnetic field disturbance on 22 October was observed during the day at all latitudes [16]. At the time of the magnetic storm onset, the electron concentration of the ionosphere began to increase again with a delay of several hours at the Wakkanai station. Furthermore, during the main phase of the magnetic storm, the electron concentration decreased and was within the background. During the recovery phase, a negative ionospheric anomaly appeared, similar to the events above (Fig. 2b, c, e, f). At the Paratunka station, the negative ionospheric anomaly had higher intensity (more than 200 conventional units, Fig. 2c) than that at the Wakkanai station (did not exceed 100 conventional units, Fig. 2f).

4 Analysis of Results and Conclusions

The obtained results confirm the complex dynamics of ionospheric parameters during increased solar activity and magnetic storms. Comparison of the results of data processing at Paratunka (53.0° N and 158.7° W) and Wakkanai (45.16° N and 141.75° W) stations showed significant dependence of ionospheric dynamics on the parameters of interplanetary medium and the magnetosphere, and on local factors.

The effects of anomalous increase and anomalous decrease of the ionospheric electron concentration of different intensity were identified before the considered events. The nature of anomalies formation and the comparison with the parameters of the interplanetary medium indicate their possible solar nature. If this hypothesis is confirmed, the study of such events is of great practical interest in the field of space weather.

During magnetic storms, general dynamics of the ionospheric parameters was observed according to the foF2 processing data at the stations. The initial phase of the storm is characterized by an anomalous increase of electron concentration (positive

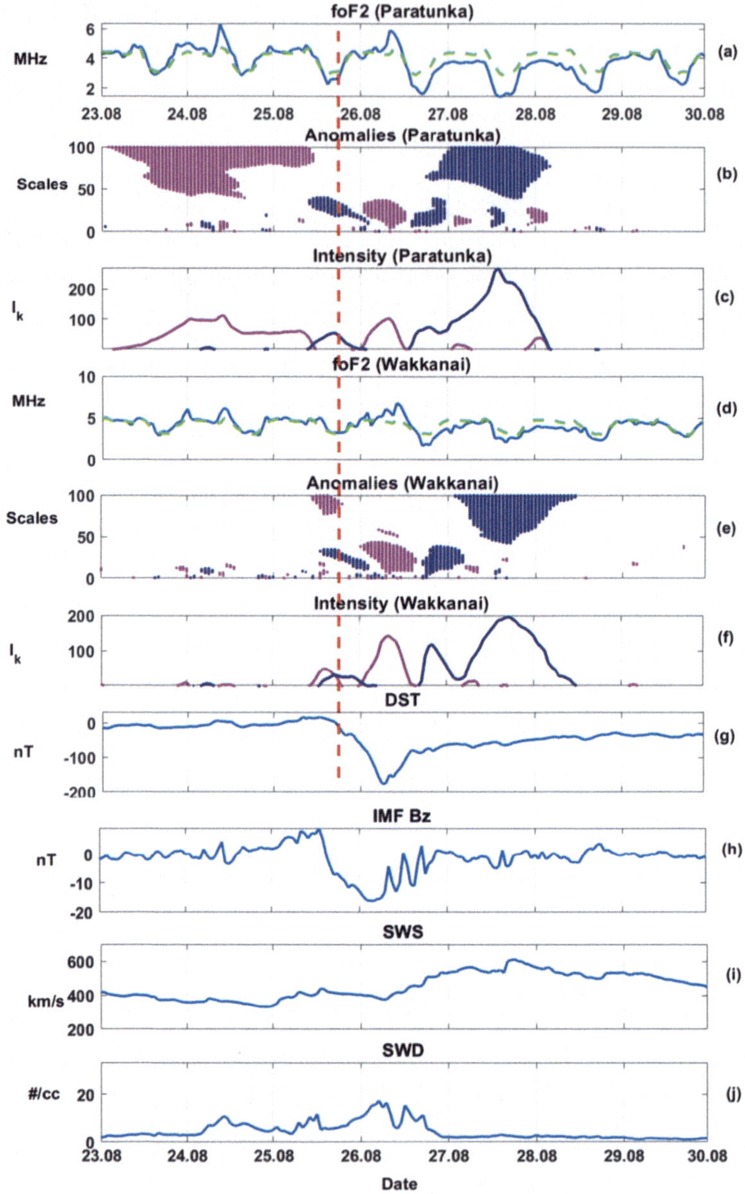

Fig. 1. Results of data processing for the period of April 18–26, 2018. In (a) and (d), blue color shows the registered foF2 data, green color shows the foF2 median values. The red vertical dashed line is the magnetic storm onset.

anomaly). It is well known that this phase is accompanied by a strong induced electric field, which penetrates to middle and low latitudes (PPEF effect). The PPEF leads to a positive ionospheric storm.

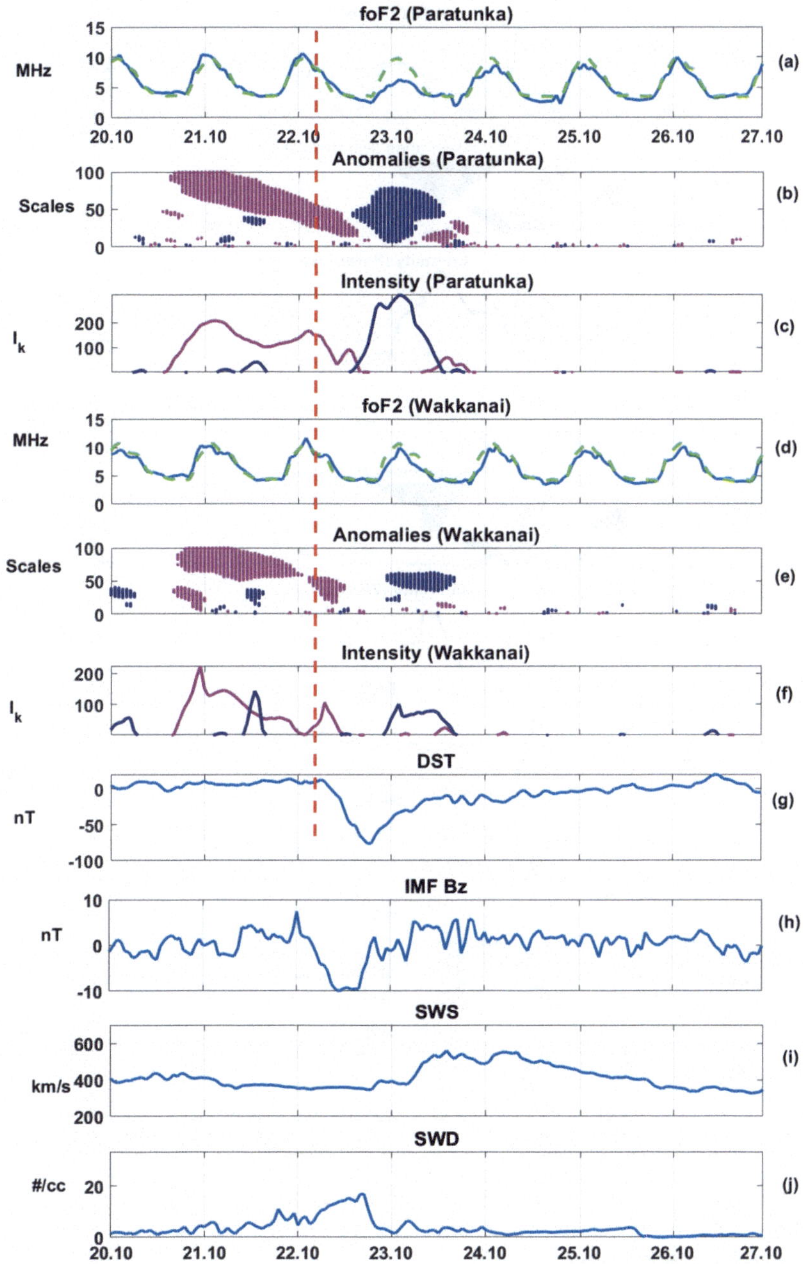

Fig. 2. Results of data processing for the period of August 23–29, 2018. In (a) and (d), blue color shows the recorded foF2 data, green color shows the foF2 median values. The red vertical dashed line is the magnetic storm onset.

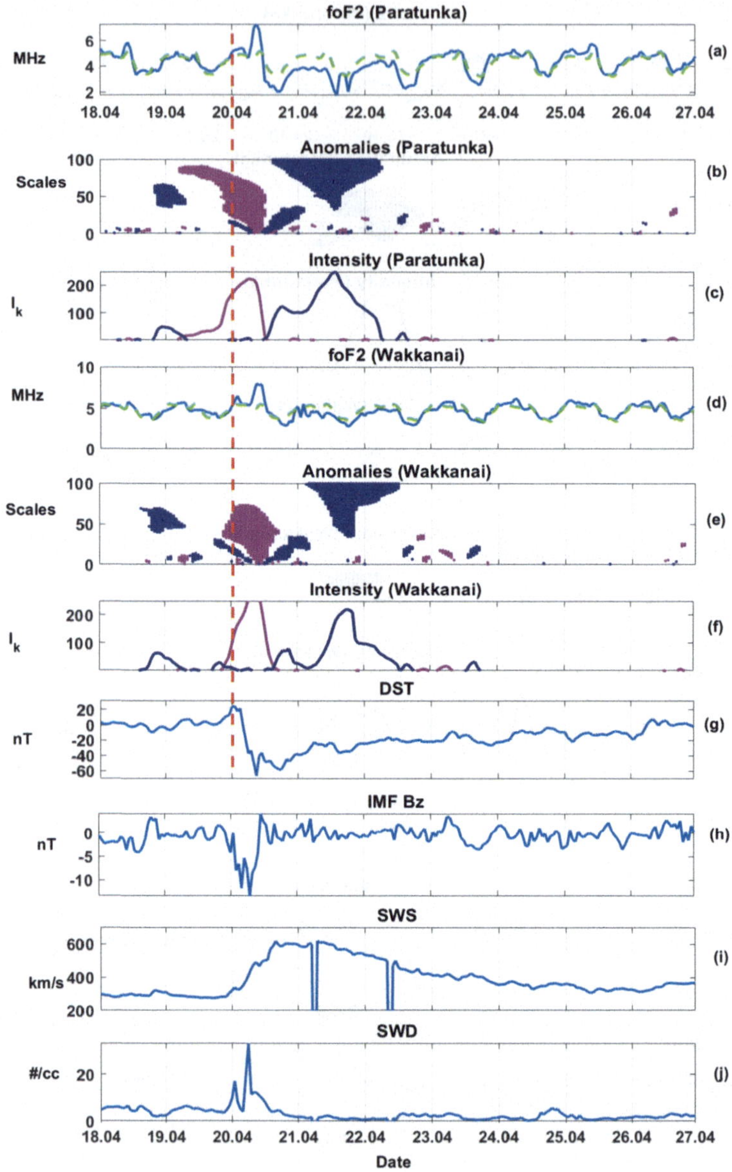

Fig. 3. Results of data processing for the period of October 20–26, 2022. In (a) and (d), blue color shows the recorded foF2 data, green color shows the foF2 median values. The red vertical dashed line is the magnetic storm onset.

During the recovery phase, long-lasting negative ionospheric storms occurred at the stations. The occurrences of negative ionospheric storms are probably related to the increases in the N2/O ratio due to the heating and elevation of the thermosphere and, consequently, to the increase in the recombination rate and the decrease in ionization.

Acknowledgements. The work was carried out within the framework of the State task on the topic (2021–2023) "Physical processes in the system of near space and geospheres under solar and lithospheric impact", registration number is AAAA-A21-121011290003-0. This Research draws upon data provided by WDC for Ionosphere and Space Weather, Tokyo, National Institute of Information and Communications Technology, WDC Kyoto and Coordinated Data Analysis (Workshop) Web (CDAWeb).

References

1. Danilov, A.D., Konstantinova, A.V.: Long-Term Variations In The Parameters Of The Middle And Upper Atmosphere And Ionosphere (Review). Geomagn. Aeron. **60**, 397–420 (2020). https://doi.org/10.1134/S0016793220040040
2. Cnossen, I., Franzke, C.: The role of the Sun in long-term change in the F_2 peak ionosphere: new insights from EEMD and numerical modeling. J. Geophys. Res. Space Physics **119**, 8610–8623 (2014). https://doi.org/10.1002/2014JA020048
3. Damboldt, T., Suessmann, P.: Consolidated database of worldwide measured monthly medians of ionospheric characteristics foF2 and M(3000)F2. INAG Bulletin on the Web, INAG-73 (2012)
4. Danilov, A.D.: Ionospheric F-region response to geomagnetic disturbances. Adv. Space Res. **52**, 343–366 (2013). https://doi.org/10.1016/j.asr.2013.04.019
5. Danilov, A.D., Konstantinova, A.V.: Ionospheric precursors of geomagnetic storms. 1. A review of the problem. Geomagn. Aeron. **59**, 554–566 (2019). https://doi.org/10.1134/S0016793219050025
6. Blagoveshchensky, D.V., Sergeeva, M.A., Kozlovsky, A.: Ionospheric parameters as the precursors of disturbed geomagnetic conditions. Adv. Space Res. **60**, 2437–2451 (2017). https://doi.org/10.1016/j.asr.2017.09.013
7. Mandrikova, O., Polozov, Y., Fetisova, N., Zalyaev, T.: Analysis of the dynamics of ionospheric parameters during periods of increased solar activity and magnetic storms. J. Atmos. Solar-Terr. Phys. **181**, 116–126 (2018). https://doi.org/10.1016/j.jastp.2018.10.019
8. Danilov, A.D., Konstantinova, A.V.: A detailed analysis of the behavior of the F_2-layer critical frequency prior to magnetic storms. Geomagn. Aeron. **62**, 590–606 (2022). https://doi.org/10.1134/S0016793222050036
9. Mallat, S.G.: A Wavelet Tour of Signal Processing: The Sparse Way. Academic Press, Boston (2009)
10. Mandrikova, O., Fetisova, N., Polozov, Y.: Hybrid model for time series of complex structure with ARIMA components. Mathematics **9**, 1122 (2021). https://doi.org/10.3390/math9101122
11. Mandrikova, O., Polozov, Y., Khomutov, S.: Wavelet model of geomagnetic field variations and its application to detect short-period geomagnetic anomalies. Appl. Sci. **12**, 2072 (2022). https://doi.org/10.3390/app12042072
12. ICSU World Data System. https://wdc.nict.go.jp/. Accessed 28 August 2023
13. Berger, J.O.: Statistical Decision Theory and Bayesian Analysis. Springer, New York (1993)
14. Coordinated Data Analysis Web. https://cdaweb.gsfc.nasa.gov/index.html. Accessed 28 August 2023
15. World Data Center for Geomagnetism, Kyoto. https://wdc.kugi.kyoto-u.ac.jp/. Accessed 28 August 2023
16. Fedorov Institute of Applied Geophysics. http://ipg.geospace.ru/. Accessed 28 August 2023

Seismoacoustic and Seismoelectric Responses of Near-Surface Sedimentary Rocks in Kamchatka

Mikhail M. Mishchenko[1(✉)], Oleg P. Rulenko[2], and Yuriy V. Marapulets[1]

[1] Laboratory of Acoustic Researches, Institute of Cosmophysical Research and Radio Wave Propagation FEB RAS, Mirnaya, 7, Paratunka 684034, Kamchatka, Russia
{micle,marpl}@ikir.ru

[2] Laboratory of Active Tectonics and Paleoseismology, Institute of Volcanology and Seismology FEB RAS, bul'var Piypa, 9, Petropavlovsk-Kamchatsky 683006, Kamchatka, Russia
rulenko@kscnet.ru

Abstract. Seismoacoustic and seismoelectric responses of near-surface sedimentary rocks, detected earlier in Kamchatka by the authors, are under investigation. Cases of these responses, recorded from November 2012 to May 2023 at Karymshina site, are considered. Highly significant correlation relation between earthquake energies and distances from hypocenters was discovered in the cases of only seismoacoustic responses, as well as both seismoacoustic and seismoelectric responses of rocks. It indicates that manifestation of seismoacoustic and seismoelectric responses on near-surface sedimentary rocks is determined by the energy emitted from an earthquake source in the form of seismic waves and delivered to an observation site by these waves. Rock acoustic responses during some earthquakes were detected not only at seismic wave frequencies but also at higher kilohertz frequencies. Such high-frequency responses were also recorded at Krutoberegovo site located 480 km from Karymshina site. Observation of high-frequency seismoacoustic signals at two sites indicates the fact that their generation is manifested in sedimentary rocks in different near-surface zones of the earth crust in Kamchatka.

Keywords: near-surface sedimentary rocks · seismoacoustic and seismoelectric responses · manifestation features

1 Relevance of Research

Geodynamic phenomena and processes, occurring at the earth crust - atmosphere boundary, are the subject of investigation of an intensively developing new direction in Geophysics, Near-Surface Geophysics [1,2]. It investigates phenomena and processes developing, in particular, in the Earth near-surface layer.

O. P. Rulenko and Y. V. Marapulets—These authors contributed equally to this work.

A. Dmitriev et al. (Eds.): STRPEP 2023, SPEES, pp. 150–158, 2023.
https://doi.org/10.1007/978-3-031-50248-4_16

Rocks are constantly deformed here under the effect of different causes (moon-solar tides, weather parameter variations, Earth eigen movements etc.). Rock deformation in seismically active regions also occurs under the impact of seismic waves from earthquakes of different energy.

Sedimentary rocks are widely spread among the near-surface rocks. They cover about 80% of the continents surface [3]. Sedimentary rocks are a complicated polydisperse water- and gas-saturated porous medium of low rigidity and, thus, are easily deformed. Of great interest is the transformation of energy of these rocks deformations by seismic waves from earthquakes into acoustic signal energy and electric field variations. Investigation of near-surface sedimentary rock seismoacoustic and seismoelectric responses, different in genesis but having common deformation nature, characterizes the deformation process better and is topical for investigation of mechanic and electric properties of these rocks. Moreover, seismic waves affect sedimentary rocks, which are in stress state. That is why the intensity of transformation of elastic impact energy of seismic waves into acoustic and electric fields energy will be mainly determined not only by rocks structure and texture but their stress-strain state as well [4,5]. Thus, when we record a response of near-surface sedimentary rocks on seismic wave impact in the form of generation of acoustic and electric signals, we can indicate the changes of properties, stress state and the deformation pattern of these rocks at the observation site.

During sedimentary rock deformations by seismic waves from earthquakes, relative micro displacements of fragments and interactions of their surfaces occur. That is accompanied by acoustic signal generation in the frequency range from seismic waves up to the first tens of kHz [6]. Seismoelectric effect of the second kind, also known in Geophysics and detected on sedimentary complex rocks [7], appears under the impact of elastic oscillations of these waves. It consists in the generation of an electric field during displacement of the fluid phase relatively rock solid fragments and distortion of the equilibrium state in the double electric layer at the fragments-fluid interface. Thus, sedimentary rocks can be considered as acoustic and electric seismographs simultaneously appearing during seismic wave passage. They are formed in the result of collective reaction of rocks fragments on their elastic deformation.

Simultaneous investigations of seismoacoustic and seismoelectric responses of near-surface sedimentary rocks were first begun in Kamchatka in 2012 and are still carried out at Karymshina site (52.83° N, 158.13° E) of IKIR FEB RAS [8–11]. This site is located in the region of Verkhne-Paratunskaya hydrothermal system of the Southern Kamchatka, in the zone of intersection of different-rank tectonic faults [12]. The main results of the investigations are as follows. Joint seismoacoustic and seismoelectric response of near-surface sedimentary rocks was discovered for the first time. Moreover, acoustic signal generation was detected not only at seismic wave frequencies but also at higher kilohertz frequencies. The both effects depend on earthquake magnitude and on the distance from a hypocenter. The dependence of intensity of rock response manifestation on earthquake energy was confirmed during two earthquakes having very close location of sources but different energy. Kilohertz clusters were observed in rock acoustic response on a P-wave during a close and deep strong seismic event. They were not observed during a remote shallow event almost of the same magnitude. Appearance of rock acoustic response in kilohertz frequency range agrees well with the behavior of vector module of soil displacement velocity and its spectrum at the observation site. We detected statistically significant relation between earthquake energies and the distances from hypocenters during joint acoustic and electric responses of rocks and in cases of presence of kilohertz frequencies in acoustic responses.

In the present paper, we added 20 cases of joint responses of near-surface sedimentary rocks recorded in 2012–2023 to the 18 cases described in [11]. Results of records of acoustic signals at the frequencies higher than seismic waves at Krutoberegovo site, located 480 km to the south-east of Karymshina site, are presented.

2 Observation Method and Data Analysis

A wide-band piezoceramic hydrophone, having the sensitivity together with a pre-amplifier of about 1 V/Pa, was used as an acoustic signals receiver. It was hung in an artificial water reservoir of the size $1 \times 1 \times 1\,m^3$ at the depth of 0.5 m. The hydrophone had a directivity pattern diagram of 60°, was oriented vertically downwards and was arranged at the point of intersection of measuring lines for recording the electric field in rocks. Electric field horizontal components were recorded by two orthogonal measuring 10-meter lines oriented along the magnetic meridian (N-S) and perpendicularly to it (E-W). Lead plates with the size of $0.25 \times 1.0\ m^2$ were used as electrodes. They were dug into the ground at the depth of 1.0 m. The resistance between N-S line electrodes was 10.5 kOhm and that of the E-W line was 8.7 kOhm. The potential difference between the electrodes was applied to the pre-amplifier with input resistanse of 1 MOhm. The pre-amplifier was near the measuring lines. Acoustic and electric signals were simultaneously digitized by a general 8-cannel 16-bit professional sound card M-Audio Fast Track Ultra 8R with the sampling frequency of 48 kHz. The frequency range of the recorded signals was 0.1 Hz–11 kHz.

According to the drilling results, the near-surface rocks at Karymshina site are sedimentary with the layer thickness of about 50 m [13]. There is a well No. 99-8 with the depth of 19 m, located 170 m from the hydrophone. Its geological column is represented by boulder-cobble deposits with sand-clay filler (0–5 m), block-rubble deposits with clay filler (5–14 m) and boulder-cobble deposits with sand filler (14–19 m). Based on the estimates from the paper [6], sources of the acoustic signals, occurring at the frequencies from the first hundreds of Hz to the first tens of kHz, are located at the distances up to the first tens of meters from the hydrophone. Taking all that into the consideration and the distance between the electrode and their position in the ground, we think that seismoacoustic and seismoelectric signals were generated in the near-surface sedimentary rocks.

Seismoacoustic and seismoelectric responses of sedimentary rocks are formed as follows. In case of the applied installation of the hydrophone, the compressional P-waves are recorded without significant distortions since seismic wave refraction can be neglected when the water reservoir is small. Water does not have form elasticity that is why shear S-waves do not propagate in water reservoirs. However, when initiating horizontal and vertical motions of soil, they cause rock acoustic emission and affect, together with the emission, the hydrophone though the hanger, manifesting in the obtained data [6]. Taking that into consideration, we considered acoustic signals occurring only during P-wave passage. These signal frequencies are determined by the sizes of rock fractions interacting at the given time. Electric field variations in rocks occur only at the frequencies of seismic wave mechanic impact.

The recorded seismoacoustic and seismoelectric signals were analyzed in different frequency ranges. In the paper [8,9], spectrograms of these signals were under consideration. A new method for investigation of joint seismoacoustic and seismoelectric responses of near-surface sedimentary rocks was suggested and applied in the work [10]. It consists in simultaneous recording of these responses and comparison of the features of their occurrences with the behavior of three displacement components and soil displacement velocity during P- and S-wave passage from strong Kamchatka earthquakes. In the paper [11], we considered the correlation relation between earthquake energies and the distances to hypocenters only for seismoacoustic responses, as well as for both seismoacoustic and seismoelectric responses of rocks. Nonparametric correlation Spearman's analysis was applied.

3 Seismoacoustic and Seismoelectric Responses of Rocks at Karymshina Site

The paper investigates the correlation relations between earthquake energy and the distance to a hypocenter for a seismoacoustic response only, as well as for both seismoacoustic and seismoelecreic responses of near-surface sedimentary rocks. We considered the cases of response records for the period from November 2012 to May 2023. The energy class K_s was considered as the earthquake energy characteristics. It is the decimal logarithm of the energy emitted from a source in the form of seismic waves [14,15]. Earthquake parameters were taken from the catalogue of Kamchatka Branch of the Federal Research Center "Geophysical Survey, Russian Academy of Sciences" (KB GS RAS) (http://sdis.emsd.ru/info/earthquakes/catalogue.php). Earthquakes with $K_s > 11$ and the epicentral distances up to 600 km were under consideration from this catalogue.

We detected 196 earthquakes, during which rock acoustic responses were observed. Rock electric responses, that is both acoustic and electric responses occurred together, were detected during 38 earthquakes from them. Earthquake epicenter maps for the both cases are represented in Fig. 1a, b, respectively.

The earthquakes, during the passage of seismic waves from which only acoustic or both acoustic and electric responses were observed, are denoted differently in Fig. 1. Using the same symbols for all 196 earthquakes as in Fig. 1, Fig. 2 compares their energy classes K_s with the distances from hypocenters to Karymshina site D.

The correlation relation between K_s and D was considered for the two earthquake groups. As it was mentioned above, the first group (No. 1) includes 38 earthquakes, during which both seismoacoustic and seismoelectric responses of rocks were observed. The second group (No. 2, 158 earthquakes) contains the earthquakes during which only seismoacoustic responses of rocks without seismoelectric responses were observed. Nonparametric correlation analysis was applied in

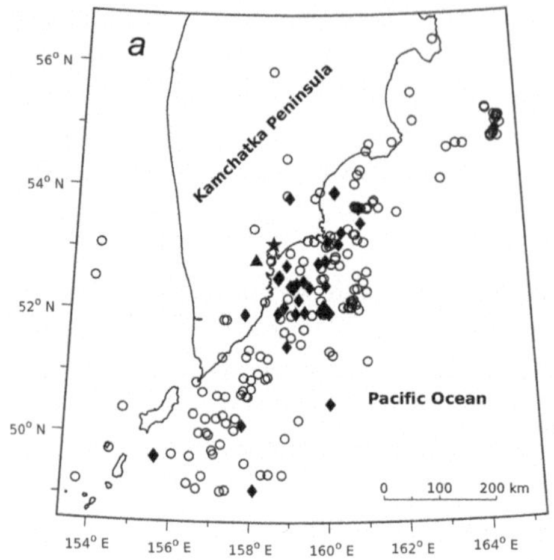

Fig. 1. Earthquake epicenter maps during which only acoustic response (a, ○) or both acoustic and electric responses of near-surface sedimentary rocks (a, b, ♦) were observed; ▲ - Karymshina site, ★ - Petropavlovsk-Kamchatskiy

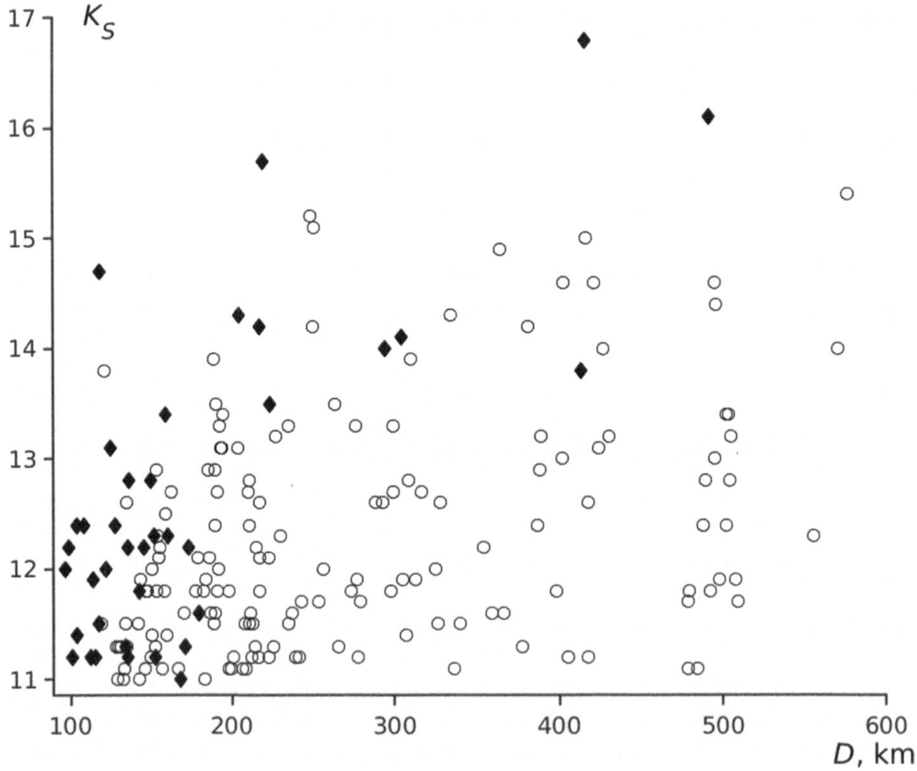

Fig. 2. Comparison of the considered earthquake energy classes K_s with distances from hypocenters to Karymshina site D. Rock response symbols (○, ♦) are the same as in Fig. 1.

Table 1. Estimates of Spearman's correlation coefficient r_s and its significance level p between earthquake energy classes and hypocentral distances to Karymshina site for the first (No. 1) and the second (No. 2) groups of earthquakes

Group/Earthquake number	r_s	p
No. 1/38	0.53	<0.001
No. 2/158	0.37	<0.001

the both cases. Spearman's rank correlation coefficient r_s was considered. It is less sensitive to spikes and errors in observation results. Moreover, it also allows us to estimate monotonous nonlinear relations between variables and, that is important in our case, can be used for sample small sizes [16].

Table 1 represents the estimates of Spearman's correlation coefficient r_s and its significance level p between K_s and D values for the two specified groups of earthquakes. According to Table 1, there is a highly significant correlation relation ($r_s = 0.53$, $p < 0.001$) between the energy classes of earthquakes, during which both seismoacoustic and seismoelecreic responses of rocks were observed, and the distances from hypocenters. In spite of the larger sample size, there is the same highly significant but weaker relation ($r_s = 0.37$) for the earthquakes, during which only seismoacoustic response of rocks was observed. The highly significant correlation relations indicate the fact that manifestation of seismoacoustic and seismoelectric responses of near-surface sedimentary rocks is determined by energy, emitted from an earthquake source in the form of seismic waves and delivered to the observation site by these waves.

4 High-Frequency Seismoacoustic Responses of Rocks at Krutoberegovo Site

As it was mentioned above, during the passages of seismic waves from some earthquakes at Karymshina site, acoustic signals appear not only at seismic wave frequencies but also at higher kilohertz frequencies. They indicate the transformation of low-frequency energy of seismic waves to significantly higher energy of acoustic radiation of sedimentary rocks close to the hydrophone. We think that such signals are generated during rock intensive deformations when relative micro displacements and interactions of smallest fragment surfaces occur.

Detection of seismoacoustic signals, including kilohertz ones, at other site in Kamchatka is of interest. To do that, we organized and began geoacoustic observations at Krutoberegovo site (56.26° N, 162.71° E), located 480 km to the northeast from Karymshina site. An artificial water reservoir of the size $1 \times 1 \times 1\,\text{m}^3$ was set to install a piezoceramic hydrophone. When digging a hole for the water reservoir, it was discovered that the rocks under the soil layer are sedimentary (large and fine send, clay band). The measuring-recording complex, observation and data analysis methods were the same as at Karymshina site.

Three close earthquakes occurred from January 2022 to July 2023. Their hypocenters were located close to each other. During the passage of seismic waves from these earthquakes, acoustic signals were recorded in high-frequency kilohertz frequency range besides the low-frequency one. Earthquake parameters, taken from the catalogue of KB GS RAS (http://sdis.emsd.ru/info/earthquakes/catalogue.php), and the distances from hypocenters to Krutoberegovo site are shown in Table 2.

As an example, Fig. 3 shows a record and a spectrogram of the acoustic signal which was observed during the passage of waves from the earthquake No. 1 (Table 2). It is clearly seen on the spectrogram of this signal that acoustic energy was emitted in different frequency clusters during the passage of seismic waves just like during some earthquakes recorded at Karymshina site [8–10]. The first cluster manifested at the frequencies up to about 100 Hz, the second one manifested

Table 2. Parameters of the earthquakes, during which high-frequency acoustic signals were recorded besides the low-frequency ones at Krutoberegovo site. D is the distance from a hypocenter.

No.	Date, UTC	Time, UTC	Lat.° N	Long.° E	H, km	K_s	D, km
1	2022.01.18	22:54:59.2	56.26	163.00	11.3	10.0	21.1
2	2023.06.29	02:24:17.6	56.24	162.97	6.3	10.5	17.3
3	2023.07.04	02:48:19.8	56.25	162.96	7.9	10.6	17.3

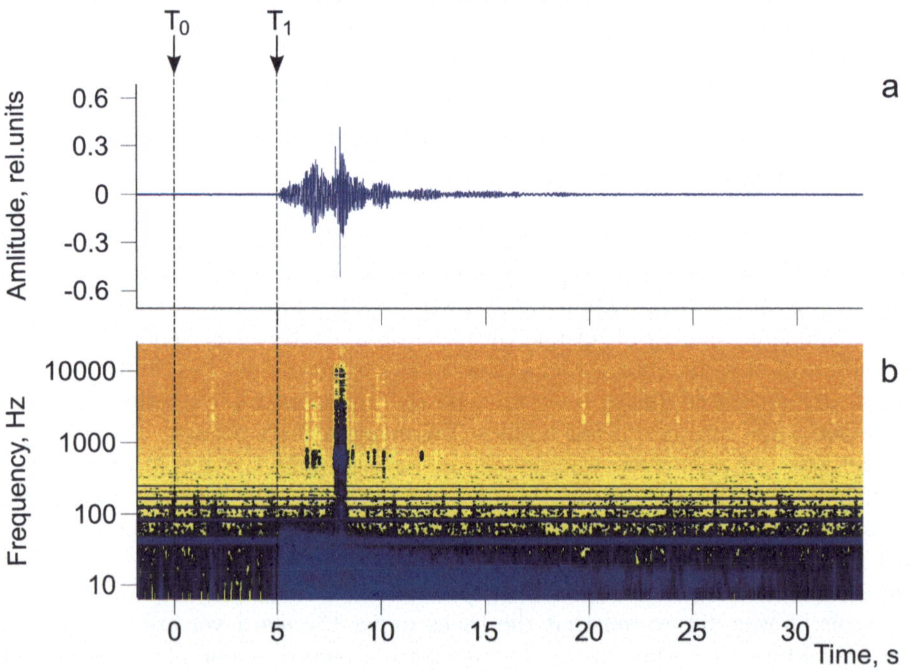

Fig. 3. Record (a) and spectrogram (b) of acoustic signal at Krutoberegovo site during the passage of seismic waves from earthquake No. 1 (Table 2). Arrows indicate the times of earthquake occurrence T_0 and P-wave onset T_1.

at the frequencies from 100 Hz to about 10 kHz. The same clusters are observed on the spectrograms of acoustic signals recorded during the earthquakes No. 2, 3 (Table 2). That all indicates the fact that generation of high-frequency seismoacoustic signals manifests in sedimentary rocks in different near-surface zones of the earth crust in Kamchatka.

5 Conclusions

We considered the cases of records of seismoacoustic and seismoelectric responses of near-surface sedimentary rocks at Karymshina site in Kamchatka from November 2012 to May 2023. The relation between the energy classes of earthquakes K_s, during which there was only seismoacoutic or both seismoacoustic and seismoelectric responses, and the distances from earthquake hypocenters D to this site were under investigation. In both cases, highly significant correlation between K_s and D was discovered. It indicates that manifestation of seismoacoustic and seismoelectric responses on near-surface sedimentary rocks is determined by the energy emitted from an earthquake source in the form of seismic waves and delivered to an observation site by these waves. The high-frequency seismoacoustic signals were also detected at Krutoberegovo site located 480 km to the northeast from Karymshina site. They were observed during three close earthquakes, the hypocenters of which were located close to each other. Observation of high-frequency signals at two sites shows that their generation manifests in sedimentary rocks in different near-surface zones of the earth crust in Kamchatka. Observation of the high-frequency seismoacoustic responses of rocks at Karymshina and Krutoberegovo sites confirms the fact that these responses occur during strong and moderate in energy remote earthquakes and during weak in energy close earthquakes.

Acknowledgments. The research was carried out in the framework of realization of the State Task AAAA-A21-121011290003-0.

References

1. Slater, L., Knight, R., Singha, K., Binley, A., Atekwana, E.: Near-surface geophysics: a new focus group. EOS Trans. Am. Geophys. Union **87**(25), 248–249 (2006)
2. Adushkin, V.V., Spivak, A.A.: Problems related to the interaction of geospheres and physical fields in near-surface geophysics. Izv. Phys. Solid Earth **55**(1), 1–11 (2019). https://doi.org/10.1134/S1069351319010014
3. Garrels, R.M., Mackenzie, F.T.: Evolution of Sedimentary Rocks. Norton, New York (1971)
4. Migunov, N.I.: Near-surface geophysics: a new focus group. Izv. Akad. Nauk SSSR Fiz. Zemli (9), 20–28 (1984). in Russian
5. Marapulets, Y.V., Shevtsov, B.M., Larionov, I.A., Mishchenko, M.A., Shcherbina, A.O., Solodchuk, A.A.: Geoacoustic emission response to deformation processes activation during earthquake preparation. Russ. J. Pac. Geol. **6**, 457–464 (2012). in Russian

6. Marapulets, Y.V., Shevtsov, B.M.: Mesoscale Acoustic Emission. Dal'nauka, Vladivostok (2012). in Russian
7. Ivanov, A.G.: Seismic and electrical effect of the second kind. Proc. Acad. Sci. USSR Ser. Geograph. Geophys. **5**, 18–21 (1940). in Russian
8. Muratov, P.V., Rulenko, O.P., Marapulets, Y.V., Solodchuk, A.A.: Electrical and acoustic response of the near-surface sedimentary rocks to seismic earthquake waves passing. Vest. KRAUNC Fiziko-Mat. Nauki **25**(5), 62–73 (2018). in Russian
9. Muratov, P.V., Rulenko, O.P., Marapulets, Y.V.: Manifestation features of nearsurface sedimentary rock electric and acoustic responses on seismic waves from earthquakes. In: E3S Web of Conferences, vol. 127, p. 02015 (2019). https://doi.org/10.1051/e3sconf/201912702015
10. Mishchenko, M.A., Rulenko, O.P., Marapulets, Y.V.: Some features of acoustic and electric responses of near-surface sedimentary rocks during deformations by seismic waves from strong earthquakes (mw > 6.5). In: E3S Web of Conferences, vol. 196, p. 02028 (2020) https://doi.org/10.1051/e3sconf/202019602028
11. Mishchenko, M.A., Rulenko, O.P., Marapulets, Y.V.: Analysis of manifestation of joint acoustic and electric responses of near-surface sedimentary rocks on the deformation by earthquake seismic waves in the south of Kamchatka. In: EPJ Web of Conferences, vol. 254, p. 02002 (2021) https://doi.org/10.1051/epjconf/202125402002
12. Serezhnikov, A.I., Zimin, V.M.: Geological structure of the Paratunka geothermal region, the influence of some geological factors on the current hydrothermal activity (hydrothermal systems and thernal fields of Kamchatka). DVNTs AN SSSR, Vladivostok (1976). in Russian
13. Kuptsov, A.V., Larionov, I.A., Shevtsov, B.M.: Geoacoustic emission during the precursory periods of Kamchatka earthquakes. Volcanol. Seismol. **5**, 45–59 (2005)
14. Fedotov, S.A.: On energetic classification of Kuril-Kamchatka Earthquakes, Q-frequency dependence and magnitude problem. Program and Abstracts, General Scientific Assembly JASPEI-IAGA. Madrid (1969)
15. Fedotov, S.A.: On Energetic Classification of Kuril-Kamchatka Earthquakes and Magnitude Problem. Nauka, Moscow (1972). in Russian
16. Borovikov, V.: STATISTICA. The Art of Computerized Data Analysis: For Professionals. Piter, St. Petersburg (2003). in Russian

Monitoring of Radiation Fields in Near-Earth Space with the Use of Kodiz Instrument On-Board Cubesat Monitor-1

G. I. Antonyuk[1], V. V. Benghin[1,2], I. A. Zolotarev[1], A. V. Sazonova[1,3], V. I. Osedlo[1(✉)], Vas V. Sazonov[4], O. Yu. Nechaev[1], and S. I. Svertilov[1,3]

[1] D.V. Skobeltsyn Institute of Nuclear Physics, Lomonosov Moscow State University, 1(2), Leninskie Gory, GSP-1, Moscow 119991, Russian Federation
osedlo@mail.ru

[2] Institute for Biomedical Problems of the Russian Academy of Sciences, 76A Khoroshevskoye Shosse, 123007 Moscow, Russian Federation

[3] Physics Department, Lomonosov Moscow State University, Leninskie Gory, Moscow 119991, Russian Federation

[4] Faculty of Space Research, Lomonosov Moscow State University, GSP-1, 1-52, Leninskiye Gory, 119991 Moscow, Russian Federation

Abstract. As part of the Constellation-270 project of Moscow University, on August 9, 2022, the Monitor-1 spacecraft was launched, on which the KODIZ device, a combined cosmic radiation detector, was installed as a payload. The KODIZ device is designed to test equipment for detecting radiation-dangerous fluxes of solar cosmic rays, as well as high-energy particles in precipitations from the Earth's magnetosphere. These particles can create additional ionization in the ionosphere. The article describes the system of detectors of the device and the procedure for processing the signals coming from them, as well as the subsequent transmission of the received information to the Earth.

Keywords: CubeSat · Cherenkov radiation detector · Semiconductor detector · Neutron detector

1 Introduction

It is well known that powerful solar flares can have a significant impact on a number of aspects of human activity [1]. One of the possible negative consequences of such solar flares is the increased radiation exposure of people on spacecraft and high-altitude aircraft located in high-latitude regions of the Earth [2–4]. To take measures to reduce increased exposure, timely information about the appearance in the low Earth orbit (LEO) of significant fluxes of high-energy protons generated by a powerful solar flare is needed. In near-Earth space, systems of patrol satellites have been successfully operating for a long time, monitoring the parameters of radiation conditions [5]. In connection with the wide development in recent years of the technology of microsatellites of the Cubesat type, during the development of the Universat-Socrates program, Mikhail Igorevich

A. Dmitriev et al. (Eds.): STRPEP 2023, SPEES, pp. 159–165, 2023.
https://doi.org/10.1007/978-3-031-50248-4_17

Panasyuk, director of the SINP MGU, proposed to create a device that could further monitor radiation conditions in the LEO using microsatellites of this type. The task was set to ensure the registration of particle flux capable of penetrating the Earth's atmosphere to the altitudes of aircraft flights. A device was developed that makes it possible to record fluxes of high-energy particles of both solar and magnetospheric origin, capable of penetrating to altitudes of 100–20 km and creating additional ionization at these altitudes. Estimates have shown that for a flight altitude of 15 km, the thickness of the residual atmosphere is 124 g/cm^2 [6], and the energy of protons capable of penetrating to such a height is about 450 meV. This made it expedient to use a Cherenkov detector as one of the sensors in order to ensure the registration of particles of sufficiently high energies [7]. To work out the methodological issues of creating a means for monitoring radiation conditions in the LEO on the basis of a microsatellite, the KODIZ device was developed - a combined detector of cosmic radiation, which is described in this article. The KODIZ device was integrated into the CubeSat 3U OrbiCraft-Pro, developed by SPUTNIX [8]. This microsatellite, named Monitor-1, was launched on August 9, 2022 by a Soyuz-2.1b launch vehicle with a Fregat upper stage as part of the SPACE-π project and successfully launched into a polar orbit with an altitude of 500 km and an inclination of 97.4° [9].

2 KODIZ Design

2.1 Detectors of the Device KODIZ

In accordance with the main task, the Cherenkov detector was chosen as the main detector of the device. A detector in the form of a Plexiglas cylinder with a diameter of 38 mm and a height of 20 mm, viewed by a photomultiplier, was used. The detection threshold of such a detector for protons is about 330 meV. It is technically difficult to obtain a detector with a higher detection threshold due to the lack of transparent substances suitable for use on Cubesat with a suitable refractive index. Therefore, to estimate the flux of protons with higher energies, we decided to use the spectrum extrapolation method. One point on the spectrum is given by a Cherenkov detector, and to obtain the second point of the spectrum, it was decided to use a semiconductor detector. This makes it possible to obtain a point for the energy threshold of 40–50 meV (for protons), which is necessary for estimating the spectrum parameters. In addition, the semiconductor detector made it possible to implement the method of dosimetric measurements, which is widely used on spacecraft for various purposes [10, 11].

Thus, to solve the main task of the device, it should have included a Cherenkov and a semiconductor detector. The Cherenkov detector was viewed by a photomultiplier, the operation of which was provided by a high-voltage converter. A bias voltage of 80 V was applied to operate the semiconductor detector. Signal processing from the detectors was provided using analog circuits and subsequent conversion to digital form. The Milandr 1986BE93U microcontroller was chosen for digital processing.

Elaboration of the device layout and its placement in the microsatellite showed that the resources of the Cubesat and the selected microcontroller made it possible to install additional detectors in the device. A thermal neutron detector, a fast neutron detector, and a scintillation detector based on a CsI crystal were chosen as such detectors.

The thermal neutron detector is based on a plastic scintillator BC-702 in the form of a disk with a diameter of 38 mm and a thickness of 6.35 mm, which is made of a lithium compound matrix enriched up to 95% with lithium-6 (^{6}Li) dispersed in a fine zinc sulfide phosphor powder ZnS(Ag). The BC-702 detector makes it possible to detect thermal neutrons with energies up to 1 eV due to the reaction (^{6} Li (n, α) ^{3} H + 4.79 meV). The fast neutron detector is based on a plastic scintillator BC-720 in the form of a disk with a diameter of 38 mm and a thickness of 15.9 mm, which is made of transparent hydrogen plastic with a ZnS(Ag) zinc sulfide phosphor. The BC-720 detector makes it possible to detect fast neutrons with energies above 1 meV by detecting recoil protons with zinc sulfide. Each of the neutron detectors, as well as the Cherenkov detector, were viewed using the same HAMAMATSU R5611A PMT with a tube diameter of 19 mm and a circular photocathode diameter of 15 mm. The radiator of the Cherenkov detector is additionally viewed by a silicon photomultiplier SiPM, which was installed opposite the PMT for simultaneous light collection of radiation in order to subsequently evaluate the possibility of using SiPM instead of the PMT.

The Cherenkov and neutron detectors are arranged in a separate unit, with its own analog signal processing board and a high-voltage voltage converter.

The second detector assembly contains two semiconductor silicon detectors and an experimental scintillation detector based on a CsI crystal. Semiconductor detectors have sensitive elements based on ion-implanted silicon with an area of 1 cm^2 and a thickness of 0.3 mm. These detectors are made in the form of a classical scheme of a telescope of detectors installed coaxially under each other at a distance of 18 mm, which provides an opening angle of about 30°. The detector assembly also includes a detector bias filter and an Amptek A225F charge sensitive preamplifier. Two signals are formed at the outputs of each preamplifier: the first one, the S-signal, has an amplitude proportional to the charge formed in the detector and a duration of about 5–10 μs, and the second, the t-signal, has a short delay time of less than 0.5 μs from the moment of arrival of the signal from the detector to the maximum amplitude. The first of the signals is fed to the analog input of the microcontroller, which allows you to convert its amplitude into digital form, and the second is used to start the process of digital processing of the incoming pulse.

A scintillation detector based on a CsI crystal is a disk 3 mm thick and 20 mm in diameter, which is viewed from both sides by SiPM, the signals from which are fed to the analog processing board. The detector is included in the instrument in order to evaluate the possibility of using in the future a small-sized solid-state cosmic radiation sensor based on such a scheme for detecting radiation.

The analog processing boards of both detector units are built in the same way and contain signal amplifiers and comparators. Each of the signals has its own comparator, in which two signal discrimination thresholds are set: low and high. The signals from the analog processing boards are fed to the digital processing unit. For each of the signals from the detector, a pulse is formed, the amplitude of which depends on the amplitude of the signal received from the detector, and two pulses of standard amplitude, coming from the comparator.

2.2 Digital Signal Processing Unit

The third node of the device contains a Milandr 1986BE93U microcontroller and a chipset for correct signal processing from detectors. This microcontroller manufactured by the Russian company Milandr is based on a 32-bit ARM Cortex-M3 RISC core [12]. Also on the board are 2AND-NOT logic circuits, circuits for matching the electrical parameters of signals received for processing, elements needed for the operation of the microcontroller, including a quartz resonator and a DC-DC power converter, and transceivers for UART and CAN channels, connectors for information exchange and microcontroller programming.

The 1986BE93U microcontroller has two built-in high-speed ADCs that can be programmatically connected to any of the eight analog signals coming from the detector nodes.

During the operation of the device, the microcontroller generates two different types of frames.

The first one - "fluxes and doses" is formed once every 10 min and is a standby frame containing the following information accumulated during this time for all detectors:

1. the number of rises above two thresholds for each of the eight detectors;
2. the number of coincidences for pairs of semiconductor, Cherenkov, neutron detectors and photodetectors of the scintillation detector;
3. sums of codes of two ADCs with and without coincidences;
4. the total number of interrupts for each pair of detectors.

The second one, "an array of ADC codes" is formed as the array fills up, which accumulates when interrupts are triggered, caused by registration in particle detectors. This array is a measurement frame containing data on the signal amplitudes from the detectors and remains unchanged when switching detectors. This information, during subsequent ground processing, allows not only to build detailed spectra of energy releases for each of the ADCs and compare them with comparator thresholds, but also to obtain two-dimensional distributions of coincidence events.

During the operation of the KODIZ device, it is possible to flexibly adjust the intervals for storing detector measurements and request measurement data at any time on command. The KODIZ device is made in the form of 3 separate nodes, which are interconnected by loops. The mass of the device is 0.747 kg. The layout of the units of the KODIZ device is shown in Fig. 1.

The exchange of information between the KODIZ device and the on-board computer of the satellite is based on the CAN (Controller Area Network) protocol. Through the exchange channel built on this protocol, control commands are sent to the device, and the information generated by it is transmitted from the KODIZ device. The transmission of measurement data is carried out in two stages. At the first stage, the data arrays generated by the KODIZ device are transferred to the Raspberry-Pi on-board computer, where it is written as a file in the file system of this computer. At the second stage, during communication sessions with the Monitor-1 microsatellite, the accumulated data files are transferred using the means of interaction with Cubsats of the OrbiCraft-Pro type, developed by SPUTNIX.

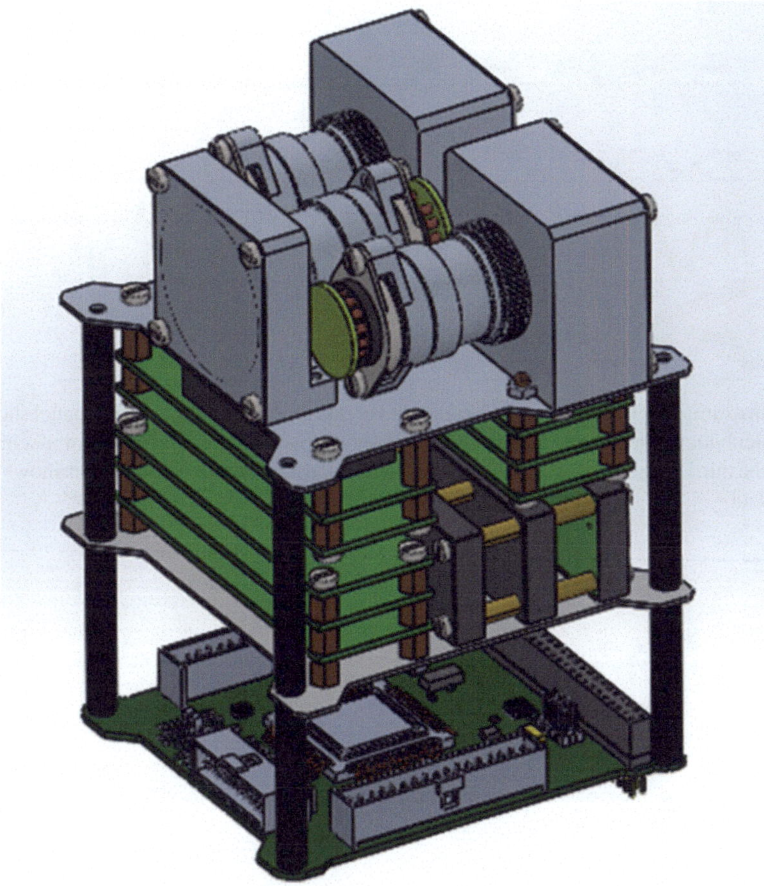

Fig. 1. Scheme of arrangement of units of the KODIZ device. The top board is for PMT detectors, the middle board is for semiconductor detectors, and the bottom board is for digital processing.

3 Results

3.1 Evaluation of the Performance of the KODIZ Device in Flight Conditions

After launching and turning on Cubesat Monitor-1, data were received from the KODIZ device. To process the received information, appropriate software tools were developed, which we will not dwell on in this article. Figure 2 shows an example of the results in the form of a change in the count rate of the instrument's detectors in the period January 17–18, 2023.

On all the graphs presented in Fig. 2, one can see the count rate variations due to the orbital motion of the satellite.

Figure 3 shows an example of the amplitude spectrum of signals coming from the Cherenkov detector.

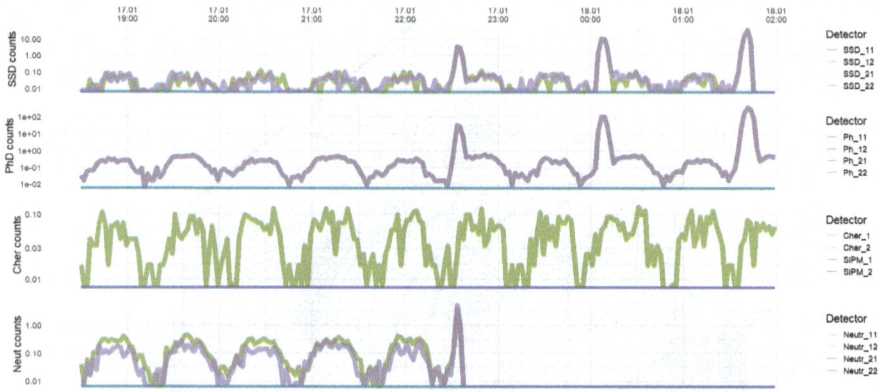

Fig. 2. An example of the change in the count rate of detectors over time. The top panel shows data from semiconductor detectors, the second panel shows photoconverter data from a scintillation detector, the third panel shows PMT Cherenkov detector data, and the bottom panel shows neutron detector data.

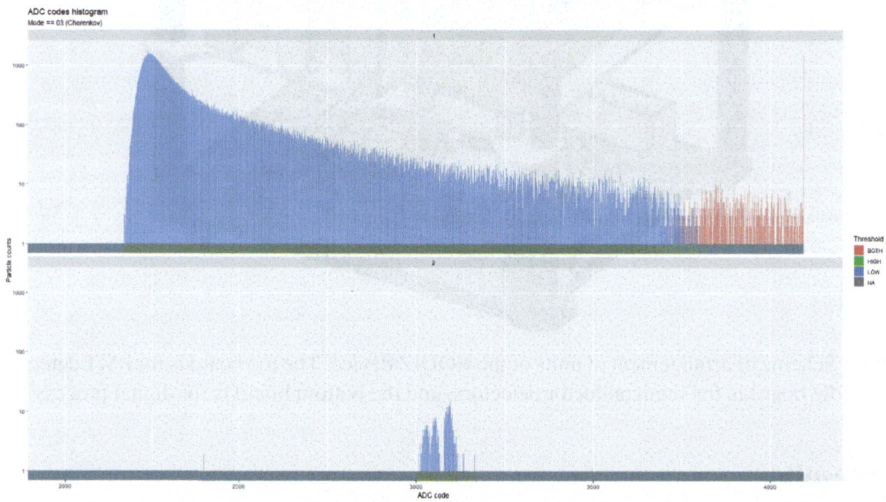

Fig. 3. Amplitude spectrum of signals from the Cherenkov detector. The top panel is the spectrum of signals from PMT, the bottom panel is the spectrum of signals from SiPM.

The presented data indicate that all detectors are functioning and the signal processing system of the device allows obtaining measurement data. A more detailed analysis of the information received will be the subject of subsequent publications.

4 Conclusion

A device was developed that makes it possible to record fluxes of high-energy particles of both solar and magnetospheric origin, capable of penetrating to altitudes of 100–20 km and creating additional ionization at these altitudes. Developed, manufactured and integrated into the microsatellite of the Cubesat type, the KODIZ device, designed to develop methods for monitoring the radiation situation in low Earth orbit. After the launch on August 9, 2022, the device successfully operates as part of the Monitor-1 microsatellite.

Acknowledgements. This work was done with the support of MSU Program of Development, Project No 23-SCH01-02.

References

1. Panasyuk, M.I., Novikov, L.S.: Physical conditions in outer space. In: Panasyuk, M.I., Novikov, L.S. (eds.) Space Model KDU, Moscow (2007)
2. Dorman, L.I., Miroshnichenko, L.I.: Solar Cosmic Rays. Nauka, Moscow (1968)
3. Petrov, V.M., Miroshnichenko, L.I.: Dynamics of Radiation Conditions in Space. Energoatomizdat, Moscow (1985)
4. Mertens, C.J., et al.: Geomagnetic influence on aircraft radiation exposure during a solar energetic particle event in October 2003. Space Weather **8**(3) (2010)
5. Cliver, E.: History of research on solar energetic particle (SEP) events: the evolving paradigm. Proc. Int. Astron. Union **4**(S257), 401–412 (2008). https://doi.org/10.1017/S1743921309029639
6. GOST 4401-81 Interstate Standard, Standard Atmosphere, Parameters, Official edition, IPK Standards Publishing House, Moscow, Introduction date 01.07.82
7. Ginzburg, V.L.: Radiation of uniformly moving sources (Vavilov-Cherenkov effect, transition radiation and some other phenomena). Uspekhi fizicheskikh nauk **166**(10), 1033–1042 (1996)
8. SPUTNIX Company, CUBESAT-PLATFORM LINE. https://sputnix.ru/ru/platformyi/linejka-cubesat-platform
9. Space-π satellites in space. https://spacepi.space/news/16-sputnikov-space-p-v-kosmose/
10. Lishnevsky, A.E., Benghin, V.V.: Results of separating the contribution of galactic cosmic rays and the Earth's inner radiation belt to the daily dose recorded by dosimeters DB-8 of the radiation monitoring system on board the International Space Station for the period from 2001 to 2014. Space Res. **58**(4), 344–352 (2020)
11. Dachev, T., Plock, O., Tomov, B., Spurny, F.: Analysis of the GCR dose rate increase onboard spacecraft and aircraft in the declining phase of the 23rd solar cycle. Fundamental Space Research, Supplement of Comptes Rend. Acad. Bulg. Sci. (2010)
12. Specification for microcircuits of the 1986BE9hu, K1986BE9hu, K1986BE9huK, K1986BE91H4, K1986BE94H4 series. PKK Milandr JSC TSKYA.431296.001SP Version 3.21.1 dated 22 July 2021

Selkov Dynamic System with Variable Heredity for Describing Microseismic Regimes

Roman Parovik(✉)

Laboratory for Simulation of Physical Processes, Institute of Cosmophysical Research and Radio Wave Propagation Far Eastern Branch of the Russian Academy of Sciences, Mirnaya, 7, Paratunka 684034, Kamchatskiy Krai, Russia
parovik@ikir.ru

Abstract. The paper proposes a new mathematical model for describing the microseismic modes of interaction of two types of fractures – seed and fractures that directly generate microseisms. The mathematical model is a system of two non-linear differential equations with derivatives in the sense of Gerasimov-Caputo of fractional variable orders. The dynamical system under study is a generalization of the Selkov fractional dynamical system studied earlier. The purpose of the study of the proposed model is to establish self-oscillatory modes using numerical calculations. The algorithm based on the theory of finite-difference schemes serves as a numerical algorithm for studying the problem. The paper shows that the proposed dynamic system can have different oscillatory modes depending on the choice of the dependence of the orders of fractional derivatives on time, as well as on the values of the model parameters. Using the numerical method of Adams-Bashfort-Moulton, oscillograms and phase trajectories are constructed depending on various types of functions - orders of fractional derivatives. Regularities of oscillatory regimes are established.

Keywords: Microseisms · Self-Oscillations · Phase Trajectories · Oscillograms · Selkov Fractional Dynamic System

JEL Classification: C3
MSC Classification: 26A33 · 34D45 · 37D45

1 Introduction

Microseisms are vibrations of the earth's surface of small amplitude, which occur as a result of natural and man-made processes. For example, natural processes can be associated with the processes of formation and passage of cyclones in the atmosphere, with the processes of the impact of seas and oceans on the

A. Dmitriev et al. (Eds.): STRPEP 2023, SPEES, pp. 166–178, 2023.
https://doi.org/10.1007/978-3-031-50248-4_18

coast. Technogenic processes associated with human activities, for example, the construction of buildings and structures, traffic, the operation of power plant engines, etc.

The condition for the occurrence of microseisms is determined by the presence of elasticity of the medium, the ability to accumulate stress to a certain critical value, and on the other hand, the presence of fragility - the ability to collapse under the action of forces, the level of which is noticeably lower than the yield point [1]. The fragility of the medium is realized at the macroscopic level due to the appearance and growth of the length of cracks. The process of multiple crack formation ends with the destruction of the medium. In this article, we study the process of crack formation at an early stage of failure.

Note that the growth of the crack length may slow down with time, and if the medium has viscosity, then the crack length can also decrease until it is completely closed. In [2], the authors believe that such a process can occur in the earth's crust at great depths at high pressure and temperature. Under such conditions, a diffusion process occurs in the crack tips, which leads to their tightening.

According to the terminology of [2], there are two types of microseisms: regular weak oscillations with periods from 2 to 10 s and less regular ones with long periods of oscillations up to 30 s. The first type of microseisms are excited by cracks of small length, which cannot be recorded by seismic equipment. We will call such cracks trigger or tr-cracks, which are triggers for larger cracks. Microseismic signals of the second type are generated by longer fractures, which are already registered by seismic equipment. We will call such cracks gs-cracks.

Next, we will investigate the mechanism of self-oscillations microseismic sources or fluctuations in the concentration of gs-fractures, by analogy with [2]. The self-oscillatory process here consists in the interaction of tr-cracks and gs-cracks, which is shown in Fig. 1.

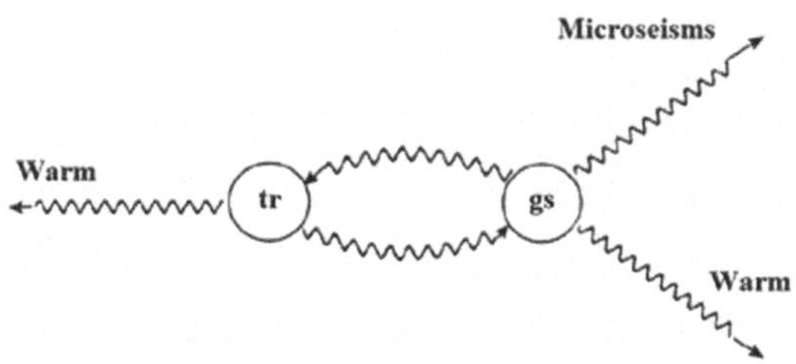

Fig. 1. Self-oscillatory process in the interaction between tr-cracks and gs-cracks [2]

Trigger tr-cracks are seed cracks with lower energy, and when the critical level is reached, the concentration transforms into gs cracks. Further, according to the Le Chatelier-Brown principle [3], gs-cracks that generate microseismic signals partially disappear after energy release, and partially turn into tr-cracks. Further, the process of interaction is repeated, and a self-oscillatory regime occurs.

In the article [2], the authors proposed to use the Selkov nonlinear dynamic system to describe the self-oscillating process of interaction of tr and gs-cracks in an elastic-brittle medium. The Selkov dynamic system studied in biology when modeling self-oscillatory modes of substrates and glycolysis products [4]. The article [2] studies the equilibrium points of the Selkov dynamic system in order to determine the conditions for the existence of undamped oscillations. Spectra of microseismic oscillations are also studied.

A generalization of the results in the article [2] is a series of works by the author [5,6], in which the memory effects in a self-oscillatory system were taken into account. The memory effect determines the dependence of the current state of an oscillatory (dynamic) system on its previous states. From the point of view of mathematics, the memory effect can be described using integro-differential equations, in particular, using fractional derivatives. A dynamical system with derivatives of fractional orders will be called a fractional dynamical system. In [6] the Selkov fractional dynamical system was studied, a numerical solution algorithm based on the Adams-Bashfort-Multon method was proposed, the stability and convergence of the method were studied, and test examples were given. In [5] the qualitative properties of the Selkov fractional dynamical system were studied, regular and chaotic regimes were studied. It is shown that a dynamic system can have different regimes, including chaotic ones.

In this article, we continue to study the fractional Selkov dynamical system in the case when the derivatives have fractional variable orders in time. A numerical algorithm for solving the solution is proposed, the simulation results were visualized in the Maple computer mathematics environment, oscillograms and phase trajectories were constructed for cases where the orders of fractional derivatives are monotonic functions of time, as well as periodic functions of time.

2 Problem Statement and Solution Technique

Consider the following nonlinear dynamical system:

$$\begin{cases} \partial_{0t}^{\alpha_1(t)} x(t) = -x(t) + ay(t) + bx^2(t)\, y(t)\,, x(0) = x_0 \\ \partial_{0t}^{\alpha_2(t)} y(t) = v - ay(t) - bx^2(t)\, y(t)\,, y(0) = y_0. \end{cases} \tag{1}$$

where $x(t) \in C^1[0,T]$ is a function that determines the concentration of tr-cracks; $y(t) \in C^1[0,T]$ is a function that determines the concentration of gs-cracks that generate microseisms, $t \in [0,T]$ is the coordinate responsible for the current time of the process, T is a constant, simulation time; x_0, y_0, v, a, b - given positive constants; fractional differentiation operators are understood in the sense of the Gerasimov-Caputo orders $0 < \alpha_1(t), \alpha_2(t) < 1$ and are determined according to (2) [7].

$$\begin{aligned} \partial_{0t}^{\alpha_1(t)}x(t) &= \frac{1}{\Gamma(1-\alpha_1(t))}\int\limits_0^t \frac{\dot{x}(\tau)\,d\tau}{(t-\tau)^{\alpha_1(t)}}, \\ \partial_{0t}^{\alpha_2(t)}y(t) &= \frac{1}{\Gamma(1-\alpha_2(t))}\int\limits_0^t \frac{\dot{y}(\tau)\,d\tau}{(t-\tau)^{\alpha_2(t)}}. \end{aligned} \tag{2}$$

We will assume that the functions $\alpha_1(t), \alpha_2(t) \in C(0,1)$.

The dynamical system (1) will be called a fractional Selkov dynamical system with variable memory. In the works of the author [5,6] the fractional Selkov dynamical system was considered, when the fractional orders of the derivatives are constant: $\alpha_1(t) = \alpha_1, \alpha_2(t) = \alpha_2$. The properties of fractional derivatives of constant order are described in detail in the books [8–10]. In the case when $\alpha_1 = \alpha_2 = 1$, then we get the classical Sel'kov dynamical system [4].

To study the Selkov fractional dynamical system (SFDS) (1), we use the Adams-Bashforth-Moulton (ABM) numerical method from the family of predictor-corrector methods. The ABM method has been studied and discussed in detail in [11,12]. We adapt this method to solve SFDS (1). To do this, we use definitions (2), and on a uniform grid with step we introduce the functions x_{n+1}^p, y_{n+1}^p, $n = 0, ..., N-1$, which will be determined by the Adams-Bashforth formula (predictor):

$$\begin{cases} x_{n+1}^p = x_0 + \dfrac{\tau^{\alpha_{1,n}}}{\Gamma(\alpha_{1,n}+1)}\sum\limits_{j=0}^{n}\theta_{j,n+1}^1\left(-x_j + ay_j + bx_j^2y_j\right), \\ y_{n+1}^p = y_0 + \dfrac{\tau^{\alpha_{2,n}}}{\Gamma(\alpha_{2,n}+1)}\sum\limits_{j=0}^{n}\theta_{j,n+1}^2\left(v - ay_j - bx_j^2y_j\right), \\ \theta_{j,n+1}^i = (n-j+1)^{\alpha_{i,n}} - (n-j)^{\alpha_{i,n}}, i = 1,2, \end{cases} \tag{3}$$

as well as functions x_{n+1}, y_{n+1}, which will be determined by the Adams-Moulton formula for the corrector:

$$\begin{cases} x_{n+1} = x_0 + \dfrac{\tau^{\alpha_{1,n}}}{\Gamma(\alpha_{1,n}+2)} \\ \times\left(\left(-x_{n+1}^P + ay_{n+1}^P + b\left(x_{n+1}^P\right)^2 y_{n+1}^P\right) + \sum\limits_{j=0}^{n}\rho_{j,n+1}^1\left(-x_j + ay_j + bx_j^2y_j\right)\right), \\ y_{n+1} = y_0 + \dfrac{\tau^{\alpha_{2,n}}}{\Gamma(\alpha_{2,n}+2)} \\ \times\left(v - ay_{n+1}^P - b\left(x_{n+1}^P\right)^2 y_{n+1}^P + \sum\limits_{j=0}^{n}\rho_{j,n+1}^2\left(v - ay_j - bx_j^2y_j\right)\right), \end{cases} \tag{4}$$

where the weight coefficients in (4) are determined by the formula:

$$\rho_{j,n+1}^i = \begin{cases} n^{\alpha_{i,n}+1} - (n-\alpha_{i,n})(n+1)^{\alpha_{i,n}} & , j = 0, \\ (n-j+2)^{\alpha_{i,n}+1} + (n-j)^{\alpha_{i,n}+1} - 2(n-j+1)^{\alpha_{i,n}+1}, & 1 \le j \le n, \\ 1, & , j = n+1, \\ i = 1,2. \end{cases}$$

Remark 1. *It is known that for the ABM method the error estimate is valid* [13]*:* $\max_{1 \le j \le k} |x_i(t_j) - x_{i,j}| = O\left(\tau^{\min(2, 1+\alpha_{i,m})}\right)$, $\alpha_i = \min_{0<\tau<T} \alpha_i(\tau)$, $x_1(t_j) = x(t_j), x_2(t_j) = y(t_j), i = 1, 2.$

Remark 2. *Note that in the classical case* $\alpha_i = 1$*, we obtain the classical ABM method of the second order of accuracy.*

Next, using specific examples, we consider various cases when the functions $\alpha_i(t)$ are monotonic, as well as periodic.

3 Results

The implementation of the numerical algorithm (3)–(4) was carried out using the Maple package.

Example 1. *The case of linearly monotonically decreasing functions* $\alpha_i(t)$ *(Fig. 2):*

$$\alpha_1(t) = 0.8 - 0.01\frac{t}{T}, \alpha_2(t) = 0.9 - 0.02\frac{t}{T}.$$

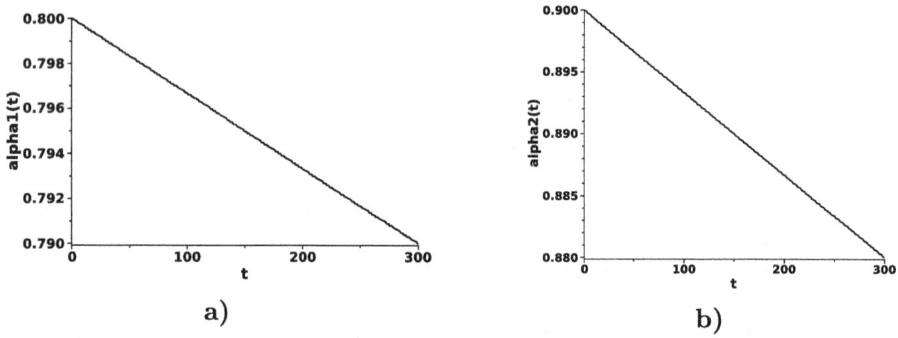

Fig. 2. a) $\alpha_1(t)$; b) $\alpha_2(t)$ for Example 1

We take the remaining parameters for system (1) as follows: $v = 0.6, a = 0.03, b = 1.3, x(0) = 1, y(0) = 0.5$. *For the numerical algorithm (3)–(4), the parameters have the following meanings:* $N = 2000, T = 300, \tau = 0.15$.

Figure 3 shows that in the case when the functions $\alpha_1(t), \alpha_2(t)$ are monotonically decreasing, then the concentration of trigger cracks and the concentration of cracks generating microseisms eventually reach a constant level. The phase trajectory is a trajectory characteristic of the limit cycle.

An important task is the study of limit cycles, in particular, their stability. According to the definition of limit cycle stability, there must exist an ε neighborhood such that all phase trajectories that start in this neighborhood approach the limit cycle asymptotically at $t \to \infty$.

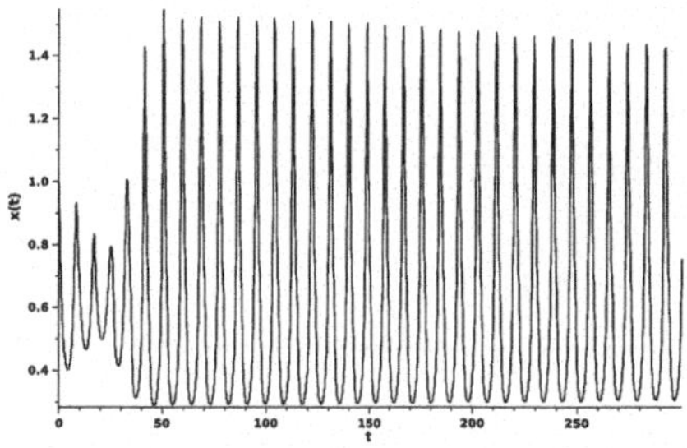

a)

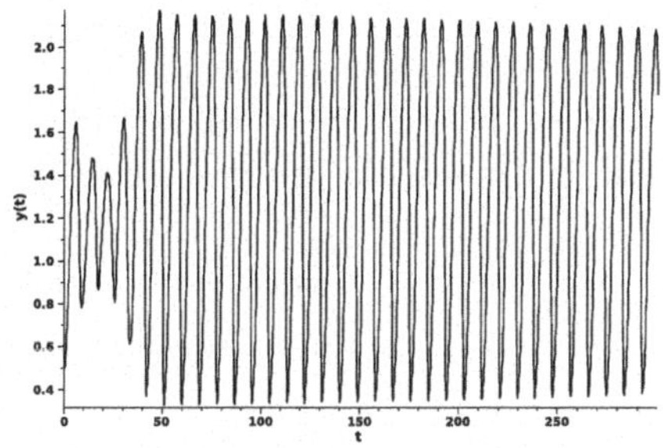

b)

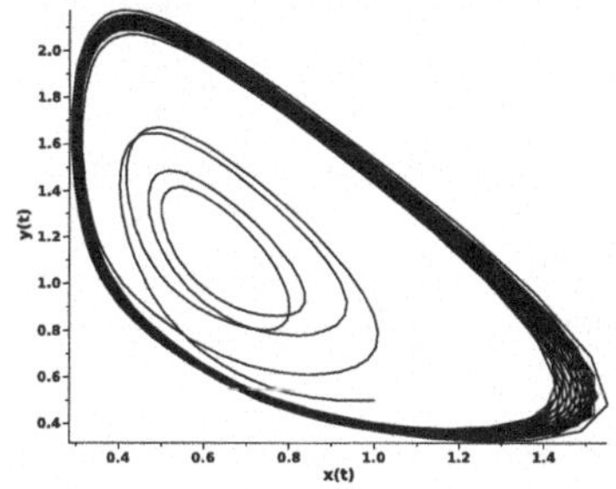

c)

Fig. 3. Osillograms: a) fluctuations in the concentration of tr-cracks; b) fluctuations in the concentration of gs-cracks. Phase trajectory c)

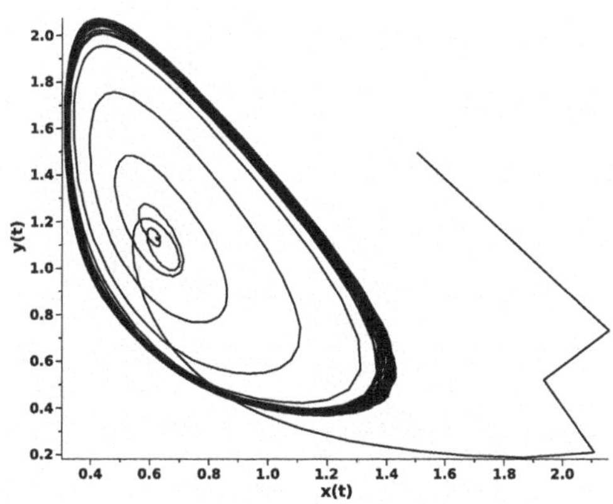

Fig. 4. Phase trajectory for $x\,(0) = y\,(0) = 1.5$

Based on the above, we will construct a phase trajectory for Example 1 in the case when the initial conditions have other values (Fig. 4): $x\,(0) = y\,(0) = 1.5$.

We see that the phase trajectory is approaching a limit cycle to the same as in Fig. 4. We see on that the phase trajectory is approaching the limit cycle to the same one as in Fig. 3c. Therefore, we can say about the possible stability of the limit cycle. For an affirmative answer to this question, it is necessary to substantiate a theorem analogous to the Poincare-Bendixson theorem for the case of fractional dynamical systems.

The author's work [6] shows that the limit cycle disappears as $\alpha_1 = 0.5$ tends. Further, the oscillatory process disappears. Let (Fig. 5)

$$\alpha_1(t) = 0.8 - 0.3\frac{t}{T}, \alpha_2(t) = 0.9 - 0.4\frac{t}{T}.$$

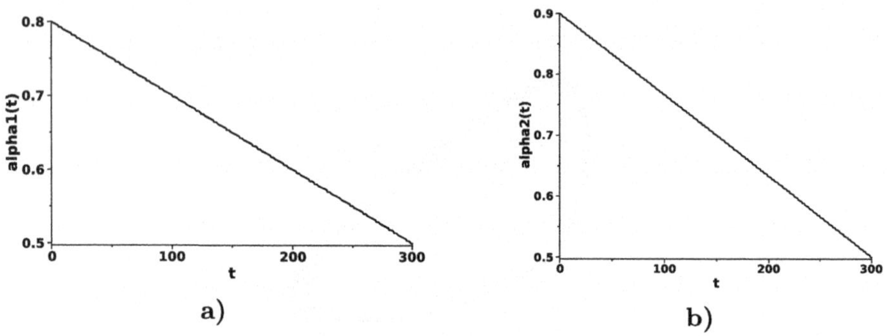

Fig. 5. a) $\alpha_1(t) \in [0.5, 0.8]$; b) $\alpha_2(t) \in [0.5, 0.9]$ for Example 1

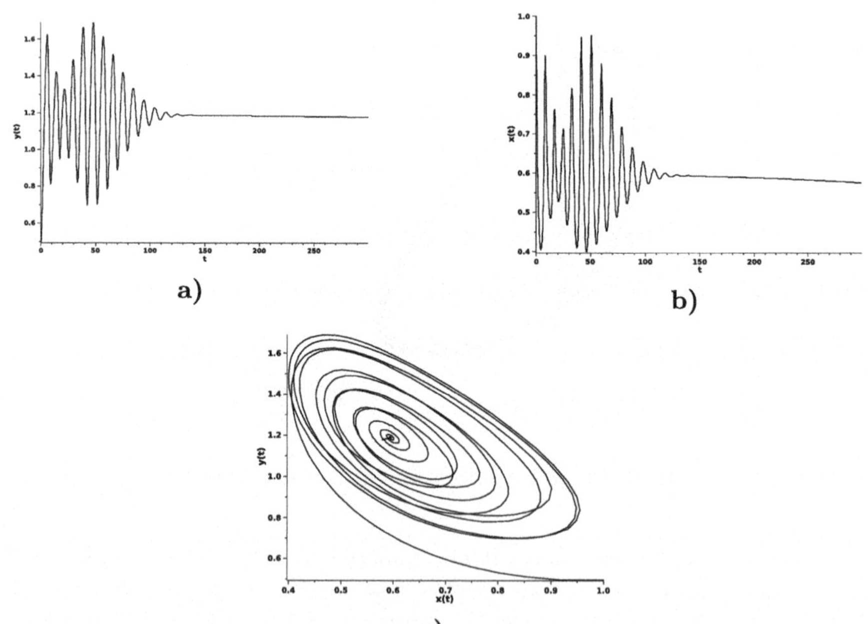

Fig. 6. Osillograms: a) fluctuations in the concentration of tr-cracks; b) fluctuations in the concentration of gs-cracks. Phase trajectory c)

The rest of the parameters will be left unchanged.

From Fig. 6 it can be concluded that, as the values of the functions $\alpha_1(t), \alpha_2(t)$ decrease to 0.5, the oscillations decay. It can be given from Fig. 5 an approximate estimate from which oscillations begin to decay: $\alpha_1 \approx 0.675, \alpha_2 \approx 0.675$.

Let us expand the range of variation of the functions $\alpha_1(t) \in [0.2, 1], \alpha_2(t) \in [0.3, 1]$. The rest of the parameters will remain unchanged.

$$\alpha_1(t) = 1 - 0.8\frac{t}{T}, \alpha_2(t) = 1 - 0.7\frac{t}{T}.$$

The simulation results are shown in Fig. 7.

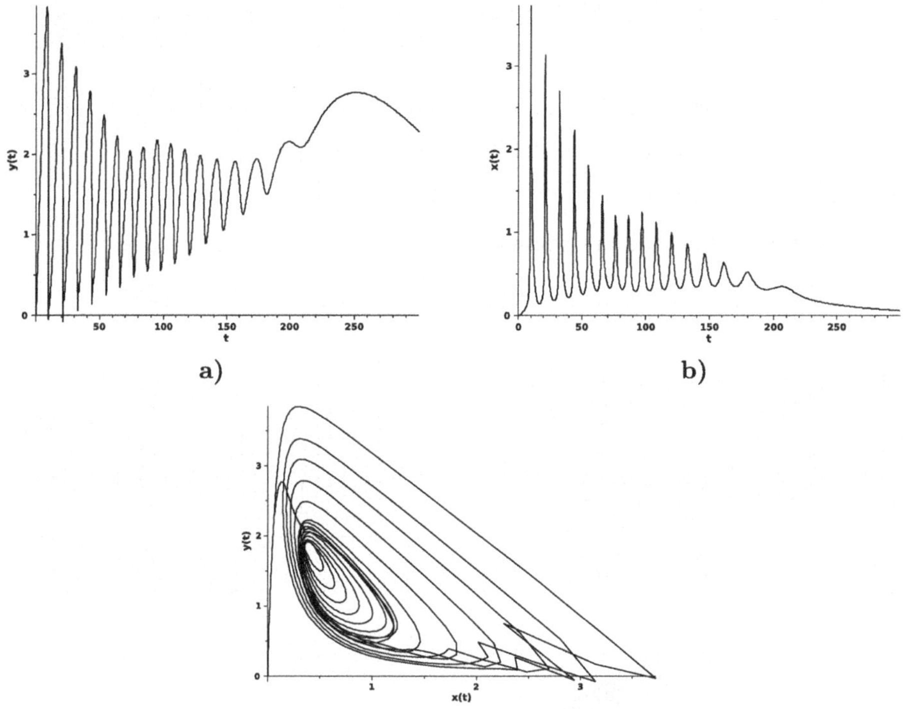

Fig. 7. Osillograms: a) fluctuations in the concentration of tr-cracks; b) fluctuations in the concentration of gs-cracks. Phase trajectory c)

Here we can see several modes: reaching the limit cycle, damped mode and aperiodic mode.

Example 2. *The case of periodic functions $\alpha_i(t)$ (Fig. 8):*

$$\alpha_1(t) = 0.9 - \frac{5\cos(0.3\pi t)}{T}, \alpha_2(t) = 0.9 - \frac{7\sin(0.2\pi t)}{T}.$$

It can be seen here (Fig. 11c) that a more complex form of the functions $\alpha_1(t)$ and $\alpha_2(t)$ can lead to limit cycles of a different form than those previously known.

Let's expand the range of functions $\alpha_1(t) \in [0.2, 0.8], \alpha_2(t) \in [0.2, 0.8]$ as in the previous examples:

$$\alpha_1(t) = 1 - \frac{30\exp(\cos(0.4\pi t))}{T}, \alpha_2(t) = 1 - \frac{30\exp(\sin(0.4\pi t))}{T}.$$

Let's see how the phase trajectory and oscillograms change compared to Fig. 11.

We see that the phase trajectory in Fig. 12c differs from the phase trajectory in Fig. 11c. Here the phase trajectory and the limit cycle have a more complex form. It should also be noted that the phase trajectory reaches the limit cycle faster.

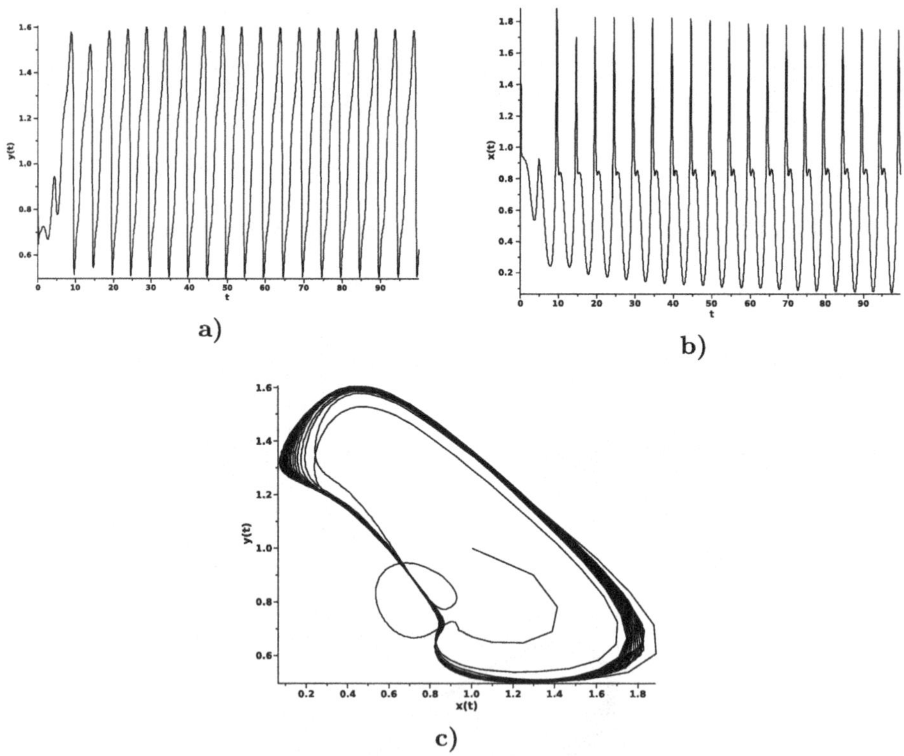

Fig. 12. Osillograms: a) fluctuations in the concentration of tr-cracks; b) fluctuations in the concentration of gs-cracks. Phase trajectory c)

4 Conclusion

The results of the work show the following taking into account the effects of memory in oscillatory systems using derivatives of fractional variable orders allows for more flexible mathematical modeling. If the orders of fractional derivatives are linearly decreasing functions, then we can observe several modes: reaching the limit cycle, damped and aperiodic. Approximate estimates of the transition or bifurcation from one regime to another were given in the work. If the orders are periodic functions, then multiperiodic regimes, limit cycles of various shapes can arise, which indicates a very rich dynamics of the process. Due to the fact that it is important for us to study the self-oscillatory regime, which is characterized by a limit cycle, further continuation of the work is connected with the study of its stability.

Another line of work is the study of chaotic regimes. Due to the fact that the methodology of works [14] is not applicable to the study of dynamic systems of fractional variable order, it is necessary to use approaches related to the construction of maps of dynamic modes [15]. This approach requires large computational resources, as well as the methodology of parallelization of numerical algorithms.

Acknowledgments. This research was funded by Russian Science Foundation grant number 22-11-00064 "Modeling of dynamic processes in geospheres taking into account hereditarity" https://rscf.ru/project/22-11-00064/.

References

1. Kearey, P.: The Encyclopedia of Solid Earth Sciences. Blackwell Science, Hoboken (1993)
2. Makovetsky, V.I., Dudchenko, I.P., Zakupin, A.S.: Auto oscillation model of microseism's sources. Geosist. perehodnykh zon **4**(1), 37–46 (2017)
3. Shpielberg, O., Akkermans, E.: Le Chatelier principle for out-of-equilibrium and boundary-driven systems: application to dynamical phase transitions. Phys. Rev. Lett. **116**(24), 240603 (2016). https://doi.org/10.1103/PhysRevLett.116.240603
4. Selkov, E.E.: Self-oscillations in glycolysis. I. A simple kinetic model. Eur. J. Biochem. **4**, 79–86 (1968)
5. Parovik, R.I., Rakhmonov, Z., Zunnunov, R.: Study of chaotic and regular modes of the fractional dynamic system of Selkov. In: EPJ Web of Conferences, vol. 254, p. 02014 (2021). https://doi.org/10.1051/epjconf/202125402014
6. Parovik, R.I.: Studies of the fractional Selkov dynamical system for describing the self-oscillatory regime of microseisms. Mathematics **10**(22), 4208 (2022). https://doi.org/10.3390/math10224208
7. Patnaik, S., Hollkamp, J.P., Semperlotti, F.: Applications of variable-order fractional operators: a review. R. Soc. A R. Soc. Publ. **476**, 20190498 (2020). https://doi.org/10.1098/rspa.2019.0498
8. Kilbas, A.A., Srivastava, H.M., Trujillo, J.J.: Theory and Applications of Fractional Differential Equations. Elsevier, Amsterdam (2006)
9. Oldham, K.B., Spanier, J.: Theory and applications of differentiation and

10. Miller, K.S., Ross, B.: An Introduction to the Fractional Calculus and Fractional Differntial Equations. A Wiley-Interscience publication, New York (1993)
11. Garrappa, R.: Numerical solution of fractional differential equations: a survey and a software tutorial. Mathematics **6**(2), 16 (2018). https://doi.org/10.3390/math6020016
12. Parovik, R.: Application of the Adams-Bashforth-Moulton method to the numerical study of linear fractional oscillators models. In: AIP Conference Proceedings, vol. 2365, p. 020006 (2021). https://doi.org/10.1063/5.0056846
13. Ma, S., Yufeng, X., Wei, Y.: Numerical solutions of a variable-order fractional financial system. J. Appl. Math. **2012**, 417942 (2012). https://doi.org/10.1155/2012/417942
14. Petras, I.: Fractional Order Nonlinear Systems. Modeling, Analysis and Simulation. Springer, Beijing, Heidelberg (2011). https://doi.org/10.1007/978-3-642-18101-6
15. Parovik, R.I., Yakovleva, T.P.: Construction of maps for dynamic modes and bifurcation diagrams in nonlinear dynamics using the maple computer mathematics software package. J. Phys: Conf. Ser. **2373**, 052022 (2022). https://doi.org/10.1088/1742-6596/2373/5/052022

Variations in the Parameters of the Signal from the A1F3 VLF Transmitter Received at the Mikhnevo Geophysical Observatory During a Series of Earthquakes in Turkey in February 2023

Yu. V. Poklad(✉), V. M. Ermak, B. G. Gavrilov, and I. A. Ryakhovskiy

Sadovsky Institute of Geosphere Dynamics of the Russian Academy of Sciences, Leninsky prospect 38 build 1, Moscow 119334, Russia
poklad@mail.ru

Abstract. The propagation path of the signal from the A1F3 VLF transmitter received at the Mikhnevo GFO passed in the area of earthquake epicenters in Turkey in February 2023. The paper considers variations in the amplitude and phase of signals before and during seismic events. It is shown that before the earthquake, there were changes in the amplitude and phase at the morning terminator. At night, fluctuations in the amplitude and phase of the signal were observed with characteristic periods of tens of minutes.

Keywords: Earthquake · Propagation · VLF · Ionosphere

1 Introduction

Electromagnetic radiation of the VLF range propagates in the earth-ionosphere waveguide with very low attenuation (3 dB per 1000 km). The upper boundary of this waveguide is the D/E region of the ionosphere (70–90 km) [1]. Any heliogeophysical disturbances (solar flares and eclipses, magnetic storms, lightning discharges, meteorites, volcanic eruptions, earthquakes) leading to changes in the lower ionosphere affect the amplitude-phase variations of VLF signals. Thus, VLF electromagnetic radiation is currently the most effective tool for studying processes occurring in the lower ionosphere.

The first attempt to use VLF/LF radio sounding to detect seismo-ionospheric effects was made by [2] and [3]. They managed to detect a significant propagation anomaly on two long distance paths from Réunion (Omega transmitter) to Moscow and Omsk a few days before the famous Spitak earthquake [3].

In [4], significant shifts in the terminator time before an earthquake were found. The morning terminator time (tm) is shifted to the early hours, and te

V. M. Ermak, B. G. Gavrilov and I. A. Ryakhovskiy—These authors contributed equally to this work.

A. Dmitriev et al. (Eds.): STRPEP 2023, SPEES, pp. 179–185, 2023.
https://doi.org/10.1007/978-3-031-50248-4_19

is shifted to the later hours. Theoretical estimates [4–6] of this effect show that the ionosphere descends before an earthquake.

A later study [5] based on 13 years of observations (11 events with a magnitude greater than 6.0) showed a relatively high probability of an anomaly in VLF propagation (in the form of a terminator time anomaly) of the order of 70–80% for larger (magnitude more than 6.0) earthquakes located relatively close to the path.

2 Results

In February 2023, a series of earthquakes hit southeastern Turkey. The first strongest shock with a magnitude of 7.8 occurred on February 6 at 01:17:34 UT. In total, on February 6, 112 aftershocks with a magnitude greater than 4.5 were recorded. Of these, 6 have a magnitude greater than 6. Further, seismic activity began to decrease. On 07.02 there were 19 aftershocks with $M > 4.5$, the most powerful of them with $M = 5.5$. 08.02 - there were 7 aftershocks with $M > 4.5$, the most powerful of them with $M = 5.4$. Later, until the end of February, up to 3 shocks per day were recorded in this area.

Since 2014, the Mikhnevo geophysical observatory (54°57′N, 37°46′51 E) has been monitoring the signals of VLF transmitters [7]. One of these transmitters is located in Israel and has the designation A1F3. It operates at a frequency of 29.7 kHz. On Fig. 1 shows a map with the location of the A1F3 transmitter, the receiver in the Mikhnevo GFO, and the epicenters of the strongest earthquakes. As can be seen from the figure, the VLF signal propagation path passes at a distance of 44–212 km from the earthquake (Fig. 1).

Table 1 shows the time, coordinates, magnitude of earthquakes greater than 6, as well as the distance from their epicenters to the A1F3 - "Mikhnevo" path.

Table 1. List of the Earthquakes

Date	Time, UT	Latitude, Degrees	Longitude, Degrees	Magnitude	Distance, km
06.02.2023	01:17:34.342	37.2256	37.0143	7.8	127
06.02.2023	01:28:15.784	37.1893	36.8929	6.7	114
06.02.2023	10:24:48.811	38.0106	37.1962	7.5	132
06.02.2023	10:26:46.742	38.0315	38.0984	6	212
06.02.2023	12:02:11.275	38.0582	36.5114	6	74
20.02.2023	17:04:29.747	36.1616	36.0251	6.3	44

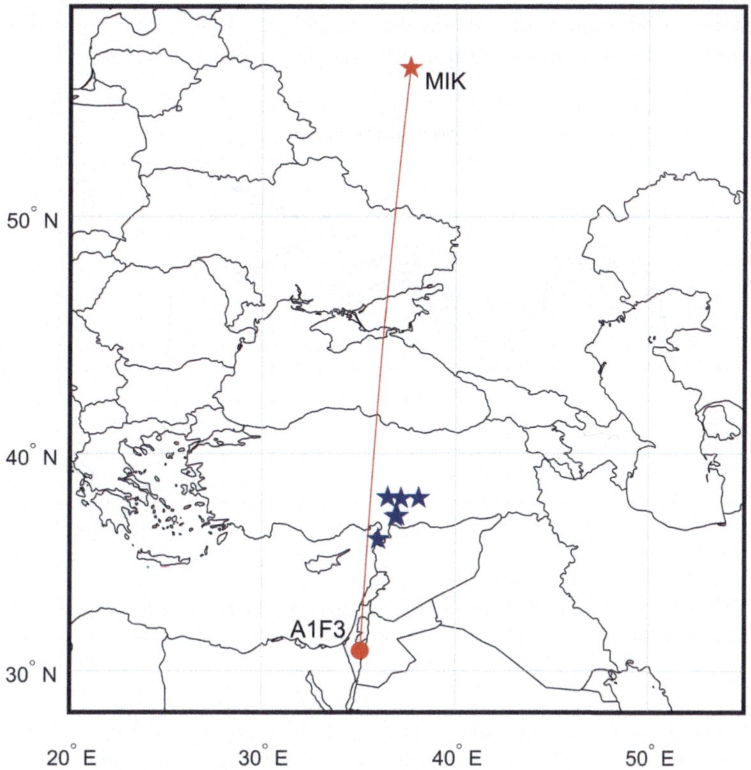

Fig. 1. Map with marks of the A1F3 transmitter (red circle), the Mikhnevo geophysical observatory (red star) and earthquake epicenters (blue stars).

On Fig. 2 shows the amplitude of the signal from the transmitter at night and during the morning terminator from February 4 to 8. It can be seen that on the night of February 5–6, approximately from 21:00 UT on February 5, strong variations in the signal amplitude begin with a characteristic period of 40–50 min. The next night, on the evening of February 6, they continue, but closer to midnight they begin to decrease and disappear by 04:00 UT on February 7. We also note that on the morning terminator on February 5 at 05:24 the signal level drops to 0.1 pT, i.e. becomes comparable to the level of natural noise.

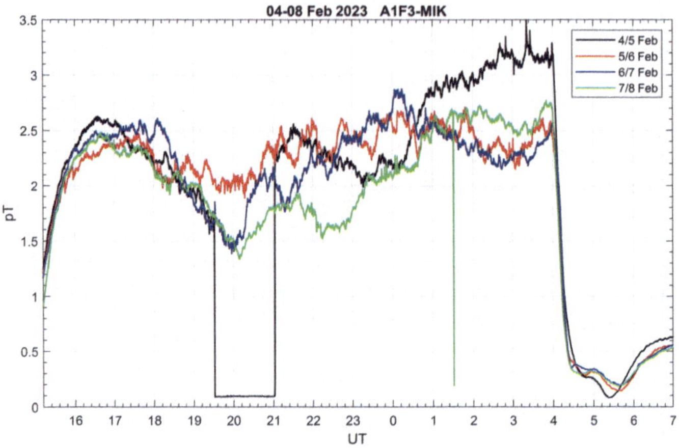

Fig. 2. Variation of the signal amplitude of the A1F3 transmitter during the night hours from 4 to 8 February.

Figure 3 shows the amplitude variation spectra of the A1F3 transmitter signal received at the "Mikhnevo" geophysical observatory at night. Signals from 16:30 UT to 03:55 UT of the next day were used for spectral processing. The black curve shows the averaged spectrum of nighttime amplitude variations for February 1–4. The red curve shows the spectrum of the signal on the night of February 5–6, i.e. just before the earthquake. We see a characteristic increase in the signal amplitude at frequencies from 0.26 to 0.35 mHz. The following night, from February 5 to February 6, when the seismic activity decreased, this maximum decreased in amplitude and shifted to the range of 0.22–0.26 mHz.

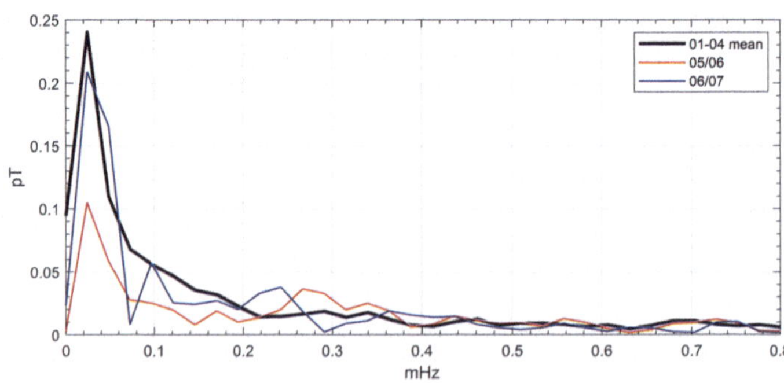

Fig. 3. Spectra of variations in the amplitude of the nighttime signals of the A1F3 transmitter from February 1 to 8.

Further, up to February 19, no such features were observed in the amplitude variations. On February 20, there was an increase in seismic activity. At 17:04:29 there was a push with a magnitude of M6.3. The distance from the epicenter to the highway was minimal - 44 km. Immediately after this shock, with an interval of 3–5 min, another 2 points occurred with a magnitude of M5.5 and M4.6. Figure 4 shows the spectra of nighttime variations in the signal amplitude from February 18 to February 21. It can be seen that on the night before these earthquakes, we also observe a characteristic increase in the signal amplitude in the frequency range from 0.2 to 0.3 mHz.

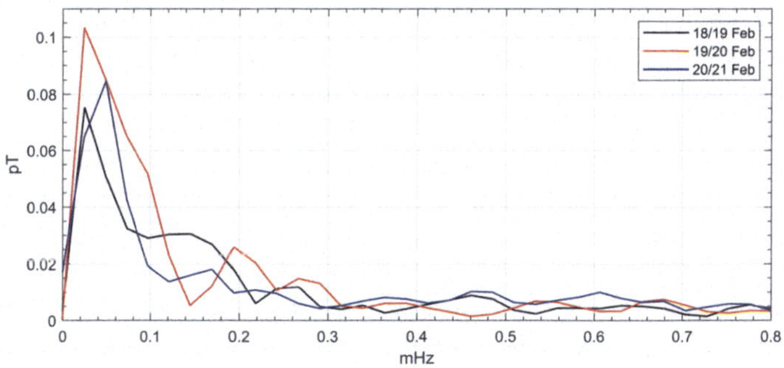

Fig. 4. Spectra of variations in the amplitude of the nighttime signals of the A1F3 transmitter from February 18 to 21.

Figure 5 shows the daily phase variations of the A1F3 transmitter signal received at the Mikhnevo geophysical observatory.

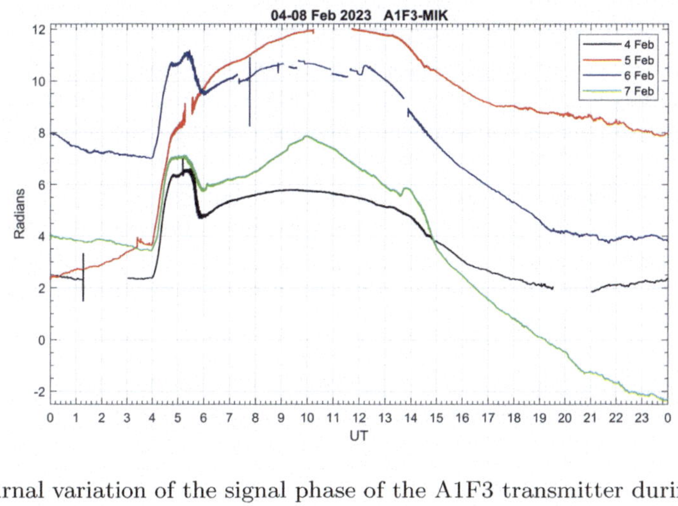

Fig. 5. Diurnal variation of the signal phase of the A1F3 transmitter during the night hours from 4 to 7 February.

It can be seen that on February 5 (red curve) at the morning terminator (from 04 to 06 UT) there is no "outlier", which is present almost unchanged at the morning terminator on September 4, 6, and 7. In addition, on February 5, during the night hours from 18 to 23 UT, signal fluctuations similar to nighttime amplitude fluctuations are observed. The same phase fluctuations, but with a smaller amplitude, are also observed on February 6 at the same time.

3 Discussion and Conclusions

To date, there are several possible hypotheses describing the mechanisms of communication between the activity of the lithosphere and the ionosphere. Three main types of interrelation of lithospheric-ionospheric processes can be distinguished (Hayakawa, 2007) (1) chemical channel, (2) acoustic channel, and (3) electromagnetic channel.

The first type - geochemical quantities (such as surface temperature, radon radiation, etc.) cause perturbation of the atmospheric conductivity, which then leads to modification of the ionosphere through the atmospheric electric field [8,9].

The second type is based on the key role of atmospheric oscillations in the lithosphere-atmosphere-ionosphere connection, and disturbances of the earth's surface (such as temperature, pressure) in a seismically active region excite atmospheric oscillations that propagate to the ionosphere [10–12].

The third mechanism is that radio emissions (in any frequency range) generated in the lithosphere propagate to the ionosphere and change the ionosphere there by heating and/or ionization. However, this mechanism turns out to be insufficient due to the low intensity of lithospheric radio emission [13]. Thus, the 1st and 2nd mechanisms are most likely candidates for this relationship [14].

In our case, the most noticeable anomalies were observed at the morning terminator on February 5th - i.e. 20 h before the first shock and the night before the first shock - i.e. 7 h before the seismic event. These facts make the acoustic mechanism of transmission of perturbations from the lithosphere to the ionosphere unlikely. The most probable mechanism for the transmission of disturbances to the ionosphere during the preparation of this series of earthquakes is the "chemical" channel.

About a day before the start of a series of earthquakes in Turkey on February 6, 2023, anomalous variations in the amplitude and phase of the signal from the A1F3 transmitter, received at the Mikhnevo geophysical observatory, began. The signal propagation path ran 44–132 km from the epicenters of the strongest shocks.

Acknowledgments. This work was supported by the Ministry of Science and Higher Education of the Russian Federation, project no. 122032900175-6.

References

1. Wait, J.R., Spies, K.P.: Characteristics of the Earthionosphere waveguide for VLF radio waves. Technical Note 300, National Bureau of Standards, Boulder, Colo (1964)
2. Gokhberg, M., Gufeld, I., Rozhnoy, A., Marenko, V., Yampolsky, V., Ponomarev, E.: Study of seismic influence on the ionosphere by super long-wave probing of the earth-ionosphere waveguide. Phys. Earth Planet. Inter. **57**(1–2), 64–67 (1989)
3. Gufeld, I.: Radiowave disturbances in period to Rudber and Rachinsk earthquakes. Phys. Solid Earth **28**, 267–270 (1992)
4. Hayakawa, M.: The precursory signature effect of the Kobe earthquake on VLF subionospheric signals. J. Comm. Res. Lab. Tokyo **43**(2), 169–180 (1996)
5. Molchanov, O., Hayakawa, M., Oudoh, T., Kawai, E.: Precursory effects in the subionospheric VLF signals for the Kobe earthquake. Phys. Earth Planet. Inter. **105**(3–4), 239–248 (1998)
6. Yamauchi, T., Maekawa, S., Horie, T., Hayakawa, M., Soloviev, O.: Subionospheric VLF/LF monitoring of ionospheric perturbations for the 2004 mid-niigata earthquake and their structure and dynamics. J. Atmos. Solar Terr. Phys. **69**(7), 793–802 (2007)
7. Ryakhovskii, I.A., Gavrilov, B.G., Poklad, Y.V., Bekker, S.Z., Ermak, V.M.: The state and dynamics of the ionosphere from synchronous records of ULF/VLF and HF/VHF radio signals at geophysical observatory "Mikhnevo". Izv. Phys. Solid Earth **57**(5), 718–730 (2021). https://doi.org/10.1134/s1069351321050177
8. Pulinets, S., Boyarchuk, K.: Ionospheric Precursors of Earthquakes, p. 315. Springer, Berlin (2004). https://doi.org/10.1007/b137616
9. Sorokin, V.M., Yaschenko, A.K., Chmyrev, V.M., Hayakawa, M.: Dc electric field formation in the mid-latitude ionosphere over typhoon and earthquake regions. Phys. Chem. Earth Parts A/B/C **31**(4), 454–461 (2006). https://doi.org/10.1016/j.pce.2005.09.001
10. Molchanov, O., Hayakawa, M., Miyaki, K., et al.: VLF/LF sounding of the lower ionosphere to study the role of atmospheric oscillations in the lithosphere-ionosphere coupling (2001)
11. Miyaki, K., Hayakawa, M., Molchanov, O.A.: The role of gravity waves in the lithosphere - ionosphere coupling, as revealed from the subionospheric LF propagation data. In: Seismo Electromagnetics: Lithosphere - Atmosphere - Ionosphere Coupling, pp. 647–653 (2002)
12. Shvets, A.V., Hayakawa, M., Maekawa, S.: Results of subionospheric radio LF monitoring prior to the Tokachi (m = 8, Hokkaido, 25 September 2003) earthquake. Nat. Hazard. Earth Syst. Sci. **4**(5/6), 647–653 (2004). https://doi.org/10.5194/nhess-4-647-2004
13. Molchanov, O., Hayakawa, M., Rafalsky, V.: Penetration characteristics of electromagnetic emissions from an underground seismic source into the atmosphere, ionosphere, and magnetosphere. J. Geophys. Res. Space Phys. **100**(A2), 1691–1712 (1995)
14. Molchanov, O.A., Hayakawa, M.: Seismo-electromagnetics and related phenomena: History and latest results. (No Title) (2008)

SME Geomagnetic Index Data Forecast Based on Wavelet Transform and LSTM Neural Networks

Yurii Polozov(✉)

Institute of Cosmophysical Research and Radio Wave Propagation FEB RAS, Kamchatsky Krai, Paratunka, Mirnaya Str., 7, 684034 Paratunka, Russian Federation
up_agent@mail.ru

Abstract. Solar-Earth relations are the sequences of transformations and energy impact transfers. The inflow of large amounts of energy into the geospheric shells leads to the development of magnetic storms. The influence of the southward component of the interplanetary magnetic field Bz causes increased geomagnetic activity. This paper proposes to use the SME index obtained in the SuperMAG project to analyze geomagnetic activity. In this paper, we propose an approach to forecast the SME geomagnetic activity index based on interplanetary magnetic field data and joint application of wavelet transform and LSTM neural network architecture. The effectiveness of the approach for different sets of input data can be evaluated using the generated neural network models. The constructed neural network model makes it possible to forecast the SME for the coming hours. The analysis of the results during calm and disturbed periods of geomagnetic activity showed the dependence of data approximation quality on the input data set for the neural network model and the forecast depth.

Keywords: SME · IMF Bz · neural networks · wavelets · data forecast

1 Introduction

Solar-Earth relations are the sequences of transformations and energy impact transfers. Large-scale processes on the Sun influence the dynamic parameters of the geospheric shells. These processes have direct and indirect impacts on anthropogenic infrastructure. Communication and navigation systems, power systems, satellites and high altitude aircraft, complex electronic devices are among the objects and systems that depend on space weather.

The inflow of large amounts of energy into the geospheric shells is associated with the Sun impact. This leads to the development of magnetic storms. The influence of the southward component of the interplanetary magnetic field (IMF) Bz causes increased geomagnetic activity [1–5]. The southern IMF Bz leads to the coupling of the IMF and the geomagnetic field lines. That results in plasma penetration into the inner magnetosphere. These processes are characterized by increased magnetosphere-ionosphere currents, which are estimated using the measurable geomagnetic activity indices Dst,

A. Dmitriev et al. (Eds.): STRPEP 2023, SPEES, pp. 186–196, 2023.
https://doi.org/10.1007/978-3-031-50248-4_20

AE, Kp. Application of IMF Bz in various approaches to the analysis and assessment of geoeffective effects is shown in the papers [6–8].

The task of selecting the magnetic index arises due to the possibility of forecasting and estimating the expected effects of increased geomagnetic activity on anthropogenic systems. As it was shown in the papers [9–10], the variations of the AE and Kp indices are influenced by the position of the auroral current relative to the stations for which the indices are calculated. This paper proposes to use the SME index obtained in the SuperMAG project [11] to analyze geomagnetic activity. The geomagnetic stations, the data of which are used for the calculation of the SME index, are located in the band of geomagnetic latitudes from 40° to 80°. That makes it possible to take into account the spatial drift of the auroral electrojet towards the equator. The SME index provides a more accurate assessment of auroral currents and does not decrease values when the auroral electrojet shifts to lower latitudes that is typical for the AE index due to the limited number of geomagnetic stations and the narrow band of geomagnetic latitudes of 65°–70° [12].

In this paper, we propose an approach based on the joint application of wavelet transform and Long Short-Term Memory (LSTM) neural network architecture to solve the problem of forecasting the SME index data using IMF Bz data. The flexible wavelet transform apparatus has been successfully used to analyze data of different nature [13–14]. Various wavelet decomposition schemes allow one to select the required components in the data. A large number of basic wavelets provides the possibility to adapt the method to the current research objectives and to a variety of data structures. The effectiveness of wavelets in the analysis of geomagnetic indices, including the neural networks (NN), has been demonstrated in the previous work [15, 16]. NN have become widely used in many fields [17–20]. There is a large number of architectures designed to solve different problems. The effectiveness of NN depends on the representativeness of the data to be trained and its nature. Despite their high ability to approximate complex internal dependencies, NN have certain limitations when applied. For example, working with noisy and non-stationary data reduces the quality of NN training and performance. The input vectors can be simplified and the quality of NN approximation can be improved by using wavelet-based data pre-processing. To avoid such problems as long-term dependencies when dealing with data with long-term relationships, it is necessary to use appropriate architectures. The author uses the LSTM architecture in order to significantly reduce the impact of long-term dependencies. Various studies have influenced the development of the architecture and it is now used for many tasks [19–20].

The approach proposed in this paper allows us to forecast SME data 1–5 h ahead based on hourly SME index and IMF Bz data. The effectiveness of the approach for different sets of input data can be evaluated using the generated neural network models. Application of wavelet transform improves the quality of the forecast and reduces NN errors. The trained neural networks were used to approximate natural SME data and showed the possibility to make forecast for the coming hours with a certain error. The dependence of NN errors on forecast depth was shown by evaluating the approximation quality of neural network models.

2 Methodology

Data

This work used IMF Bz and SME hourly data for the period 1999–2022. IMF Bz data (GSM coordinates) were obtained from the Advanced Composition Explorer spacecraft and accessed at https://omniweb.gsfc.nasa.gov. SME index data were obtained from the SuperMAG project at https://supermag.jhuapl.edu.

SME Index and IMF Bz Intensity Estimation

The continuous wavelet transform allows us to decompose the initial data into wavelet coefficients distributed over scaling levels [13]:

$$\left(W_{\Psi}x_{b,a}\right) := |a|^{-\frac{1}{2}} \int_{-\infty}^{+\infty} x(t)\Psi\left(\frac{t-b}{a}\right)dt, x \in L^2(R), a, b \in R, a \neq 0, \quad (1)$$

where Ψ is the basis wavelet, a is the scale, b is time. The change in the wavelet coefficients amplitudes at different scale levels depends on the appearance of local features of the signal corresponding to that scale, and outside the local features the absolute values of the wavelet coefficients are close to zero [13]. When decreasing the scale a, the neighbourhood $t = b$ that affects the original function x also decreases. This allows us to make a multi-scale analysis of the original function. To approximate the analyzed neighbourhood with the minimum number of signal samples, a minimum carrier wavelet basis is chosen. In [21], selection of the best wavelet basis for the AE index is shown. We use the Daubechies wavelet 3, taking into account that the SME index is based on a similar calculation [11, 12] from geomagnetic station data as the AE index. Using (1), we obtain the separation of positive and negative values of the wavelet coefficients:

$$W_{\Psi}x_{b,a} = \begin{cases} (W_{\Psi}x_{b,a})^{+}, \textit{if } W_{\Psi}x_{b,a} > 0 \\ (W_{\Psi}x_{b,a})^{-}, \textit{if } W_{\Psi}x_{b,a} \leq 0 \end{cases} \quad (2)$$

The intensity of positive and negative changes at the time b can be determined by evaluating the values of the wavelet coefficients on the ranges of scale levels [21], where A is the range of scale levels for which the intensity estimation is performed:

$$I_b^{\pm} = \sum\nolimits_{A} (W_{\Psi}x_{b,a})^{\pm}. \quad (3)$$

Construction of a Neural Network Model of the SME Index

SME index time series have a complex non-linear structure, containing local features of different amplitude and duration. NN allows one to approximate such data [19, 20, 22] and model complex temporal dependencies. The LSTM architecture is an updated version of the RNN and was originally designed to solve long-term dependency problems. It solves the vanishing gradient problem and, with additional control mechanisms for the information propagating across blocks, makes it possible to store it over long periods [23]. The LSTM architecture is a set of memory blocks that organize recurrently connected subnets (Fig. 1).

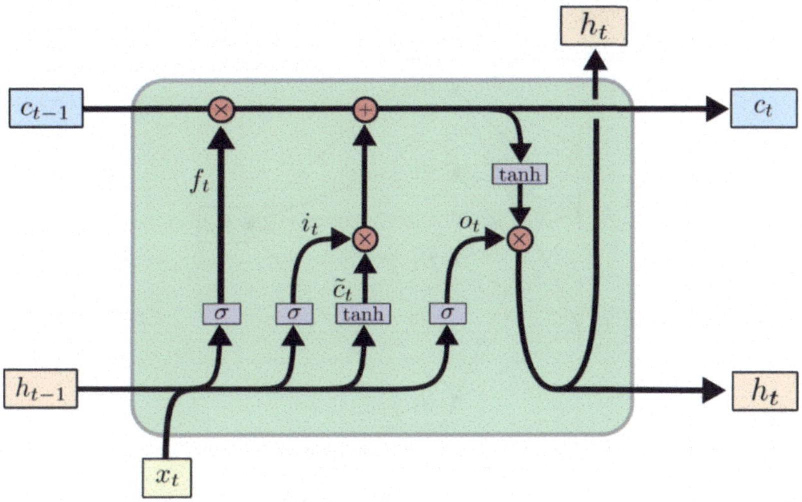

Fig. 1. Basic architecture of a long short-term memory network according to Olah [http://colah.github.io/posts/2015-08-Understanding-LSTMs].

An analytical view of the memory update implementation in the LSTM cell from Fig. 1 is shown below:

$$\begin{aligned} f_t &= f\left(w_f\left[h_{t-1}, x_t\right] + b_f\right) \\ i_t &= f\left(w_i\left[h_{t-1}, x_t\right] + b_i\right) \\ \tilde{c}_t &= f\left(w_c\left[h_{t-1}, x_t\right] + b_c\right) \\ c_t &= c_{t-1}f_t + i_t\tilde{c}_t \\ o_t &= f\left(w_o\left[h_{t-1}, x_t\right] + b_o\right) \\ h_t &= o_t f(c_t). \end{aligned}$$

where x_t is the input data; w_f, w_i, w_c, w_o are the states of weight matrices; b_f, b_i, b_c, b_o are the biases; h_t is the hidden state of the LSTM; f_t, i_t, o_t are the values of the forgetting gate, input gate and output gate, respectively; c_t, $\tilde{c}_t$ are the current and time states of a cell; $f(\cdot)$ is the activation function.

The LSTM structure consists of an input layer, one or more hidden layers and an output layer. Figure 2 shows the architecture of the network. In the work we used the single hidden layer architecture and the number of input vectors varied depending on the amount of input data. Thus, several neural network models were constructed and their performance quality was evaluated by the following indicators: Root Mean Square Error (RMSE) and Mean Absolute Percentage Error (MAPE).

$$RMSE = \sqrt{\frac{1}{N}\sum_{i=1}^{N}\left(y_i - \hat{y}_i\right)^2}$$

$$MAPE = 100\% \frac{1}{N}\sum_{i=1}^{N}\left|\frac{y_i - \hat{y}_i}{y_i}\right|$$

where N is the sample length, y_i is the SME data, $\hat{y}_i$ is the NN forecast.

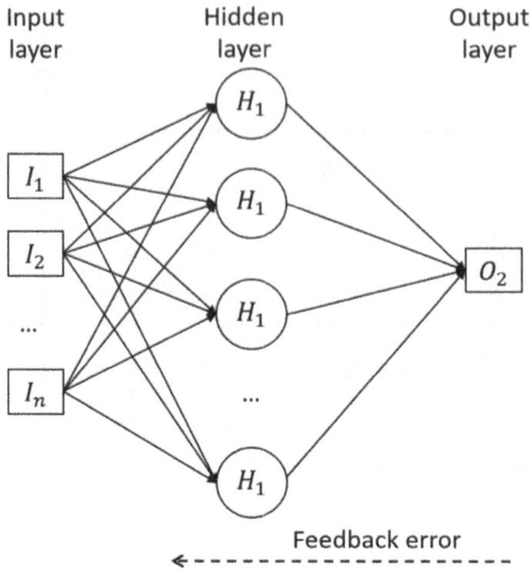

Fig. 2. Architecture of the LSTM model.

3 Experimental Results

The correlation between the IMF Bz variations and the AE geomagnetic activity index was shown in [21]. At the same time, the SME is an analogue of the AE index, but obtained at a larger number of stations. It also depends on the changes in the interplanetary environment parameters [11, 12]. The joint application of SME and IMF Bz data in neural network models can be used to forecast SME variations. In order to obtain adequate results, input data pre-processing is required. Using wavelet transform, we represented input data as ranges of wavelet coefficient scaling levels [21] and divided them into frequency periods corresponding to 1–4, 5–8, 9–16 and 17–24 h variations. The division into frequency periods is based on the results of [21], which describe the different durations of the southward turn of the IMF Bz and its correlation with changes in the AE index. Each frequency period contains information about the duration of the influence of the negative IMF Bz. This information reflects the processes of energy accumulation in the electrojet. Furthermore, the obtained scale ranges were fed to neural network inputs by separate vectors. Figure 3 shows the results of wavelet processing and calculation of the corresponding intensity ranges (the period from 08:00 UT 06 August 2000 to 08:00 UT 15 August 2000 is shown).

Energy penetrates to the Earth's magnetic field mainly at the southward turn of the IMF Bz that corresponds to the negative values on the graph in Fig. 3 b. Only the negative values of the initial time series, reduced to absolute values, were used in the wavelet processing of IMF Bz. A local change in IMF parameters and their shift towards negative values is shown by the increasing positive intensity I_b^+ for IMF Bz (Fig. 3d, f, i, k). The significant changes in the amplitude of the magnetic index variations occur are close in time, as can be seen from the graphs of the initial IMF Bz and SME data (Fig. 3a, b green ovals). The representation in the form of the wavelet components (Fig. 3c–k) shows early increase of the IMF Bz to the maximum values compared to the SME, especially for the significant changes in the intensity amplitude (Fig. 3g–k, green arrows).

The neural network models were constructed using four sets of input data: SME; SME and IMF Bz; SME, IMF Bz and intensity of IMF Bz I_b^+ wavelet coefficients; SME, IMF Bz and intensity of IMF Bz I_b^+ and SME $I_b^\pm$ wavelet coefficients. As output data of the NN, the forecast of SME was given for 2 and 5 readings ahead. The performance of the obtained models was verified on the test data, which were not used in the training process. The obtained estimates for RMSE and MAPE are shown in Tables 1 and 2.

The estimation results show that, when adding the IMF Bz data, the network quality improves, the RMSE values decrease that indicates a decrease in the amplitude of anomalous spikes in errors (Table 1). Meanwhile, application of the initial IMF Bz almost does not influence the MAPE percentage errors (Table 2). For both RMSE and MAPE estimates, the addition of the IMF Bz I_b^+ intensity improves the forecast quality. In its turn, the additional representation of the SME as the intensity of SME $I_b^\pm$ wavelet coefficients significantly reduces the NN errors in the forecast (Tables 1 and 2).

Analyses of the NN estimates for the two and five forecast readings show that the RSME for the first two readings is close in both types of models until SME $I_b^\pm$ is added to the input data. When the full input data set is used, the total RSME of networks for the two forecast readings is lower than that for 5 readings. The analysis of the MAPE estimates shows a similar picture, but MAPE performance already improves when IMF Bz I_b^+ is added to the network with two readings.

The results of data processing using the constructed NN are shown in Fig. 4 (the period from 19:00 UT 11 June 2004 to 18:00 UT 19 June 2004 is shown). Neural networks were trained on data from 1999–2003 for the period shown in Fig. 4. The complete set of SME, SME $I_b^\pm$, Bz, IMF Bz I_b^+ vectors were used. The NN performance graphs (Fig. 4b–h) show the current SME data for the forecast and the NN forecast. For the NN with the forecast for 2 readings ahead for the first and second readings, the graphs are introduced in Fig. 4b, c, respectively. For the NN with the forecast for 5 readings ahead, the graphs are introduced in Fig. 4d–h. Figure 4 shows the period containing calm and perturbed states of the SME. During the SME calm state from reading 24 to reading 48, there was an increase in the index amplitude towards the positive (northward) direction in the IMF Bz data (Fig. 4a). Typically, such changes have no noticeable effect on geomagnetic environment that is confirmed by the SME data in Fig. 4 (graphs in blue).

After a calm period, disturbances are observed in the SME data. They are divided into blocks 1–3 (Fig. 4, purple dashed line). The SME changes occur with a small delay after the IMF Bz turns to the negative (southward) direction. The moments of the SME amplitude maxima are in agreement with the forecast of the neural network, but the performance of the neural network models depends on the forecast depth and disturbances duration. For the forecasting depth of one reading, the NN allow us to build a graph close to the SME real values, and the NN with the forecast depth of 2 readings shows better results than the NN with the forecast depth of 5 readings. At the

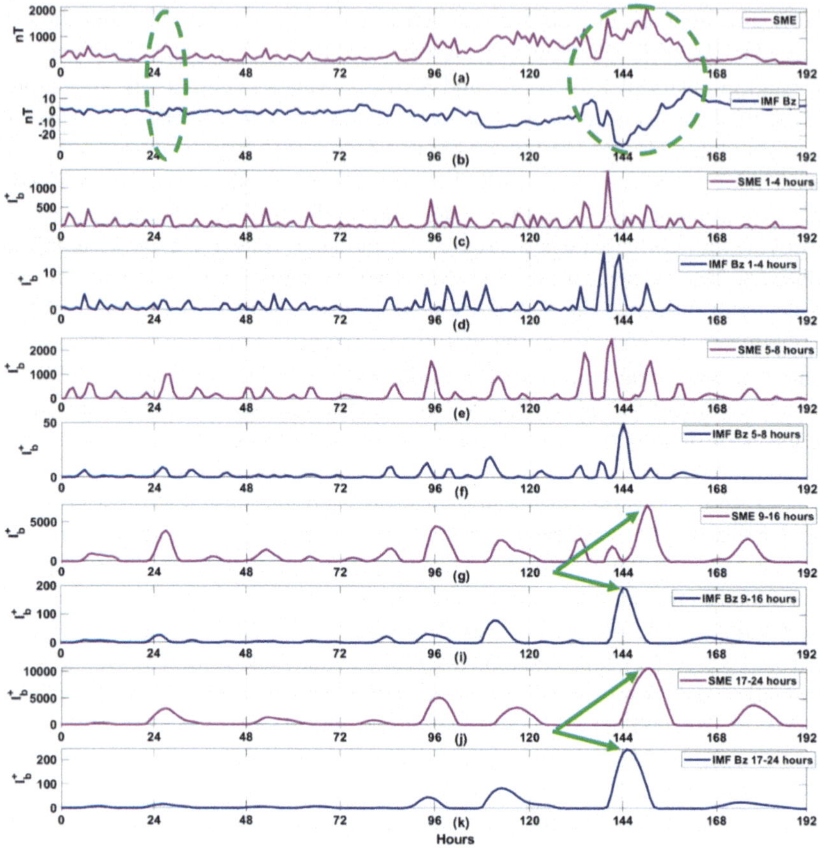

Fig. 3. Positive intensities of IMF Bz and SME wavelets ranges.

maximum forecast depth of 5 readings, there is a significant smoothing of the graph. Single short-term disturbances are not detected by the NN (block 1 in Fig. 4h). As the disturbance duration increases, the forecast quality for the 5-th reading improves, but single short-term spikes are not forecasted (block 2, Fig. 4h). Short-term disturbances of significant amplitude are the most difficult to be forecasted for the maximum forecast depth (block 3, Fig. 4h). They also contribute significantly to the increase of the MAPE and especially of the RSME.

Based on the data in Tables 1 and 2 and Fig. 4, it can be seen that application of preliminary wavelet processing of data makes it possible to reduce the error values of neural network models, while increase of the forecast depth affects the quality of the networks performance and worsens the ability to forecast short-term changes in SME.

Table 1. RMSE estimation of NN.

	For 2 forecast readings			
Forecast reading number	SME	SME, Bz	SME, Bz, Bz I_b^+	SME, SME $I_b^\pm$, Bz, Bz I_b^+
1	176	139	134	47
2	200	172	158	78
Total	189	157	146	59
	For 5 forecast readings			
1	179	143	138	63
2	201	175	161	89
3	209	198	170	98
4	214	208	175	112
5	219	213	183	118
Total	205	189	166	98

Table 2. MAPE estimation of NN.

	For 2 forecast readings			
Forecast reading number	SME	SME, Bz	SME, Bz, Bz I_b^+	SME, SME $I_b^\pm$, Bz, Bz I_b^+
1	46	45	36	14
2	52	47	44	19
Total	49	46	40	15
	For 5 forecast readings			
1	47	48	38	19
2	52	47	50	22
3	54	51	49	27
4	56	54	56	30
5	57	55	56	31
Total	53	51	50	26

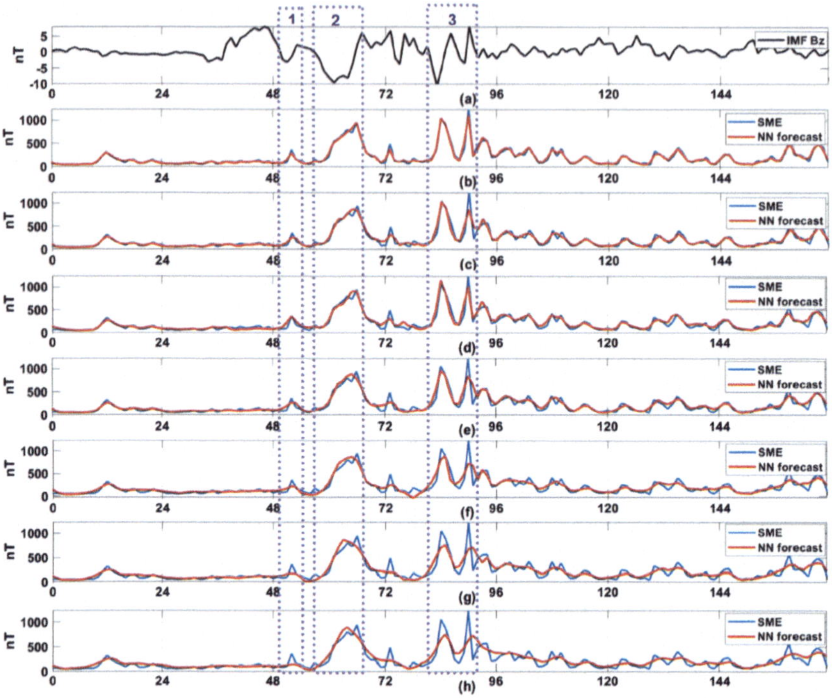

Fig. 4. Results of processing of IMF Bz and SME data with NN. (a) IMF Bz data; (b)–(c) NN forecast for 2 h ahead from the 1-st to the 2-nd, respectively; (d)–(h) NN forecast for 5 h ahead from the 1-st to the 5-th, respectively.

4 Conclusions

The analysis of the obtained results showed the possibility of constructing neural network models of the SME index based on wavelets and NN of the LSTM architecture. The use of wavelet pre-processing of the input data improves the quality of NN approximation. The modelling of the SME index during calm and disturbed periods showed the dependence of NN errors on the forecast depth and the level of local disturbances in SME data. The proposed approach to forecast the SME can be used to analyze space weather effects on the geomagnetic field. The development of a real-time forecasting system (as data become available) will be the subject of further studies.

Acknowledgements. The work was carried out within the framework of the State task on the topic (2021–2023) "Physical processes in the system of near space and geospheres under solar and lithospheric impact", registration number AAAA-A21-121011290003-0.

References

1. Dungey, J.W.: Interplanetary magnetic field and the auroral zones. Phys. Rev. Lett. **6**, 47–48 (1961). https://doi.org/10.1103/PhysRevLett.6.47

2. Cowley, S.W.H.: The earth's magnetosphere: a brief beginner's guide. Eos. Trans. AGU **76**, 525 (1995). https://doi.org/10.1029/95EO00322
3. Burton, R.K., McPherron, R.L., Russell, C.T.: An empirical relationship between interplanetary conditions and *Dst*. J. Geophys. Res. **80**, 4204–4214 (1975). https://doi.org/10.1029/JA080i031p04204
4. Gonzalez, W.D., et al.: What is a geomagnetic storm? J. Geophys. Res. **99**, 5771 (1994). https://doi.org/10.1029/93JA02867
5. Kane, R.P.: How good is the relationship of solar and interplanetary plasma parameters with geomagnetic storms? J. Geophys. Res. **110**, A02213 (2005). https://doi.org/10.1029/2004JA010799
6. Marques De Souza, A., Echer, E., Bolzan, M.J.A., Hajra, R.: Cross-correlation and cross-wavelet analyses of the solar wind IMF B_z and auroral electrojet index AE coupling during HILDCAAs. Ann. Geophys. **36**, 205–211 (2018) https://doi.org/10.5194/angeo-36-205-2018
7. Guarnieri, F.L., et al.: A correlation study regarding the AE index and ACE solar wind data for Alfvénic intervals using wavelet decomposition and reconstruction. Nonlin. Processes Geophys. **25**, 67–76 (2018). https://doi.org/10.5194/npg-25-67-2018
8. Souza, A.M., Echer, E., Bolzan, M.J.A., Hajra, R.: A study on the main periodicities in interplanetary magnetic field Bz component and geomagnetic AE index during HILDCAA events using wavelet analysis. J. Atmos. Solar Terr. Phys. **149**, 81–86 (2016). https://doi.org/10.1016/j.jastp.2016.09.006
9. Boroyev, R.N., Vasiliev, M.S.: Substorm activity during the main phase of magnetic storms induced by the CIR and ICME events. Adv. Space Res. **61**, 348–354 (2018). https://doi.org/10.1016/j.asr.2017.10.031
10. Boroyev, R.N., Vasiliev, M.S., Baishev, D.G.: The relationship between geomagnetic indices and the interplanetary medium parameters in magnetic storm main phases during CIR and ICME events. J. Atmos. Solar-Terr. Phys. **204**, 105290 (2020). https://doi.org/10.1016/j.jastp.2020.105290
11. Newell, P.T., Gjerloev, J.W.: Substorm and magnetosphere characteristic scales inferred from the SuperMAG auroral electrojet indices. J. Geophys. Res. **116**, 2011JA016936 (2011). https://doi.org/10.1029/2011JA016936
12. Boroev, R., Vasiliev, M.: Studying auroral activity using the SME index at the magnetic storm main phase during CIR and ICME events. Solnechno-Zemnaya Fizika **7**, 19–24 (2021). https://doi.org/10.12737/szf-74202103
13. Mallat, S.G.: A Wavelet Tour of Signal Processing: The Sparse Way. Academic Press, Boston (2009)
14. Mandrikova, O., Polozov, Y., Khomutov, S.: Wavelet model of geomagnetic field variations and its application to detect short-period geomagnetic anomalies. Appl. Sci. **12**, 2072 (2022). https://doi.org/10.3390/app12042072
15. Polozov, Y.: Analysis of the data of IMF Bz and AE index for the period 1999–2018. In: E3S Web Conferences, vol. 196, p. 02005 (2020)
16. Polozov, Y.A.: Forecasting the AE index based on neural networks. In: AIP Conference Proceedings 2910. International Conference on SMART Automatics and Energy, pp. 020130-1–020130-6 (2023). https://doi.org/10.1063/5.0166561
17. Polozov, Y.A.: AE index variations during extreme space weather and its forecast. In: AIP Conference Proceedings 2910. International Conference on SMART Automatics and Energy, pp. 020131-1–020131-6 (2023). https://doi.org/10.1063/5.0166562
18. Tang, R., et al.: The comparison of predicting storm-time ionospheric TEC by three methods: ARIMA, LSTM, and Seq2Seq. Atmosphere **11**, 316 (2020). https://doi.org/10.3390/atmos11040316
19. Perol, T., Gharbi, M., Denolle, M.: Convolutional neural network for earthquake detection and location. Sci. Adv. **4**, e1700578 (2018). https://doi.org/10.1126/sciadv.1700578

20. Jiao, F., Huang, L., Song, R., Huang, H.: An improved STL-LSTM model for daily bus passenger flow prediction during the COVID-19 pandemic. Sensors **21**, 5950 (2021). https://doi.org/10.3390/s21175950
21. Polozov, Y.: Wavelet-based analysis of interplanetary magnetic field and AE-index. In: Solovev, D.B., Kyriakopoulos, G.L., Venelin, T. (eds.) SMART Automatics and Energy, vol. 272, pp. 469–476. Springer, Singapore (2022). https://doi.org/10.1007/978-981-16-8759-4_49
22. Haykin, S.S.: Neural Networks: A Comprehensive Foundation. Prentice Hall, Upper Saddle River (1999)
23. Hochreiter, S., Schmidhuber, J.: Long short-term memory. Neural Comput. **9**, 1735–1780 (1997). https://doi.org/10.1162/neco.1997.9.8.1735

Observations of Space Weather Effects from the Moscow University Nano-satellite Constellation Sozvezdie-270

Sergey Svertilov[1,2](✉), Vitaly Bogomolov[1,2], Andrey Bogomolov[1], Vladislav Osedlo[1], Victor Bengin[1], Ivan Zolotarev[1], Anatoly Iyudin[1], Vladimir Kalegaev[1], Irina Myagkova[1], Ivan Yashin[1], Georgy Antonyuk[1], Karina Zhilchenko[1], and Vasilii Sazonov[3]

[1] D.V. Skobeltsyn Institute of Nuclear Physics, M.V. Lomonosov Moscow State University, Leninskie Gory 1(2), GSP-1, 119991 Moscow, Russia
sis@coronas.ru
[2] Physics Department, M.V. Lomonosov Moscow State University, Leninskie Gory, 119991 Moscow, Russia
[3] Faculty of Space Research, M.V. Lomonosov Moscow State University, Leninskie Gory, 119991 Moscow, Russia

Abstract. Within the framework of the Moscow University space project SOZVEZDIE-270, a constellation of cubesat nano-satellites with a set of instruments is being deployed, which, among other goals, provides monitoring of the near-Earth space radiation environment, control of the geo- and heliophysical conditions. Along with the space constellation, a network of ground receiving stations is also being created. During the project implementation, 11 spacecraft of the cubesat format have been launched to date. Currently, there are 6 such spacecraft operating in near-Earth orbit, which transmit scientific and telemetric data. During 2023–2024 it is planned to launch at least 8 more such satellites into low circular polar orbits. Multi-satellite constellation has been implemented, which makes it possible to carry out simultaneous measurements of particle and quantum fluxes using the same type of instruments at different points in the near-Earth space. Such measurements provide unique information about the sub-relativistic electron flux dynamics, including variations due to precipitation, which is of great importance for understanding the mechanisms of trapped and quasi-trapped electron acceleration and losses.

1 Introduction

The term "space weather" is used to refer to a number of natural electromagnetic phenomena occurring in the solar wind - magnetosphere - ionosphere - atmosphere system and determining the dynamics of radiation fields in interplanetary and near-Earth space (NES), the state of the ionosphere and the magnetic-wave environment in the NES [1]. The root cause of the phenomena that determine "space weather" is somehow connected with the Sun, including its flare activity, as a result of which a significant amount of particles accelerated in flares, mainly electrons and protons, can be ejected into interplanetary

A. Dmitriev et al. (Eds.): STRPEP 2023, SPEES, pp. 197–210, 2023.
https://doi.org/10.1007/978-3-031-50248-4_21

space, which, penetrating into the Earth's magnetosphere in the region of the polar caps, significantly change the pattern of radiation distribution in the NES. In some cases, so-called coronal mass ejections (CME) occur, which, under certain conditions, cause strong geomagnetic disturbances, including magnetic storms, as a result of which the spatial and temporal distributions of fluxes of both trapped and quasi-trapped particles can also change significantly in the NES, as well as electromagnetic fields.

Thus, some of the main manifestations of the "space weather" effects are energetic charged particle flux variations in different regions of the NES. In this case, the particle and quantum flux variations detected by the instruments can be caused both by the satellite crossing a compact area with an increased concentration of particles (the so-called spatial effects) and by an increase (or decrease) in intensity over time (temporal effects). In turn, both spatial and temporal effects can be associated both with the acceleration of particles and with their precipitation from the regions of trapped radiation, as well as with the penetration into the magnetosphere of particles accelerated in solar flares, i.e. the so-called solar cosmic rays (SCR).

As for the increases in the fluxes of trapped and quasi-trapped relativistic electrons associated with their acceleration, they are divided into several types. The first type is associated with "shock" acceleration and the formation of a "new" radiation belt [2, 3]. The second type is associated with slow increases in fluxes during the magnetic storm recovery phase. The third type (particle microbursts) has been little studied so far and, at present, is usually associated with fast acceleration during wave-particle interaction.

One of the popular mechanisms for the acceleration of energetic electrons is the interaction with very low-frequency (VLF) waves [4, 5], which can be generated during the development of the Kelvin-Helmoltz instability at the magnetopause. The mechanisms of acceleration of energetic electrons by high-frequency waves are also considered [6]. Radiation belt electrons can effectively interact with various modes of plasma waves in the frequency range $0.1\ \Omega_{O+} < \omega < 0.8\ |\Omega_e|$, where Ω_{O+} is O^+ ion cyclotron frequency, Ω_e is electron cyclotron frequency, including chorus emissions outside the plasmasphere, whistling hisses and electromagnetic ion cyclotron (EMIC) waves inside the plasmasphere, auroral kilometer radio emission (AKR). As a result of the resonant wave-particle interaction, both diffusion along pitch angles and diffusion in the momentum space arise. The non-linear wave-particle interaction can lead to a non-diffusion character of the occurring processes.

All experimental confirmations of the validity of the proposed mechanisms of relativistic electron acceleration are indirect, therefore, despite the progress made in solving the problem of describing the relativistic electron acceleration, a number of key problems can be identified, the solution of which requires further research. First of all, it should be noted that the problem of changing the geometry of the magnetosphere during a magnetic storm has not been largely resolved, which makes it difficult to create an adequate magnetic field model that describes the magnetic fields of a storm. The performed analysis shows that, despite the available results, the problem of transport, acceleration, and losses of relativistic electrons is still far from its final solution. The solution of these problems requires complex studies, including measurements of both the fluxes of the particles themselves and the magnetic field in various regions of the magnetosphere, as well as simultaneous observations of ULF-VLF waves in a wide frequency range.

The Earth's radiation belt dynamics is determined not only by acceleration processes, but also by particle loss processes. The mechanisms for the loss of energetic electrons can be divided into three groups. The first group includes shadowing by the magnetopause, which is typical for magnetic storms, i.e., a change in the configuration of drift trajectories (due to the Dst effect or radial diffusion directed outward from the Earth), as a result of which closed trajectories become open [7–9]. The second type includes ionization losses and Coulomb scattering, which play a significant role and determine the lifetime of electrons in the inner belt by $L < 1.5$ [10]. As for the outer belt, it is usually believed that the main mechanism of electron leakage from it is the development of cyclotron instability due to the interaction of electrons from the radiation belts with VLF/ELF/EMIC waves [11–16]. Losses of particles from the outer radiation belt can be quite intense [17], 2002). Electron fluxes can decrease by several orders of magnitude. Thus, the dynamics of precipitation of relativistic electrons from radiation belts can be considered as an independent physical problem, directly related to the processes of formation of the space-time structure of electron flows in the region of stable capture, i.e. the radiation belts themselves.

Precipitating electrons deserve more detailed consideration, since they are regularly found on drift shells corresponding to regions of trapped radiation ($L \sim 1.6$–1.9, 3.0–5.0), but at altitudes significantly lower than those at which the main "population" of the radiation belts is located, and, respectively, may pose a threat to spacecraft that were not originally designed to operate in the radiation belts. In addition, precipitation is often observed near the boundaries of the radiation belts, in particular, in the region close to the inner boundary of the outer radiation belt, in the so-called "slot", while observed as short-lived (observed during one orbit of the spacecraft) irregular increases in the electron flux, and more time-stable particle fluxes. The most compact regions with an increased electron concentration can have typical dimensions of the order of tens to hundreds of meters, i.e., like an electron beams. When the satellite crosses such areas, variations in the electron detector count rate with duration of tens of milliseconds will be observed. The temporal effects resulting from precipitation can also be very short-lived, down to sub-millisecond values of characteristic durations. In this case, precipitation can lead both to an increase in fluxes, at relatively low altitudes "under the belts" due to the appearance of precipitating particles on them, and to a decrease in fluxes trapped radiation areas due to their depletion (drop out) due to precipitation.

Precipitation of relativistic electrons can affect the properties of the upper Atmosphere, change the electrical and chemical properties of the stratosphere and mesosphere, leading, among other things, to ozone destruction.

As noted above, in addition to energetic protons and electrons of the radiation belts, SCRs are an important factor in "space weather". In near-Earth orbits, fluxes of SCR protons with energies > 5–10 meV are detected in high latitude regions, where there are no fluxes of protons from radiation belts with such energies. Therefore, SCR proton fluxes from powerful solar flares should be taken into account when constructing a 3D picture of the current spatial distribution of energetic charged particle fluxes in the NES.

To construct an adequate dynamic picture of the 3D distribution of energetic radiation fluxes in the NES, it is necessary to take into account their temporal changes, including short-term variations in high-energy electron fluxes. In this case, the central task is to identify the detected variations in particle fluxes as a consequence of spatial or temporal effects. This problem is difficult to solve using a single spacecraft. This requires simultaneous measurements on several spacecraft, which can be implemented using a multi-satellite constellation. Depending on the geomagnetic activity, such problems can be most effectively solved by a comparative analysis of data from several spacecraft operating both in close and different orbits. The implementation of such a strategy may involve options such as:

1) sequential passage of the same area by closely spaced satellites, which will allow the most reliable separation of spatial and temporal effects;
2) simultaneous measurements on different *L*-shells, which is necessary to restore the dynamic picture of the trapped particle flux distribution in a wide range of orbits, which, in particular, will make it possible to observe the shift of the maxima of the radiation belts during geomagnetic disturbances;
3) simultaneous measurements at the same height with the same type of instruments located on several satellites, shifted in longitude relative to each other, which will allow us to estimate the influence of the local time factor on the particle flux dynamics.

Some of these tasks can be solved by a constellation of several simultaneously launched spacecraft with identical detectors on board, the other part by joint analysis of data from one or more nanosatellites together with data from a larger spacecraft equipped with a complex of instruments that provides detailed measurements of particle fluxes, electromagnetic field and other parameters of the magnetospheric plasma. In any case, such simultaneous multipoint measurements should be carried out on several spacecraft using the same type of instruments.

It seems that micro and nano satellites of standard format, including cubesats, are the most suitable spacecraft for implementing the multi-satellite measurement strategy under consideration. Such devices are relatively cheap, do not require development stages and special tests. As a rule, they are launched into orbit as by-pass mission, which also seriously reduces the cost of the mission. At the same time, there is no need to completely duplicate the functionality of large spacecraft. The advantage of the constellation of micro and nano satellites lies precisely in the possibility of a more detailed study of the time variations of different space radiation components in different regions of near-Earth space.

In recent years, Moscow University has been implementing its own space program, during which particle and quantum fluxes in near-Earth space are monitored and electromagnetic transients of various nature are observed. In this regard, the successful launch of such satellites as Universitetsky-Tatiana [18], Universitetsky-Tatyana-2 [18], Vernov [19, 20], Lomonosov [21] should be noted. The next step in this direction will be a new project of the Moscow University “Constellation-270”. Within the framework of this project, a constellation of cubesat nano-satellites with a set of instruments is being deployed, which, among other goals, provides monitoring of the near-Earth space radiation environment, control of the geo- and heliophysical conditions.

2 Multi-satellite Mission of Moscow University

2.1 Scientific and Educational Experiments On-Board Cubesats

The first experience of space radiation monitoring using measurements on a cubesat satellites was obtained during the implementation of the SiriusSat educational project, during which, on August 15, 2018, cosmonauts from the International Space Station launched two 1U cubesats ("SiriusSat-1" and "SiriusSat-2"), which have successfully worked in orbit for more than 2 years. Schoolchildren were directly involved in the creation of these satellites payload during the project shift at the Sirius Educational Center. The deployment of the multi-satellite constellation was started on July 5, 2019, when 3 spacecraft of the 3U cubesat format were successfully launched from the Vostochny Cosmodrome. Those satellites were SOKRAT, VDNKh-80 and AmurSat (together with Amur State University). All these satellites were equipped with instruments developed at SINP MSU for monitoring space radiation and observing transient phenomena in the Earth's atmosphere. These instruments are the *De*tector of *Co*smic *Ra*diation (DeCoR) and the AURA ultraviolet photometer [22]. The MSU space program continued with the successful launch on September 28, 2020 of three spacecraft, i.e. 2 cubesats of 6U format DEKART and Norbi, one cubesat of 1.5U format Yarilo-2. Three identical DeCoR-type instruments were installed on the DEKART satellite in such a way that their axes are mutually normal to each other, which makes it possible to estimate the angular distributions of the detected particle and quantum fluxes. The DeCoR instruments were also installed on the Norby satellite of the Novosibirsk State University and on the Yarilo-2 spacecraft of the N.E. Bauman Moscow State Technical University.

As noted above, an important advantage of multi-satellite experiments is the possibility of simultaneous measurements at different points of the near-Earth space, on the one hand, and measurements in the same areas with successive passage of spacecraft through them, on the other. The experience of joint operation of the same type instruments on different spacecraft has shown that it is principally possible to carry out the same type of successive measurements in the same regions of the near-Earth space on a relatively short time base, which makes it possible to separate temporal and spatial effects in the detected count rate variations. An example of such measurements is data on the distribution of electron fluxes of sub-relativistic energies in the outer radiation belt. Comparison of the output data of the same type instruments obtained on similar orbits in close time intervals showed the similarity of the main typical features, i.e. counting rate maxima corresponding to the passage of the outer belt, narrow maxima observed symmetrically near the inner belt boundary, corresponding to precipitation zones.

In addition, during the experiments on the mentioned spacecraft, information was obtained on the dynamics of sub-relativistic electron fluxes, including the nature of short-term (with typical times of the order of 1 ms) flux variations due to precipitation, which is of great importance for understanding the mechanisms of acceleration and losses trapped and quasi-trapped electrons [23].

In Sect. 3, the results of observations of various space weather effects, including the filling of the polar caps with SCR particles and the dynamics of the particle spatial distribution in the outer radiation belt during a geomagnetic storm, will be considered in more detail. Events of similar type, also observed using the Moscow University satellite constellation, were previously considered in the article [24].

With the use of the multi-satellite constellation of Moscow University, flashes of UV radiation from the Earth's atmosphere, as well as candidates for astrophysical gamma-ray bursts, were also detected.

The results of the first multi-satellite measurements on spacecraft of the cubsat type confirm the suitability of small-sized low-power instruments of the DeCoR type for monitoring space radiation, detecting fast changes in the charged particle and gamma ray fluxes. Thus, the implementation of a unique multi-satellite constellation has begun, which makes it possible to carry out simultaneous measurements of particle and quantum fluxes, as well as transient atmospheric phenomena using the same type of instruments at different points in near-Earth space.

2.2 The Project "Constellation-270"

The next step in the development of the multi-satellite constellation of Moscow University will be the new project "Constellation-270". During the implementation of this project, it is planned to create a spacecraft system that allows, in a mode close to real time, to determine the levels of radiation loads caused by ionizing radiation not only in the region of the orbits of the spacecraft (SC) themselves, but also to determine the radiation environment in a significant part of the radiation belts, up to the orbits of global navigation satellite systems (GNSS) or geostationary (GSO). A separate goal is to study bursts in various ranges of the electromagnetic spectrum of astrophysical and atmospheric origin.

The main difference between this project and the meteorological and geophysical spacecraft currently operating on the orbit, one of the goals of which is to control the radiation environment in the near-Earth space, is the determination of radiation loads and notification of hazardous situations in near real time. This implies detection of energetic particle flux changes with times typical for geomagnetic disturbances, from tens of minutes and hours to several days or even weeks. As observations show, on such time scales during the main phase of a magnetic storm or substorm, the radiation situation in the region of the outer edge of the Earth's radiation belts can change, which corresponds to the orbits of the GNSS and GSO. In this case, radiation doses, the probability of internal and external electrization of the satellite, etc. can significantly change. It should be noted that the variations in radiation fluxes will be different in lower and higher orbits. For example, in lower orbits, changes in radiation conditions occur more slowly and with a delay relative to the onset of a geomagnetic disturbance than in higher orbits. It is this goal to determine the radiation loads in different spacecraft orbits at time intervals comparable to the typical times of radiation flux changings in the near-Earth space, that is central to this project, and this is its difference from existing and planned space systems, which allow, in essence, to identify only local changes in radiation loads and which are not able to estimate radiation levels on other spacecraft from these data, on which radiation monitoring devices are not installed.

It is assumed that within the framework of the "Constellation-270" project, at least 20 cubesat spacecraft will be launched into low circular orbits (altitude ~ 500–800 km), on which it is planned to install various devices for detecting energetic charged particles, gamma quanta, as well as detector-photometers for studying the UV emission of the atmosphere. In the future, it is planned to expand the scope of experiments, which, along with cosmophysical and geophysical problems, should cover research in the field of astrobiology, inter-satellite communications, etc.

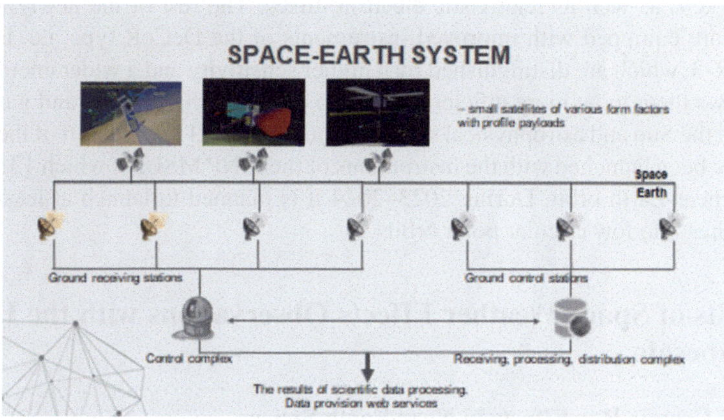

Fig. 1. The system of orbital and ground facilities providing control and data reception from spacecraft of a multi-satellite constellation.

In addition to the deployment of an orbital constellation, this project involves the creation of a network of receiving stations distributed along the meridians. At the first stage of the project implementation, it is planned to deploy at least 5 ground receiving points using antennas operating in the VHF, S and X bands. Thus, a unified system of spacecraft and ground receiving stations will be created, which should be installed in the regions from the Kaliningrad region to Kamchatka in such a way as to ensure the control of the spacecraft, as well as the regular reception of data from constellation satellites located at different points of the near-Earth space, which will allow significantly increase the amount of transmitted information (see Fig. 1).

The most important goal excluded of the project is the educational component [25, 26]. The educational component of the project is aimed at involving schoolchildren and students in space research. As part of this work, it is planned to significantly increase the number of ground reception points that can be installed in schools and other educational institutions. Schoolchildren and students are expected to participate in the preparation and conduct of space experiments, the reception and processing of scientific and telemetric information obtained using antennas installed directly on the territory of the schools participating in the project. Thus, schoolchildren will be involved in the creation of electronic components of devices, in particular, in their ground-based experimental testing, in obtaining and systematizing calibration data. Schoolchildren will have to participate in updating the databases used for space weather forecasting, processing and analyzing data obtained during the implementation of the space experiment.

As part of the implementation of the "Constellation-270" project, in addition to the cubesats launched earlier and operating in orbit (VDNKh-80, AmurSat, DEKART, Norbi), on August 9, 2022, three more cubesat 3U satellites were launched into orbit:

Monitor-1, Skoltech-b1, Skoltech-b2. As a result of the launch on June 27, 2023, the MSU multi-satellite constellation was replenished with 6 more satellites: Monitor-2, Monitor-3, Monitor-4, UTMN2, SiriusSat-SINP and Avion. The Monitor-1 satellite is equipped with a new instrument developed by SINP MGU, that is the *Co*mbined *D*etector of Radiation (*Iz*luchenii in Russian) KODIZ, which monitors proton fluxes with energies above 300 keV, as well as relativistic electron fluxes. The rest of the newly launched spacecraft are equipped with improved instruments of the DeCoR type, i.e. DeCoR-2 and DeCoR-3, which are distinguished by a higher sensitivity and a wider energy range, which allows them to be more efficiently used to detect the hard X-ray and gamma ray burtss from the Sun and astrophysical sources. Thus, to date, 17 spacecraft of the cubesat format have been launched with the instruments of the SINP MSU, of which 13 satellites operate in near-Earth orbit. During 2023–2024 it is planned to launch at least 8 more such satellites into low circular polar orbits.

3 Results of Space Weather Effects Observations with the Use of Cubesats

3.1 Solar Cosmic Ray Effects in Near-Earth Space

As examples, we consider an increase in the fluxes of energetic particles in the polar cap in July 2023 due to the arrival of solar cosmic rays (SCR), as well as changes in electron fluxes in the outer radiation belt during geomagnetic activity on March 2022.

Solar proton flux increasing was detected on 2023, July 18 by the SKIF instrument on-board geostationary satellite Elektro-L2 satellite. The corresponding time dependence of flux density of protons with energies from 9 to 110 MeV is presented in the top panel of Fig. 2. The source of this proton event was a solar flare of M2.7 class according to the GOES classification. This flare occurred at 22^h41^m UT on 2023, July 17 (see the bottom panel of Fig. 2) in the active region AR13363, which was at the time of the flare at the southern hemisphere practically on the western limb (flare coordinates S25W81). The maximum proton ($E_p > 10$ MeV) flux caused by this event (614 pfu according to the GOES data) was observed on July, 18 at 06 $^h30^m$ UT.

The time dependences of outputs of DeCoR instrument on board the DEKART satellite obtained for time interval from 2023, July 15 to 2023, July 22 are presented in Fig. 3. It means the fluxes of electrons with energies >300 keV, which were measured directly and fluxes of gamma quanta with energies >100 keV caused mainly by bremsstrahlung of electrons with corresponding energies. It can be seen from this figure that before the SCR event, i.e. 2023, July 15 and July 17 the electron flux time dependences were typical for measurements along the low altitude orbit with high inclination. The very intensive peaks were observed, when satellite crossed the outer radiation belt and South Atlantic Anomaly regions. The situation was changed drastically on July, 18, when electron fluxes in polar caps increased so high, that became comparable with values measured in radiation belts. Then, as particles injected in polar caps dissolved the electron flux time dependences became the same that they were before SCR event. Although even on July, 21 they were higher than before SCR income.

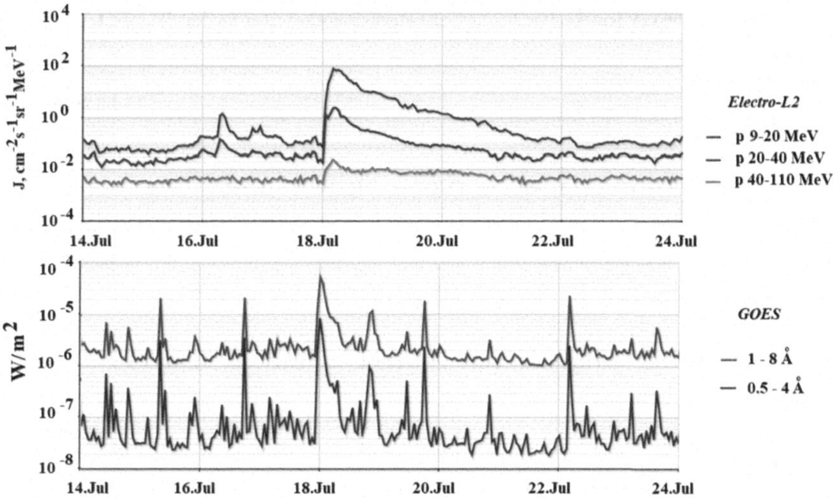

Fig. 2. Time dependences of electron flux density measured in different energy ranges by SKIF instrument on-board Elektro-L2 satellite (upper panel) and X-ray fluxes obtained from GOES data (bottom panel).

3.2 The Outer Radiation Belt Dynamics During the Magnetic Storm

The relatively high geomagnetic activity was observed on 2022, March. A solar wind high-speed stream (SW HSS) with a maximum speed of ~500 km/s and a density of ~10 particles/cm^3 came to the Earth's orbit on 2022, March 13. This stream caused a magnetic storm with a maximum $K_p = 6$ and $Dst = -83$ nT. Data on the solar wind speed and density for the time interval from 2022, March 11 to March 19, 2022 are presented in the top panel of Fig. 4. The geomagnetic activity indexes K_p and Dst for the same time are presented in the bottom panel of Fig. 4. The triangles in Fig. 4 mark the moments of measurements with the use of décor instrument on-board DEKART satellite on March 12, 14, 15, 16, and 18.

As it could be seen from Fig. 4, the clear magnetic storm occurred on 2022, March 13. In this time K_p index reached values about 6, while D_{st} variation was about -75. Approximately at the same time, the solar wind velocity increased to ~540 km/s, and solar wind density increased to 20 cm^{-3}, that indicates the arrival of a shock wave, which actually inspired the magnetic storm.

The outer belt electron flux variations before, during and after the magnetic storm are illustrated by curves in Fig. 5. There we can see dependences of electron fluxes J_e on drift shell number (McIllwine parameter) L. The top panel of the Figure shows electron flux value J_e, which were obtained from count rates of electrons with energy more than 300 keV detected in plastic scintillator of DeCoR instrument divided on its geometry factor. The bottom panel of Fig. 5 presents measurements in another channel of DeCoR instruments, I.E. gamma quanta with energy >100 keV, caused mainly by electron bremsstrahlung. Thus, really this channel presents the fluxes of electrons with corresponding energies. The curves were obtained from observational data for March 12, 14, 15, 16, and 18.

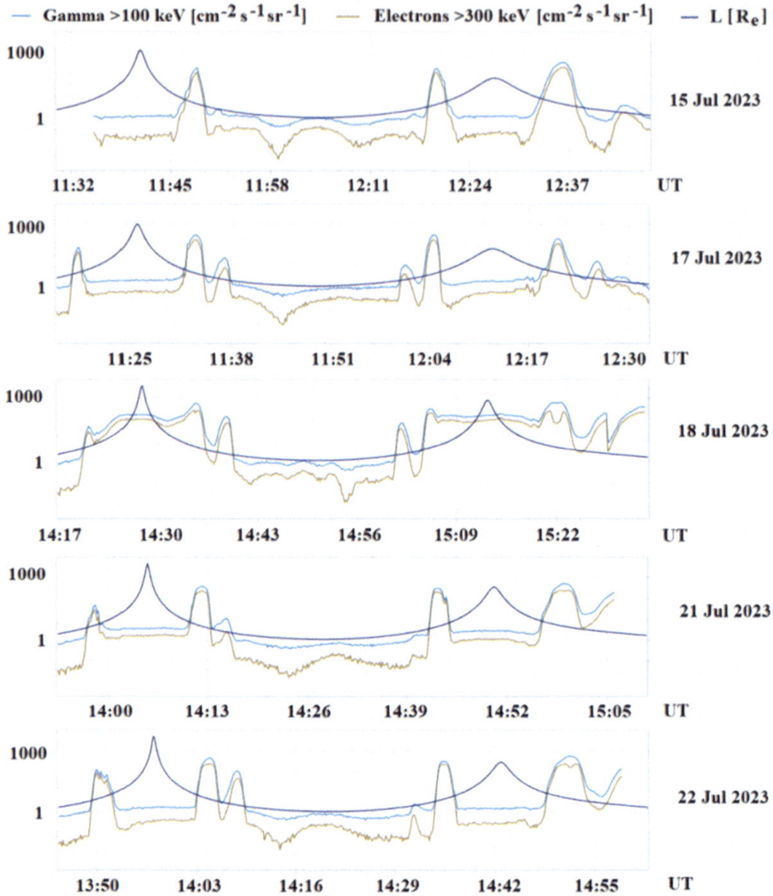

Fig. 3. Outputs of DeCoR instrument on board DEKART satellite obtained for orbits from 2023, July, 15 to July, 22 (panels from top to bottom). The time dependences of fluxes of electrons with energies higher than 300 keV are marked by blue lines, the dependences of fluxes of gamma quanta with energies higher than 100 keV are marked by brown lines. The dark blue lines indicate the values of McIllwine parameter *L*.

As it could be seen from the figure, during the recovery phase of the magnetic storm (March 14 and 15), the outer belt noticeably expanded, its polar boundary shifted from $L =$ to $L > 15$, and its maximum shifted closer to the Earth, to smaller *L*, i.e. from $L = 4.5$ to $L = 3.5$. The equatorial boundary of the belt began to return to its initial position, while the polar one remained in its primary place. On March 18, the outer belt, remaining wide (*L* values from 4 to 10), shifted further from the Earth, the maximum was observed at $L = 6.5$–7.

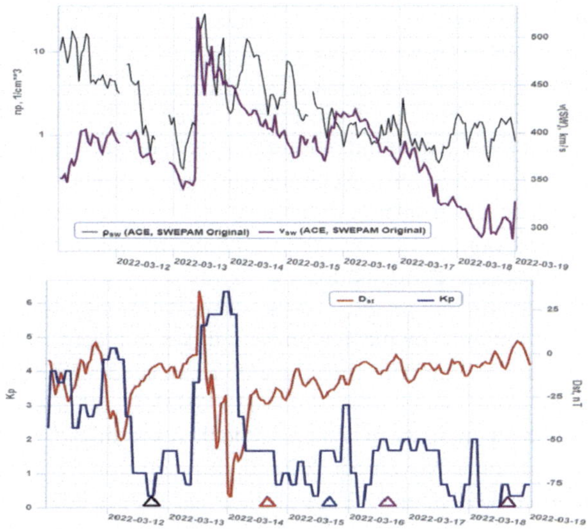

Fig. 4. Top panel: time dependences of solar wind velocity (violet line) and density (black line). Bottom panel: time dependences of geomagnetic activity indexes K_p (blue line) and Dst (red line).

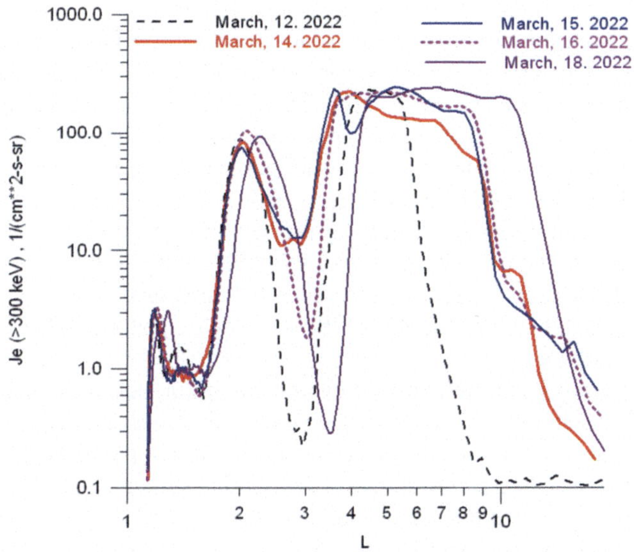

Fig. 5. Dependences on L of fluxes of electrons with energies higher than 300 keV before, during and after magnetic disturbance on 2023, March 13.

4 Conclusion

The first experience of joint switching on of identical devices on different spacecraft, implemented on satellites of Moscow University, showed the fundamental possibility of carrying out the same type of sequential measurements in the same regions of near-Earth space on a relatively short time base. Such measurements provide unique information about the sub-relativistic electron flux dynamics, including variations due to precipitation, which is of great importance for understanding the mechanisms of trapped and quasi-trapped electron acceleration and losses.

During the experiments on satellites of the cubesat format, already launched into orbit, information was obtained on the dynamics of fluxes of electrons of sub-relativistic energies. It is necessary to compare those events associated with the arrival of SCR particles in near-Earth space, which were observed in 2021, October–November [24] and in 2022, July. Although the increase of solar particle fluxes in July 2022 was significantly more intense than in October-November 2021, their resorption from the polar cap was much faster. Thus, three days after the start of the event in July 2022, the particle count rate in the polar caps dropped by almost 2 orders of magnitude (July 21), and on July 22 it reached almost the initial level. In the case of the October-November 2021 event, three days after it began, the particle count rates in the polar caps fell slightly. This indicates the influence of specific geomagnetic conditions on the dynamics of solar particle fluxes in the polar caps, which require a separate detailed analysis.

As for the spatial distribution of sub-relativistic electron fluxes in outer radiation belt during a geomagnetic storm a comparative analysis of the consequences of two geomagnetic storms, i.e. December 1, 2021 [24] and March 13, 2022 on the distribution of sub-relativistic electron fluxes in the outer radiation belt showed that in both cases there is an expansion of its boundaries, most significantly of its Arctic edge, i.e. from L ~7 to ~10–11 (in December 2021) [24], to ~20 (in March 2022), see Fig. 5. At the same time, the internal (equatorial) edge of the belt changed slightly, i.e. from L ~4 to ~3–3.2. Moreover, in the case of the event on March 13, 2023, the slot was also filled with particles, although the structure of the outer belt remained visible. Thus, it confirm well-known conclusion about stability of the outer belt equatorial edge as well during geomagnetic disturbances.

Based on the analysis of flight tests of instrument prototypes installed on satellites of the cubesat type, a method for monitoring fast variable processes in space radiation can be developed and requirements for advanced detecting instruments can be developed. Such a technique should be based on the choice of the optimal strategy for the implementation of space missions, including the determination of the required number of spacecraft, the establishment of suitable orbits and technical parameters of satellites, as well as payload parameters.

Thus, the implementation of a unique multi-satellite constellation has begun, which makes it possible to carry out simultaneous measurements of particle and quantum fluxes, as well as transient atmospheric phenomena using the same type of equipment at different points in near-Earth space.

Acknowledgements. This work was done with the support of MSU Program of Development, Project **№23-Ш01-02**.

References

1. Getzelev, I.V., Gusev, A.A., Darchieva, L.A., et al.: The model of space-energy distribution of trapped particle (electron and proton) fluxes inthe Earth radiation belts. MSU, Moscow (1991). (in Russian)
2. Li, X., et al.: Multi satellite observations of the outer zone electron variation during the November 3–4, 1993, magnetic storm. J. Geophys. Res. **102**(A7), 14123–14140 (1997)
3. Baker, D.N., et al.: A strong CME-related magnetic cloud interaction with the Earth's magnetosphere: ISTP observations of rapid relativistic electron acceleration on May 15, 1997. Geophys. Res. Lett. **25**(15), 2975–2978 (1998)
4. O'Brien, T., McPherron, R., Sornette, D., Reeves, G., Friedel, R., Singer, H.: Which magnetic storms produce relativistic electrons at geosynchronous orbit? J. Geophys. Res. **106**(A8), 15533–15544 (2001)
5. Green, J.C., Kivelson, M.G.: Relativistic electrons in the outer radiation belt: differentiating between acceleration mechanisms. J. Geophys. Res. **109**, A03213 (2004). https://doi.org/10.1029/2003JA010153
6. Trakhtengerts, VYu., Rycroft, M.J.: Whistle and Alfven Mode Cyclotron Masers in Space. Cambridge University Press, Cambridge (2008)
7. Ukhorskiy, A.Y., Anderson, B.J., Brandt, P.C., Tsyganenko, N.A.: Storm time evolution of the outer radiation belt: transport and losses. J. Geophys. Res. 111, A11S03 (2006) https://doi.org/10.1029/2006JA011690
8. Kim, K.-C., et al.: Numerical calculations of relativistic electron drift loss effect. J. Geophys. Res. **113**, A09212 (2008). https://doi.org/10.1029/2007JA013011
9. Ohtani, S., Miyoshi, Y., Singer, H.J., Weygand, J.M.: On the loss of relativistic electrons at geosynchronous altitude: Its dependence on magnetic configurations and external conditions. J. Geophys. Res. **114**, A01202 (2009). https://doi.org/10.1029/2008JA013391
10. McDonald, W.H., Walt, M.: Distribution function of magnetically confined electrons in a scattering atmosphere. Ann. Phys. **15**, 44–48 (1961)
11. Tverskoy, B.A.: The Earth's radiation belt theory. In: Proceedings of 9th ICRC, London, vol. 1, pp. 546–547 (1965)
12. Horne, R.B., Thorne, R.M.: Relativistic electron acceleration and precipitation during resonant interactions with whistler-mode chorus. Geophys. Res. Lett. **30**(10), 1527 (2003). https://doi.org/10.1029/2003GL016973
13. Summers, D., Thorne, R.M.: Relativistic electron pitch-angle scattering by electromagnetic ion cyclotron waves during geomagnetic storms. J. Geophys. Res. **108**, 1143 (2003). https://doi.org/10.1029/2002JA009489
14. Millan, R.M., Lin, R.P., Smith, D.M., McCarthy, M.P.: Observation of relativistic electron precipitation during a rapid decrease of trapped relativistic electron flux. Geophys. Res. Lett. **34**, L10101 (2007). https://doi.org/10.1029/2006GL028653
15. Thorne, R.M., et al.: Rapid local acceleration of relativistic radiation-belt electrons by magnetospheric chorus. Nature **504**, 411–414 (2013). https://doi.org/10.1038/nature12889
16. Li, W., et al.: Evidence of stronger pitch angle scattering loss caused by oblique whistler-mode waves as compared with quasi-parallelwaves. Geophys. Res. Lett. **41**, 6063–6070 (2014). https://doi.org/10.1002/2014GL061260
17. Onsager, T.G., Rostoker, G., Reeves, H.-J., Obara, G.D., Singer, H.J., Smithtro, C.: Radiation belt electron flux dropouts: local time? Radial and particle-energy dependence. J. Geophys. Res. **107** (2002). https://doi.org/10.1029/2001JA000187
18. Sadovnichii, V.A., Panasyuk, M.I., Yashin, I.V., et al.: Investigations of the space environment aboard the Universitetsky-Tat'yana and Universitetsky-Tat'yana-2 microsatellites. Solar Syst. Res. **45**(1), 3–29 (2011)

19. Panasyuk, M.I., et al.: Experiment on the *Vernov* satellite: transient energetic processes in the Earth's atmosphere and magnetosphere. Part I: description of the experiment. Cosmic Res. **54**(4), 261–269 (2016). https://doi.org/10.1134/S0010952516040043
20. Panasyuk, M.I., et al.: Experiment on the *Vernov* satellite: transient energetic processes in the Earth's atmosphere and magnetosphere. Part II. First results. Cosmic Res. **54**(5), 343–350 (2016). https://doi.org/10.1134/S0010952516050051
21. Sadovnichii, V.A., Panasyuk, M.A., Amelyushkin, A.M., et al.: Lomonosov" satellite—space observatory to study extreme phenomena in space. Space Sci. Rev. **212**(3–4), 1705–1738 (2017). https://doi.org/10.1007/s11214-017-0425-x
22. Bogomolov, V.V., et al.: A first experience of space radiation monitoring in the multi-satellite experiment of Moscow University in the framework of the Universat-SOCRAT project. Moscow Univ. Phys. Bull. **73**(6), 676–683 (2020). https://doi.org/10.3103/S00271349200 60089
23. Prokhorov, M.I., et al.: Analysis of fast variations in electron fluxes in the gap region using the normalized range method based on measurement data on the *SiriusSat-1* satellite. Cosmic Res. **60**(4), 241–253 (2022). https://doi.org/10.1134/S0010952522040062
24. Bogomolov, A.V., Bogomolov, V.V., Iyudin, A.F.: Space weather effects from observations by Moscow University Cubesat constellation. Universe **8**, 282 (2022). https://doi.org/10.3390/universe8050282
25. Osedlo, V.I., Antonyuk, G.I., Bengin, V.V., et al.: Educational project of Moscow university "Monitor" based on cubsat constellation. In: Fourth Russian Symposium on Nano-Satellites, RusNanoSat-2021, Samara, 28–30, June 2021. S.P. Korolev Samara National Research University (2021). (in Russian). http://volgaspace.org/rusnanosat/
26. Bogomolov, V.V., Antonyuk, G.I., Bengin, V.V., et al.: Scientific and educational project "Monitor" based on satellites of cubsat format. In: First International Conference on Space Education "Road to Space", Moscow, 5–8 October 2021, pp. 53–54. Institute of Space Research, RAS (2021). (in Russian). https://roadtospace.cosmos.ru/docs/2021/RoadToSpace-AbstractBook-2021-v2.pdf

Matching the Parameters of Shell Turbulence Models with the Probabilities of Interaction of Wave Shells

Gleb Vodinchar and Lyubov Feshchenko(✉)

Institute of Cosmophysical Research and Radio Wave Propagation FEB RAS, Mirnaya Str. 7, Paratunka, 684034 Kamchatka, Russia
gvodinchar@ikir.ru, feshenko.lk@yandex.ru

Abstract. The paper considers shell models of turbulence. These models make it possible to reproduce realistic distributions of spatial field spectra. Within the framework of the model, a system of quadratic nonlinear ordinary differential equations with parameters is formed.

The work proposes a computing technology that allows automating the process of constructing shell models. The technology is based on symbolic computing. The result is parametric classes of shell models that have the necessary analytical properties.

After constructing and solving the system, the coefficients of linear interactions turn out to be linear combinations of a certain set of free parameters. To numerically study models, it is necessary to fix the parameter values. This paper describes one formal method for selecting free parameters in which the interaction coefficients will be maximally consistent with the interaction probabilities.

Keywords: turbulence · shell models · computer algebra · automation of model development

1 Introduction

Shell models of turbulence are chains of ordinary differential equations that describe the processes of energy transfer along a scale spectrum in developed turbulence. Well-known shell models reproduce realistic distributions of spatial spectra of fields of real turbulent systems quite well [1–5].

To construct such models, the wave number axis is divided into geometrically expanding zones (shells). A sequence of wave numbers $k_n = q^n k_0$ appears. With a suitable choice of spatial scale when non-dimensionalizing the model, we can assume that $k_0 = 1$. The denominator of the progression q can be chosen to be arbitrarily large and greater than one. Often they take $q = 2$.

The dynamics of each field X of a turbulent system is described by real or complex variables $X_n(t)$ (collective variables) whose absolute values are interpreted as measures of field X pulsations. In scale ranges with wavenumbers from $[k_n; k_{n+1})$. This description (for integer q) corresponds to the representation of

A. Dmitriev et al. (Eds.): STRPEP 2023, SPEES, pp. 211–220, 2023.
https://doi.org/10.1007/978-3-031-50248-4_22

a turbulent system as a set of hierarchically nested interacting vortices, when each vortex of the nth scale contains q^α vortices of the $(n+1)$th scale. Here α is equal to the dimension of the physical space of the turbulent system. Each collective variable then gives a measure of the total intensity of eddies of a given scale.

For these variables, a system of equations is compiled that is qualitatively similar to the original equations of the turbulent system. Since the equations of hydrodynamics are quadratically nonlinear, we are talking about a quadratically nonlinear dynamic system. The coefficients for the nonlinear terms of such a system determine in the model the transfer of energy along the inertial scale interval in developed turbulence. In this work we will consider only complex variables.

It is known that in the non-dissipative limit in the equations of hydrodynamics there are conserved quadratic quantities (quadratic invariants). Therefore, it is necessary that the equations of shell models have quadratic invariants, which are in some sense analogues of these quantities.

The main mathematical task in deriving the equations of shell models is to calculate the coefficients for nonlinear terms that would ensure the existence of certain quadratic invariants. We are not talking about a set of specific numerical coefficients, but about some parameterized sets.

In this paper, we describe a technology for the automated construction of complex shell models, in which the calculation of nonlinear interaction coefficients is implemented using symbolic computing systems (computer algebra systems - CAS) [6–8]. We will carry out direct calculations using the Maple package, although all calculations can be carried out using any CAS.

This work generalizes the results of [9] for the case of complex models of hydrodynamic turbulence and turbulent convection.

2 Building Shell Models

The general form of the MHD equations has the form

$$\begin{aligned}
&\frac{\partial \mathbf{v}}{\partial t} + (\mathbf{v}\bigtriangledown)\mathbf{v} = -\frac{1}{\rho} + \nu \bigtriangleup \mathbf{v} + \frac{1}{\mu\mu_0\sigma} + rot\,(\mathbf{B}\times\mathbf{B}) + \mathbf{f},\\
&\frac{\partial \mathbf{B}}{\partial t} = rot\,(\mathbf{v}\times\mathbf{B}) + \mu \bigtriangleup \mathbf{B}, \qquad (1)\\
&\bigtriangledown\mathbf{v} = 0, \bigtriangledown\mathbf{B} = 0,
\end{aligned}$$

where $\mathbf{f} = \mathbf{f}(\mathbf{x}, t)$ is the mass density of external forces.

If both sides of the Eq. (1) are affected by the spatial Fourier transform, then these equations are transformed into equivalent equations in Fourier space

$$\frac{\partial \hat{v}_j}{\partial t} = i \int_{R^3} d\mathbf{q} \int_{R^3} d\mathbf{s} S_{jml}(\mathbf{k},\mathbf{q},\mathbf{s})\delta(\mathbf{k}+\mathbf{q}+\mathbf{s})\hat{v}_m^*(\mathbf{q})\hat{v}_l^*(\mathbf{s})+$$
$$\frac{1i}{\mu\mu_0\sigma} \int_{R^3} d\mathbf{q} \int_{R^3} d\mathbf{s} L_{jml}(\mathbf{k},\mathbf{q},\mathbf{s})\delta(\mathbf{k}+\mathbf{q}+\mathbf{s})\hat{B}_m^*(\mathbf{q})\hat{B}_l^*(\mathbf{s})$$
$$-\nu k^2 \hat{v}_j(\mathbf{k})_n + \hat{F}_j^*(\mathbf{k},t), \tag{2}$$
$$\frac{\partial \hat{B}_j}{\partial t} = i \int_{R^3} d\mathbf{q} \int_{R^3} d\mathbf{s} W_{jml}(\mathbf{k},\mathbf{q},\mathbf{s})\delta(\mathbf{k}+\mathbf{q}+\mathbf{s})\hat{v}_m^*(\mathbf{q})\hat{B}_l^*(\mathbf{s})-$$
$$\mu k^2 \hat{B}_j(\mathbf{k})_n.$$

where $S_{jml}(\cdot,\cdot,\cdot), L_{jml}(\cdot,\cdot,\cdot), W_{jml}(\cdot,\cdot,\cdot)$ – some real tensor function, $\hat{F}_j(\mathbf{k},t) = \mathbf{k}\times\left(\mathbf{k}\times\hat{\mathbf{f}}\right)/k^2$.

Thus, in the Eq. (2) on the left there is a change in the velocity spectrum, on the right three processes appear, the first is nonlinear interaction, the second is dissipative, and the third is the supply of energy. In the nonlinear term, a certain function is integrated over all possible wave numbers $\mathbf{q}$ and $\mathbf{s}$. That is, there is an interaction of three waves, and there is a δ-function, the argument of which is the sum $\mathbf{k}+\mathbf{q}+\mathbf{s}$, then only those waves whose vectors can be used to form a triangle.

The construction of shell models is based on this spectral analogue of the Navier-Stokes and induction Eq. (2). Shell turbulence models simulate nonlinear energy transfer between microscales.

In magnetohydrodynamic models, in the limit of zero viscosity and in the absence of external forces, some field integrals, the so-called conservation laws, will always be preserved. Parseval's equality states that there is equality of two integrals, one in physical space, the other in Fourier space. This equality allows us to write conservation laws in both spaces. So, when $\nu = 0$ and $\mathbf{f}(\mathbf{x},t) \equiv 0$, the following quantities are preserved in the Navier-Stokes and induction equations

$$E_K = \int_{\mathbb{R}^3} \frac{1}{2}\left(\mathbf{v}^2(\mathbf{x},t) + \mathbf{B}^2(\mathbf{x},t)\right) d\mathbf{x} =$$
$$\frac{1}{8\pi^3}\int_{\mathbb{R}^3} \frac{1}{2}\left(\hat{\mathbf{v}}(\mathbf{k},t)\cdot\hat{\mathbf{v}}^*(\mathbf{k},t) + \hat{\mathbf{B}}(\mathbf{k},t)\cdot\hat{\mathbf{B}}^*(\mathbf{k},t)\right) d\mathbf{k},$$
$$A^2 = \int_{\mathbb{R}^3} \mathbf{A}^2(\mathbf{x},t)\, d\mathbf{x} = \frac{1}{8\pi^3}\int_{\mathbb{R}^3} \hat{\mathbf{A}}\cdot\hat{\mathbf{A}}^*\, d\mathbf{k}, \tag{3}$$
$$H_C = \int_{\mathbb{R}^3} \mathbf{v}(\mathbf{x},t)\cdot\mathbf{B}(\mathbf{x},t)\, d\mathbf{x} = \frac{1}{8\pi^3}\int_{\mathbb{R}^3} \hat{\mathbf{v}}\cdot\hat{\mathbf{B}}^*\, d\mathbf{k},$$
$$H_B = \int_{\mathbb{R}^3} \mathbf{B}(\mathbf{x},t)\cdot\mathbf{A}(\mathbf{x},t)\, d\mathbf{x} = \frac{1}{8\pi^3}\int_{\mathbb{R}^3} \hat{\mathbf{B}}\cdot\hat{\mathbf{A}}^*\, d\mathbf{k}.$$

Each of these integrals is a function of time, but is proven to be constant. The first expression E is the total energy of the turbulent flow. The second expression is the square of the magnetic potential A^2. Third is cross-helicity H_C. And the fourth invariant is magnetic helicity H_B. Moreover, energy and cross helicity are conserved for any ideal fluid, but the conservation of magnetic helicity and the square of the vector potential are invariants that depend on the dimension of the space for which the movement occurs. For three-dimensional flows, the cross helicity is preserved; for two-dimensional flows, the square of the vector potential is preserved. All these invariants are quadratic in the velocity field and real.

In the eighties, classes of models were built that describe magnetohydrodynamic movements within the framework of shell models. Let us consider models of the GOY type of MHD turbulence (Gledzer–Okhitani–Yamada) in dimensionless form

$$\frac{dU_n}{dt} = \imath k_n \sum_{i,j} S_{ij} U^*_{n+i} U^*_{n+j} + \imath k_n \sum_{ij} L_{ij} B^*_{n+i} B^*_{n+j} -$$

$$\mathsf{Re}^{-1} k_n^2 U_n + f_n(t), \tag{4}$$

$$\frac{dB_n}{dt} = \imath k_n \sum_{i,j} W_{ij} U^*_{n+i} B^*_{n+j} - \mathsf{Re_m}^{-1} k_n^2 B_n,$$

where $k_n = q^n$ is the wave number of the n-th shell, S_{ij}, l_{ij}, W_{ij} are real coefficients (matrices of nonlinear interactions), $f_n(t)$ – models external energy supply to the n-th shell. Usually only $f_0(t) \neq 0$. $\mathsf{Re_m}$ – magnetic Reynolds number

Different models within this class differ in their nonlinear interaction matrices. Which are determined by the restrictions introduced into the model. One of the important restrictions is the limitation on the range of interactions. Because shells of different sizes cannot interact in any way. Therefore, we strictly enter $S_{ij} = 0$ if $|i| > P$ or $|j| > P$, where the indices i and j are <<distance>> by wave numbers in logarithmic scale. Here P is the maximum <<range>> of interactions. Then the summation occurs not from $-\infty$ to $+\infty$, but from $-P$ to P.

The second limitation is the fundamental possibility of interaction between shells. Previously, the presence of a δ function in the Eq. (2) was considered. That is, interaction is possible only when a triangle can be formed from wave vectors.

3 Invariants

There is a third limitation - this is the need for the existence of invariants. The time invariance of each equation of quadratic forms of collective variables means that their time derivatives are identically zero, and any such derivative will be the sum over all scales of the products of both variables and their derivatives.

Total Energy

$$E = \frac{1}{8\pi^3} \int_{\mathbb{R}^3} \left(\hat{\mathbf{v}}(\mathbf{k},t) \cdot \hat{\mathbf{v}}^*(\mathbf{k},t) + \hat{\mathbf{B}}(\mathbf{k},t) \cdot \hat{\mathbf{B}}^*(\mathbf{k},t) \right) d\mathbf{k} \sim$$

$$\frac{1}{2} \sum_{n=-\infty}^{+\infty} \left(U_n(t) U_n(t)^* + B_n(t) B_n(t)^* \right).$$

Squared magnetic potential (for two-dimensional flows). The magnetic field potential has the form $\hat{\mathbf{A}}(\mathbf{k},t) = \imath\mathbf{k} \times \hat{\mathbf{B}}(\mathbf{k},t)/k^2$.

$$A^2 = \frac{1}{8\pi^3}\int_{\mathbb{R}^3} \hat{\mathbf{A}} \cdot \hat{\mathbf{A}}^* \, d\mathbf{k} \sim \sum_{n=-\infty}^{+\infty} k_n^{-2} \left|B_n(t)\right|^2.$$

Cross helicity

$$H_C = \frac{1}{8\pi^3}\int_{\mathbb{R}^3} \hat{\mathbf{v}}(\mathbf{k},t) \cdot \hat{\mathbf{B}}^*(\mathbf{k},t) \sim \sum_{n=-\infty}^{+\infty} \left(U_n(t)B_n^*(t) + U_n^*(t)B_n(t)\right).$$

Magnetic helicity (for three-dimensional flows)

$$H = \frac{1}{8\pi^3}\int_{\mathbb{R}^3} \mathbf{B} \cdot \left(\frac{-i}{k^2}\mathbf{k} \times \hat{\mathbf{B}}^*\right) d\mathbf{k} = \frac{1}{8\pi^3}\int_{\mathbb{R}^3} \frac{1}{k^2}\left[\mathbf{B} \cdot \left(-i\mathbf{k} \times \hat{\mathbf{B}}^*\right)\right] d\mathbf{k}$$
$$\sim i \sum_{n=-\infty}^{+\infty} \frac{1}{k_n} B_n(t)B_n^*(t).$$

Let us consider the structure of equations for the coefficients S_{ij}, L_{ij} and W_{ij}. For each law, similar groups of homogeneous linear equations are formed. Formally, there are infinitely many of them in each group. All these equations contain subscripts $\pm i, \pm j, \pm(i-j)$. The pair (i,j) defines an equation within one group. Taking into account the restriction on long-range interaction, each term of the linear interaction matrices is equal to zero if $|i| > P$ or $|j| > P$. Then only a finite number of equations do not degenerate into identities.

Their identifiers (i,j) are given by the predicate

$$d(i,j) = (|i| \leq P \wedge |j| \leq P) \vee (|i| \leq P \wedge |i-j| \leq P) \vee$$
$$(|j| \leq P \wedge |i-j| \leq P),$$

This predicate identifies a finite subsystem in an infinite, generally speaking, system of equations.

Thus, using computer algebra tools, equations are automatically entered and solved. The linear system turns out to be of very large dimension, but since its construction occurs automatically, no difficulties arise. We performed symbolic transformations using Maple tools, but you can use any other shell that allows you to work with symbolic transformations.

So, for GOY type models, equations and conservation laws were considered. For other models, similar equations are constructed using computer algebra and solved.

4 Spectral Laws

After constructing and solving the models, the question of choosing free parameters arises. Additional considerations could be the introduction of power laws.

Since it is a known fact in the theory of turbulence that, in homogeneous isotropic turbulence within the iteration interval, some characteristics have power laws of distribution over scales, that is, over wave numbers. Let $S(k) \sim k^\lambda$ be a characteristic of a stationary flow. If we introduce the spectral distribution law, then the total value of this characteristic in the n-th shell will be

$$S_n = S\left(q^n \le k \le q^{n+1}\right) = \int_{q^n}^{q^{n+1}} S(k)\, dk \sim \int_{q^n}^{q^{n+1}} k^\lambda\, dk \sim q^{n(\lambda+1)}.$$

As an example, consider Kolmogorov's law for the energy $E(k) \sim k^{-5/3}$. Applying the relations (4) we find that the energy in the n-th scale shell should be $E_n \sim q^{(-2/3)n}$. Let's consider what is necessary for the shell model to have a similar spectral law. Since in the shell model the total energy in the n-th shell is $|U_n + B_n|^2$. Then it is necessary to require that there exist stationary solutions $U_n = U_0 q^{-n/3}$ and $B_n = B_0 q^{-n/3}$. After substituting the solution into the model equations, we obtain

$$\imath q^n \sum_{i,j} \left(S_{ij} |u_0|^2 q^{-(n+i)/3} q^{-(n+j)/3} + L_{ij} |U_0|^2 q^{-(n+i)/3} q^{-(n+j)/3} \right) = 0$$
$$\imath q^n \sum_{i,j} W_{ij} |B_0|^2 q^{-(n+i)/3} q^{-(n+j)/3} = 0.$$

From this equation, using algebraic transformations, we obtain

$$\sum_{i,j} \left(\frac{U_0}{B_0} S_{ij} + L_{ij} \right) q^{-(i+j)/3} = 0$$
$$\sum_{i,j} \left(W_{ij} \right) q^{-(i+j)/3} = 0.$$

By adjusting the control parameters if necessary, it is possible to ensure that U_0 is equal to B_0.

Thus, to introduce spectral laws into the model, it is necessary to add spectral equations to those equations obtained earlier.

To compile models with coefficients that guarantee conservation laws, fulfillment of the necessary spectral laws, and restrictions on the range of interactions, the authors developed algorithms based on computer algebra methods. The algorithms are implemented programmatically in the Maple package [10].

The problem of matching interaction probabilities and coefficients was solved using a specially developed application in C++. This is due to the fact that the probabilities of interactions are determined using simulation modeling, so carrying out accurate computer algebra calculations becomes pointless. In addition, floating point calculations in computer algebra systems are very slow. And the minimization problem presented above requires a very large amount of calculations.

5 Probabilities of Interaction

An algorithm for calculating the probabilities of interaction of p_{ij} waves from n, n_i and n_j shells was described earlier. Thus, a quantitative measure of the interaction of shells appears based on probabilistic considerations Fig. 1.

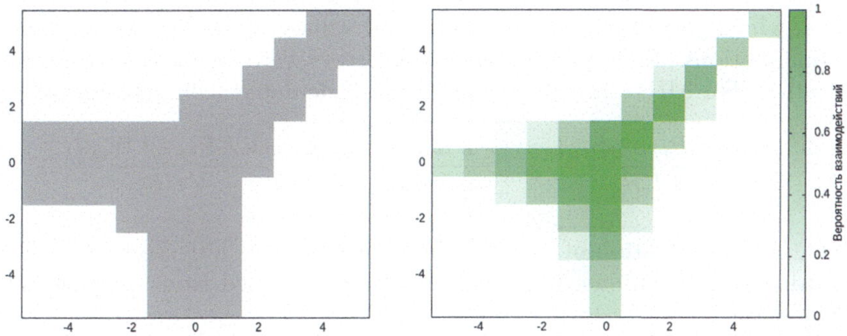

Fig. 1. Possibilities of interaction (left) and probabilities of interaction of waves p_{ij} (right) for pairs of indices i and j for $q = \left(1 + \sqrt{5}\right)/2$ – "golden ratio" and $P = 5$.

On the other hand, the coefficients S_{ij}, L_{ij} and W_{ij} can be considered as measures of the interaction of the n-th, $(n+i)$-th and $(n+j)$ shells. How large they are for fixed values of free variables is a measure of the interaction of shells. The need to reconcile interaction coefficients with interaction probabilities is obvious.

Let $S_{ij} = S_{ij}(\mathbf{s})$, $L_{ij} = L_{ij}(\mathbf{s})$ and $W_{ij} = W_{ij}(\mathbf{s})$, where $\mathbf{s} = [s_1, \ldots, s_k]$ is a vector of free parameters. One of the options for determining free coefficients is as follows. We considered non-zero coefficients that are expressed in terms of free parameters and considered sums of the following type that need to be minimized over free variables

$$\sum_{i,j} \left[\left| \frac{|S_{ij}(\mathbf{s})| - p_{ij}}{p_{ij}} \right| + \left| \frac{|L_{ij}(\mathbf{s})| - p_{ij}}{p_{ij}} \right| + \left| \frac{|W_{ij}(\mathbf{s})| - p_{ij}}{p_{ij}} \right| \right] \rightarrow \min$$

From a formal point of view, this is simply an optimization problem. But since it is not the coefficients themselves that need to be compared, but their modules, since the probabilities p_{ij} are only positive, and the coefficients of nonlinear interactions can be of any sign. Then the objective function becomes non-differentiable. In addition, I would like to stay within the framework of rational numbers. Therefore, one of the possible options is the brute force option. The choice of just such an objective function is based only on practical considerations. Since it is this expression that ultimately gives a good agreement between the interaction probabilities p_{ij} and the coefficients of nonlinear interactions.

6 Computational Experiments

Let us briefly describe the computational algorithm for this minimization problem. The following files are available: probability file, decision matrix coefficients file.

Next, various fractions of the form p/q were introduced, where p varied in the range $[-10; 10]$, and q - in the range $[1; 10]$. Thus, we have 127 different numbers, which more than cover the required values of the free parameters. So, 127^J combinations of free parameters arise. Errors were calculated for each such combination. For MHD models, the error was calculated using the formula

$$O = \sum_{i,j} \left| \frac{VER_{ij} - \mid S_{ij} \mid}{VER_{ij}} \right| + \left| \frac{VER_{ij} - \mid L_{ij} \mid}{VER_{ij}} \right| + \left| \frac{VER_{ij} - \mid W_{ij} \mid}{VER_{ij}} \right|. \quad (5)$$

Next, the combination of free variables with the smallest error was selected. Thus, using this technique, one single model is specified from multiparameter MHD models.

Let us present some calculation results for MHD models at $P = 3$, which is the minimum nonlocality.

Figure 2 shows the logarithms of the ratio of the obtained interaction coefficients to the interaction probabilities for the two-dimensional case

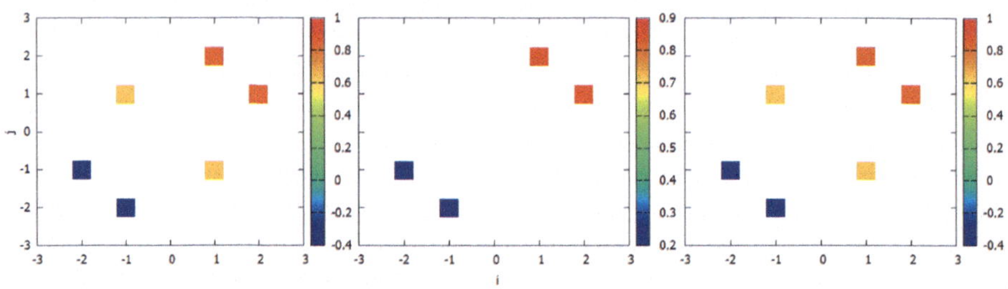

Fig. 2. Ratios of the resulting interaction coefficients to the interaction probabilities on a logarithmic scale (2D).

Figure 3 shows the logarithms of the ratio of the obtained interaction coefficients to the interaction probabilities for the three-dimensional case

The plots show a good interaction between interaction coefficients and interaction probabilities.

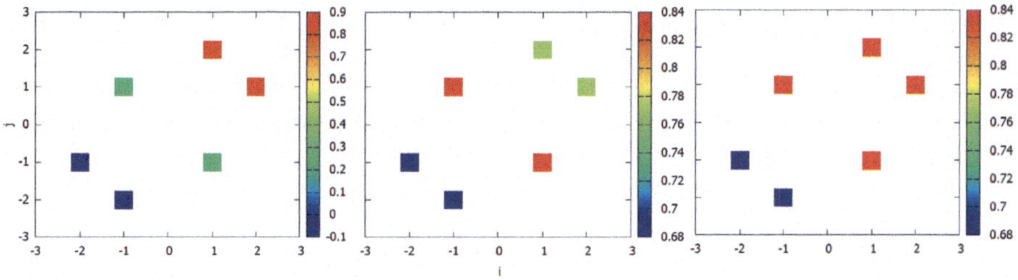

Fig. 3. Ratios of the resulting interaction coefficients to the interaction probabilities on a logarithmic scale (3D).

7 Discussion

Previously, we developed approaches using computer algebra that allow us to obtain parametric classes of models that provide models of the required conservation laws and spectral laws. In this case, the coefficients of linear interactions turn out to be linear combinations of a certain set of free parameters. To numerically study models, it is necessary to fix the parameter values. And by varying these parameters, you can provide large or small values of certain coefficients. It may turn out that when the parameters are fixed, the coefficient corresponding to unlikely interactions will be large, and vice versa. It is clear that such a choice of coefficients is unnatural.

In this paper, we describe one formal method for selecting free parameters in which the interaction coefficients will be maximally consistent with the interaction probabilities.

As a result, we received a comprehensive technology for constructing models with numerical values of parameters. The generated shell models can be directly studied further by numerical methods.

Acknowledgments. The work was carried out within the framework of realization of the State Task №AAAA-A21-121011290003-0.

References

1. Ditlevsen, P.: Turbulence and Shell Models. University Press, Cambridge (2011)
2. Ching, E.S.C., Guo, H., Cheng, W.C.: Understanding the different scaling behavior in various shell models proposed for turbulent thermal convection. Phys. D Nonlinear Phenom. **237**(4), 2009–2014 (2008). https://doi.org/10.1016/j.physd.2008.01.001
3. Hattori, Y., Rubinstein, R., Ishizawa, A.: Shell model for rotating turbulence. Phys. Rev. E **70**(4), 046311 (2004). https://doi.org/10.1103/PhysRevE.70.046311
4. Mingshun, J., Shida, L.: Scaling behavior of velocity and temperature in a shell model for thermal convective turbulence. Phys. Rev. E **56**(1), 441 (1997). https://doi.org/10.1103/PhysRevE.56.441

5. Frik, P.G.: Turbulence: approaches and models, Izhevsk, Moscow (2010). (in Russian)
6. Davenport, J., Sire, I.: Computer Algebra. Mir, Moscow (1991). (in Russian)
7. Matrosov, A.V.: Maple 6. Solving Problems in Higher Mathematics and Mechanics. BHV, St. Petersburg (2001). (in Russian)
8. Chichkarev, E.A.: Computer Mathematics with Maxima. ALT Linux, Moscow (2012). (in Russian)
9. Vodinchar, G.M., Feshchenko, L.K.: Automated generation of turbulence shell models by computer algebra methods. Comput. Technol. **26**(5), 65–80 (2021). https://doi.org/10.25743/ICT.2021.26.5.006. (in Russian)
10. Vodinchar, G.M., Feshchenko, L.K.: Symbolic computations in simulations of hydromagnetic dynamo. In: E3S Web of Conferences, vol. 20, p. 02017 (2017). https://doi.org/10.1051/e3sconf/20172002017

About Some Small Effects in Magnetic Field at Observatories Paratunka (Kamchatka, Russia) and Choutuppal (India)

Alexey Y. Gvozdarev[1](✉), Sergey Y. Khomutov[1], Kusumita Arora[2], and Phani Chandrasekhar[2]

[1] Geophysical Observatory Paratunka, IKIR FEB RAS, Mirnaya, 7, Paratunka 684034, Kamchatka, Russia
{gvozdarev,khomutov}@ikir.ru

[2] Magnetic Observatory, National Geophysical Research Institute, Uppal Road, Hyderabad 500007, Telengana, India

Abstract. Small (up to 1 nT) effects in the magnetic field recorded at the observatories Paratunka (PET, IKIR FEB RAS, Kamchatka, Russia) and Choutuppal (CPL, CSIR-NGRI, India) are considered, including diurnal and seasonal variations primarily in the total intensity field and baselines. The effects are not related to the technical features of magnetometers, electromagnetic interference or man-made noise. Inhomogeneities in the magnetization of rocks in the pavilions area and its temperature dependence, as well as induction effects, are considered as a possible mechanisms. The results of direct measurements of magnetic susceptibility are presented, the manifestations of its inhomogeneities are compared qualitatively with model calculations.

Keywords: magnetic observatory · local effects · magnetic susceptibility · temperature dependence

1 Introduction

Magnetic observatories are one of the most important elements of the Earth's magnetic field monitoring system. They provide long continuous homogeneous data series with high accuracy. The development of magnetic measurement techniques and methods at observatories leads to an increase in the quality of the results obtained, including field and time resolution, long-term stability, and resistance to external influences on the equipment [1]. However, this increases the importance of small effects of natural origin, which do not exceed 0.5–1.0 nT and which were previously unavailable for research. In addition, the magnitude of these effects generally lies within the requirements for modern magnetic data [2].

S. Y. Khomutov, K. Arora and P. Chandrasekhar—Equal Contribution.

A. Dmitriev et al. (Eds.): STRPEP 2023, SPEES, pp. 221–234, 2023.
https://doi.org/10.1007/978-3-031-50248-4_23

In this paper, we consider several examples of such effects recorded at the observatories Paratunka (IKIR FEB RAS, Kamchatka, Russia) and Choutappal (CSIR-NGRI, Hyderabad, India), which are certified magnetic observatories of the INTERMAGNET. Note that the effects under consideration are not directly related to the problems of measuring equipment, such as, for example, the direct dependence of magnetometer sensors and electronics on temperature, electromagnetic interference from other instruments used at observatories, etc. [3]. These effects are generated by local sources and manifest themselves in a magnetic field, that is, they represent a natural and objective result of magnetic measurements. Some aspects of such effects are covered in monographs [4,5].

2 Local Effects in the Magnetic Field at the Observatory Paratunka

The Observatory Paratunka of the IKIR FEB RAS is located on the Kamchatka Peninsula, 30 km from Petropavlovsk-Kamchatsky and is the easternmost operating observatory in Russia. Magnetic observations have been continuously carried out since 1967, and vast experience has been accumulated. Since the beginning of the 2000s, the equipment and infrastructure have been modernized, which allowed obtaining the status of INTERMAGNET observatory in 2013. A detailed description of the equipment, infrastructure and observation systems is provided in [6].

2.1 Spatial Gradient of the Total Field Intensity in the Pavilion for Absolute Observations

Modern standards of magnetic service require absolute observations of the components of the total field and its intensity. At the Observatory Paratunka, total field intensity F in the absolute pavilion is measured using the scalar Overhauser magnetometer POS-1 [7], continuously at a fixed place and during several minutes on the main pillar, usually twice a day. The absolute pavilion is a wooden building with an area of 3.5×3.5 m, which does not contain magnetic elements, raised on asbestos-cement pipes with a filler approximately 1.2 m above the ground. Inside there are two pillars from glass blocks up to the floor level, also standing on asbestos-cement pipes.

Figure 1 shows the values of the gradient gradF between the main pillar (point 1) and the place of permanent installation of POS-1 sensor (point 2), obtained since July 2016. The value of the gradF is estimated as the difference F(2)-F(1) between points 1 and 2 when moving the sensor there and back. The average gradF value is about 8 nT at a distance of about 3 m, the height of the points above the floor is about 1.5 m and above ground level is about 3 m. It should be noted that only a single scalar magnetometer POS-1 was used for measurements, so possible systematic errors do not affect gradF estimates. The inaccuracy of the sensor installation at points 1 and 2 does not exceed 1–2 mm, therefore the effect due to the inhomogeneity of the field is expected to be less than 0.01 nT. The temperature inside the pavilion is maintained stable throughout the year at +16°C with extreme short-term anomalies from +8 to +25°C. Thus, it can be assumed that the observed seasonal variations in a range of up to 0.5 nT and a trend of −0.06 nT/year are reliable and are not related to magnetometer problems or electromagnetic interference.

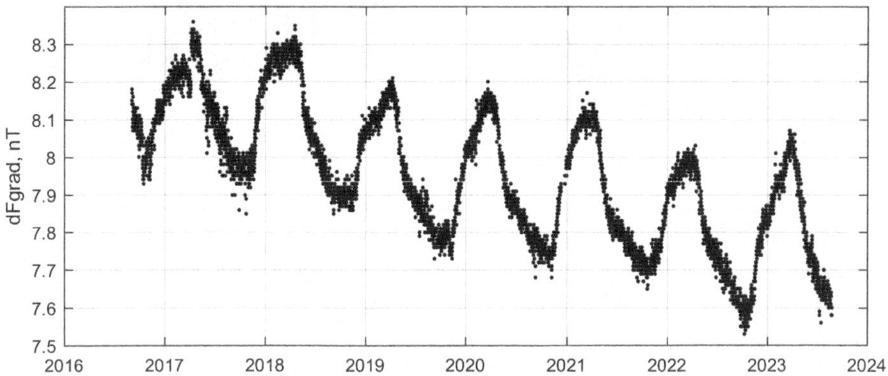

Fig. 1. Spatial gradient gradF of the total field intensity between two fixed points in the absolute pavilion at the Observatory Paratunka

2.2 Variations of the Magnetic Field in the dIdD Magnetometer Pavilion

The magnetometer dIdD GSM-19FD (GEM Systems, Canada), including the sensor, console and power supply, is installed in a small wooden pavilion measuring 2 × 1 m on a pillar about 20 cm high above ground level (see [6]). The pavilion has no heating system, only passive thermal insulation. In late winter and early spring, it can be almost completely covered with snow. DS18B20 digital temperature sensors are installed inside and outside, the temperature inside the pavilion varies from −15°C to +25°C throughout the year with a range of daily fluctuations up to 10°C in summer.

Regular measurements of magnetic field variations using dIdD have shown that the magnetometer has a temperature dependence, both in the angular components declination D and inclination I, and in the total field intensity F. In general, the temperature coefficients are close to those given in the specification for the magnetometer by the developer - up to 0.1 nT/°C [8]. Figure 2a, b shows 10-day sets of minute values of the deltaF, that is, the total field intensity F from dIdD measurements in January and August 2021 after the removing of natural geomagnetic variations using records of the magnetometer GSM-90 installed in the pavilion with a stable temperature conditions. It is clearly seen

Table 1. Magnetic susceptibility $\chi, 10^{-3}$ of soil probes

Probes	Distance from pavilion, m	at indoor temperature	frozen probes	ratio frozen/room
0	5	3.40 ± 0.60		
1	15	3.46 ± 0.14	2.62 ± 0.20	0.757
2	25	3.15 ± 0.17	2.38 ± 0.20	0.756
3	35	3.24 ± 0.12	2.41 ± 0.20	0.744
4	45	4.45 ± 0.39	2.71 ± 0.32	0.609

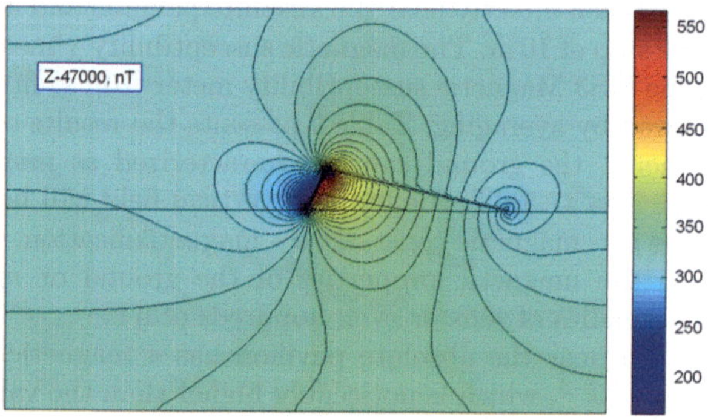

Fig. 3. Model distribution of the magnetic field vertical component Z in the vicinity of the embankment near the southern wall of the absolute pavilion. The step between the isolines is 7 nT. The size of the calculation area (in vertical section) is 2 by 3 m.

lead to the formation of a horizontal gradient of soil magnetization and the appearance of a field gradient that will depend on the soil temperature.

Structural heterogeneities in the soil can lead to the appearance of anomalies. Examples of such inhomogeneities are the pits that were dug for the installation of pillars in the new simplified pavilions for magnetometers dIdD (see Sect. 2.2) and POS-4. These pits were partially filled with the material of the pillars, but this does not remove the inhomogeneity in χ. To assess the expected effects under conditions of increased magnetical susceptibility of the surface layer of soil on the territory of the observatory, special measurements were carried out: at point No. 3 (Table 1) a pit of 0.8×1.0 m in size and 1 m in depth was dug. Previously, a magnetic survey was carried out on an area of 4×4 m at heights of 0.15 and 1.0 m, which was repeated after digging the pit. Measurements were carried out using a magnetometer GSM-19W, natural geomagnetic variations were removed using the records of the stationary scalar magnetometer GSM-90. At the same time, soil samples were taken at different depths and their magnetic susceptibility was measured (the results are presented in Table 2). These measurements showed a noticeable growth of χ with depth.

Table 2. Magnetic susceptibility χ of soil probes taken at different depths

Probes	Depth, cm	rock	$\chi, 10^{-3}$
1	0	sand	3.24 ± 0.12
2	20	sand	3.99 ± 0.10
3	40	sandy loam of ochre color	4.72 ± 0.39
4	60	sandy loam of ochre color	4.00 ± 0.18
5	80	light yellow clay	8.99 ± 0.46
6	100	light yellow clay	8.28 ± 0.83

The distribution of the change in total field intensity F before and after digging the pit is shown in Fig. 4. The results show a change in the field in range of 90 nT at a height of 0.15 m and 10 nT at a height of 1 m. The positive anomaly is located to the north of the pit, the negative one is located to the south with the center at the southern edge of the pit.

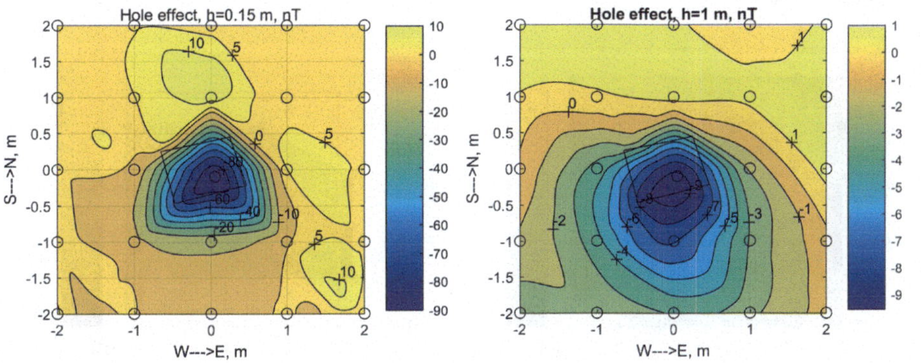

Fig. 4. Distribution of the difference of the total field intensity F before and after digging the pit: left panel is at a height of 0.15 m above the surface, right panel is at a height of 1 m.

2.3.2 The Gradient GradF Changes in the Absolute Pavilions As it was shown in Sect. 2.3.1, inhomogeneities of the magnetic properties of rocks in the area of the absolute pavilion can create noticeable anomalies. The magnitude of these anomalies may vary with temperature variations. Therefore, a magnetic survey was carried out under the pavilion and outside it at a height of about 1 m above the ground. The survey dates were close to the extremes of the gradF (see Fig. 1). The first survey in November 2022 was in conditions of the beginning of freezing of the soil under the pavilion. The second survey was in April 2023, when only the surface layer of the soil thawed while maintaining frozen soil at depth.

The snow and ice conditions allowed measurements to be made only under the pavilion. The distribution of F is shown in Fig. 5. As can be seen in Fig. 5, the field is highly heterogeneous, the range of changes over the 5×4 m area reaches up to 90 nT. The main anomalies are located at the western wall near the stairs (negative) and at the southeastern corner (positive). The positive anomaly is caused by the remains of a concrete blind area buried in the ground, left over from the old absolute pavilion. A negative anomaly may also be caused by some undetectable undemounted structural elements of the old pavilion. The load-bearing and measuring pillars of the new pavilion, apparently, are not noticeable sources of the anomalous field, although they are a little visible in the results of the survey carried out in 2012 on the floor inside the pavilion.

The gradF estimation based on the distributions in Fig. 5 at the points 1 and 2 under the pavilion gives values of +42.9 nT in November and +41.4 nT in April, compared with +7.6 nT and +8.1 nT measured directly at points 1 and 2. The total error of the survey results due to positioning and interpolation errors does not exceed 1.5 nT. Thus, our studies show the presence of seasonal variability in the distribution of the field under the pavilion, but the data collected is insufficient to prove this mechanism of changing the gradF between points 1 and 2.

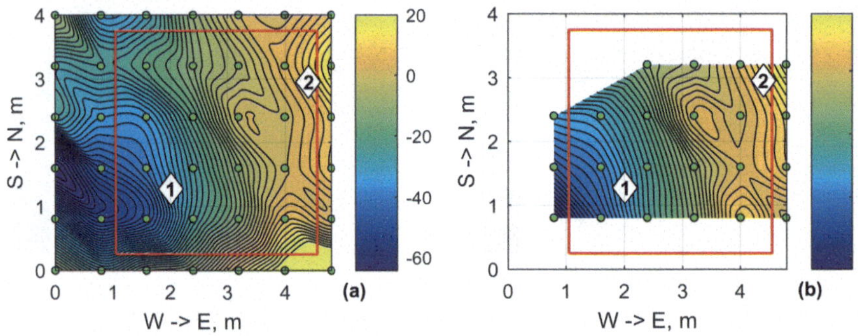

Fig. 5. The distribution of total field intensity F (in the nT) at a height of about 1 m from the ground in the vicinity and under the absolute pavilion according to the survey results on November 08, 2022 (a) and April 03, 2023 (b). The red rectangle represents the walls of the pavilion, the labels 1 and 2 show the locations of the POS-1 sensor during absolute and continuous measurements.

3 About Some Small Effects in Magnetic Field at Low Latitude Observatory Choutuppal, India

Baseline stability is the most important criterion for judging ground magnetic observatory data quality, as long as there are no faults in the absolute equipment, variometer, or observing method. In practice, temperature variations in the variometer, pier tilts, electrical component deterioration, anthropogenic noise, and other factors might affect baseline stability over time [12]. Despite the fact that the observatories are located in a magnetic-free area and are monitored by well-trained observers, temperature changes over variometers are the primary causes that result in baseline instability, in addition to other hidden factors. Magnetic variometer's sample rate and resolution continue to increase for scientific applications, while major measurement discrepancies between instruments appear. As a result, examining data by comparing observations among variometers is becoming increasingly important and crucial for assessing the data quality of every observatory [13].

Due to the presence of anthropogenic noise, the data collected at the HYB observatory at a 1-second interval is compromised. Consequently, the Choutuppal (CPL) low latitude observatory was established with the primary objective of generating high-quality 1-second data while maintaining stable baselines [14]. Both fluxgate installations employ identical instruments, including a DTU space fluxgate magnetometer, a Bartington Mag-01H, and a GSM-90F1 for measuring the total field. These installations are situated in underground vaults with a depth of around 4.5 m.

In this section, we talk about the nature of the data quality at CPL by comparing it to HYB. We also compare the baseline trends from different magnetometers by looking at how temperature affects the recording systems and the spot values from different magnetometers during the time of absolute observations at the CPL campus.

Figure 6 shows examples of 1 min dF (difference between the three-component vector and scalar data) from the CPL and HYB observatories for April 2017 and May 2020. From Fig. 6, it is evident that the dF at CPL displays a clear diurnal trend that is more pronounced in 2017 and somewhat subdued in 2020; however, this trend is not present in the dF of HYB data. In both years the range of dF is less than 1 nT. The diurnal trend may be explained as the effect of temperature on fluxgate measurements at CPL; however, in that case, the same trend should be present in the dF of HYB data as the daily/monthly/annual temperature ranges in the vaults at both locations are similar.

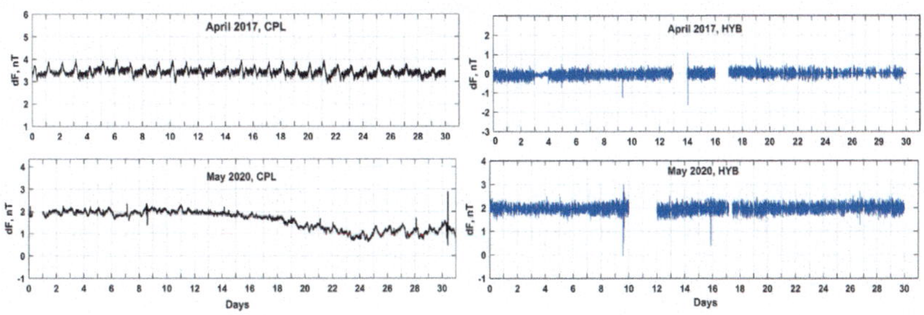

Fig. 6. Nature of dF at CPL and HYB during April 2017 and May 2020.

Therefore, to investigate the influence of temperature on variation data at CPL in greater depth, we performed an experiment in which baselines were calculated using distinct fluxgate magnetometers operated under identical environmental conditions at both CPL and HYB. At 1 sec recording, the variation data reveal significant variances from different instruments, and Fig. 7 shows an example of the H component on one quiet day (26 May 2017) and one disturbed day (29 May 2017), for the CPL observatory. From Fig. 7, it is obvious that the amplitudes of the disparities in spot values increase significantly on a disturbed day compared to a quiet day between the variometers.

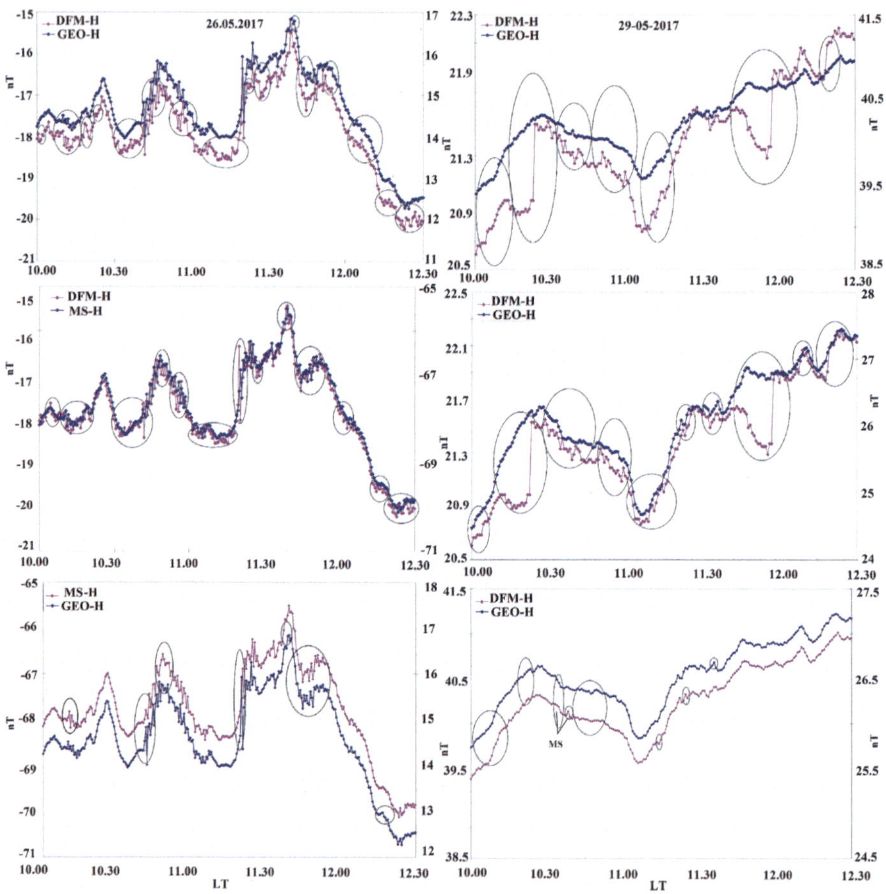

Fig. 7. Spot variations in the H-component of DFM, GEOMAG (GEO), Magneto Telluric System (MS) during 26th (quiet day) and disturbed day (29th) May 2017 at CPL Observatory.

Figure 8(a, b, d, f) shows that temperature fluctuations in DFM, GEO, and MS are found to be less than 0.2°C over a day in different variometer huts at the CPL campus. Comparing the dF changes and temperature variations in sensors (DFM, GEO, and MS), no close relationship was found on May 26th, 2017. This appears to indicate that the observed differences in the dF diurnal trend are not caused by temperature changes in the variometer huts but by other factors. Further, we have also noticed that when the Kp index is high, the magnitude of the dF goes up by 2 nT, occasionally.

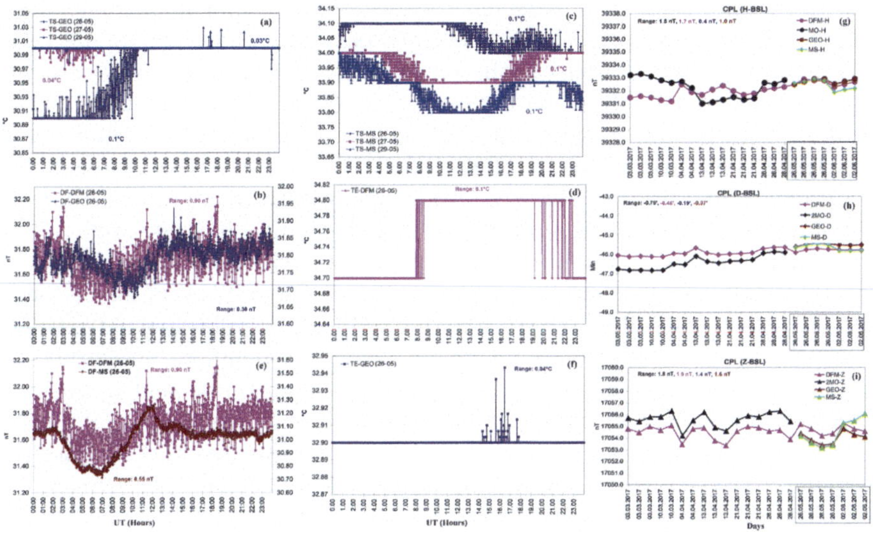

Fig. 8. Temperature fluctuations in GEOMAG (GEO) and MS sensors observed between May 26th-29th, 2017; dF variations in DFM and GEO (c); DFM and MS (e) with temperature of electronics in DFM (d) and GEO (f) during May 26th, 2017; (g-i) Observed baselines at CPL from different fluxgate variometers for H, D, and Z components for the months March-May 2017.

After removing the variations from various fluxgate variometers used at various variometer huts around campus, the baselines from D-I measurements at CPL absolute room pillar-1 are shown in Fig. 8(h–i). The highlighted box on the bottom x-axis indicates the time period of the experiment conducted at CPL, and baselines between the instruments are observed to be substantially variable on the days of observation. The spot values of the variometers during these days were found to be significantly different from the nearby values after analysis.

Our investigations with various fluxgate magnetometers revealed that the baselines between the fluxgates varied by 0.4 to 3.3 nT in H, −0.19 to −0.95 min in D, and 1.4 to 3.6 nT in Z, respectively, which is not the case at HYB. Therefore, the observed deviations in the baselines between the variometers at CPL could be because of due to other factors, as well as due to the induction effect and this phenomenon is found to be stronger when the magnetosphere is active.

4 Discussion

The selection of suitable sites for establishing magnetic observatories are undertaken with extreme care [4], yet natural and artificial influences cause contamination in the data. This may lead to long terms drifts on data [12] by accumulation of errors and increase in data uncertainty. In this work, we have documented the small effects in magnetic observatory data from a mid latitude observatory PET, 100 km west of the Kurile Kamchatka trench and a low latitude observatory CPL in the center of the Indian Peninsula.

In the former, i.e. PET, seasonal changes of the gradient of about 0.5 nT in the total field values inside the absolute pavilion, which cannot be related to temperature gradients within the pavilion, magnetometer errors or electromagnetic interference. Modeling indicates that the significant magnetization of soil under the absolute pavilion and its temperature dependence and possible unidentified structural elements of the pavilion create a sharp gradient inside the pavilion and cause its changes. Additionally total field values from dIdD measurements in a non-temperature controlled pavilion matches well with the values from a GSM90 in a temperature controlled pavilion in August, but not in January, shows seasonal variations in temperature dependence of magnetometer dIdD that can be caused by the surrounding soil and the pillar magnetization. To investigate this effect, the temperature dependence of magnetic susceptibility of ferromagnetic rocks, which may create an additional magnetic field is considered. For this purpose, magnetic susceptibility of near surface rocks under and around the absolute pavilion to a distance up to 45 m were measured and a higher value was found near the pavilion. Modeling shows that a material with increased susceptibility can create a magnetic anomaly of more than 10 nT at a height of 1 m. Furthermore measurements have revealed that susceptibility values decreased to upto 75% of its indoor value, when brought to freezing temperatures; thus it is inferred that inhomogeneous freezing can create to very localized field gradients. Similar seasonal effects of the temperature of near-surface rocks to total field intensity during the study of the volcanomagnetic signals are considered in [15].

At CPL, a diurnal pattern of the dF of range 1 nT (difference between scalar F and computed F) is noted, which is quite absent at the nearby observatory of HYB, 50 km away, although both observatories have the same equipment and identical methods of computation and very similar patterns of daily, seasonal and annual temperature patterns. Further, for days where $Kp > 3$, the dF range increases to nearly 2 nT. Subsequently, with the deployment of a variety of fluxgate magnetometers from different manufacturers, with different temperature coefficients in different pavilions in CPL, it was further clear that diurnal temperature changes were not the cause of the diurnal signature in dF. The effect of the changes in the spot values of the fluxgate magnetometers on instantaneous values of baselines could be 0.4 to 3.3 nT in H, −0.19 to −0.95 min in D, and 1.4 to 3.6 nT in Z. It may be speculated that very local variations of susceptibility may create transient fields or local variations in soil conductivity, the flow of groundwater [16,17] may lead to a variable induction effect, enhanced on disturbed days. Careful analysis of the correlation of H and Z components, may shed light on this aspect.

5 Conclusion

For many years, small variations have been observed in the magnetic data obtained at the Paratunka Observatory (PET, IKIR FEB RAS, Kamchatka, Russia) and at the Observatory Choutuppal (CPL, CSIR-NGRI, India). These variations are within the limits defined by the INTERMAGNET standards, but they are quite obvious and require explanation. Such variations at PET include, for example, seasonal variations in the gradient of the total field F inside the absolute pavilion with amplitude up to 0.5 nT and its long-term trend up to 0.1 nT per year, measured by the Overhauser magnetometer POS-1. Other example is daily variations of F obtained by the magnetometer dIdD GSM-19FD with amplitude up to 0.8 nT.

Various mechanisms are possible that can create such fields: the magnetization of rocks, soils, pillars basements, the heterogeneity of the distribution of conductivity in the underlying rocks, and even the flow of groundwater. In particular, it is shown that the magnetic susceptibility of Kamchatka soils is sufficient to create anomalies of tens and even hundreds of nT due to the heterogeneity of magnetic properties and microrelief. A decrease in magnetic susceptibility during freezing of the soil has been revealed, it is assumed that this effect can cause changes in the distribution of the field in the pavilions.

Even though a site with the least magnetic gradient is chosen for an observatory, preferably with less than 1 nT/m gradient, it is often found that the magnetic gradient patterns are not permanent and change over time. These slow gradient changes can distort the data, which can affect the accuracy of long term data and baselines of a geomagnetic observatory. Based on experimental measurements and modeling, the observed "small effects" recorded in PET and CPL could be caused by gradient changes of the local magnetic field arising from small changes in magnetic susceptibility, depending upon the crustal field contributions, as well as uneven electrical conductivity distribution in the ground, and magnetized rocks and soil in the area and/or even the flow of groundwater.

Acknowledgments. The work of A.Y. Gvozdarev and S.Y. Khomutov was carried out within the framework of realization of the State Task No. AAAA-A21-121011290003-0, the work of Kusumita Arora and Phani Chandrasekhar are supported by CSIR National Geophysical Research Institute, Ref.No: NGRI/Lib/2023/ Pre-116.

References

1. Thomson, A.W.P.: Geomagnetic observatories: monitoring the earth's magnetic and space weather environment. Weather **69**(9), 234–237 (2014)
2. St-Louis, B.: INTERMAGNET Operations Committee and Executive Council: INTERMAGNET Technical Reference Manual, 5.0.0 edn. INTERMAGNET, Edinburgh (2020). INTERMAGNET. http://intermagnet.org/docs/Technical-Manual/technical_manual.pdf

3. Khomutov, S.Y., Mandrikova, O.V., Budilova, E.A., Arora, K., Manjula, L.: Noise in raw data from magnetic observatories. Geosci. Instrum. Method. Data Syst. **6**(2), 329–343 (2017). https://doi.org/10.5194/gi-6-329-2017
4. Jankowski, J., Sucksdorff, C.: IAGA Guide for Magnetic Measurements and Observatory Practice, Warsaw (1996). https://iaga-aiga.org/data/uploads/pdf/guides/iaga-guide-observatories.pdf
5. Nechaev, S.A.: Guide for Stationary Geomagnetic Observations. Institute of Geography B. B. Sochava SB RAS, Irkutsk (2006). in Russian
6. Khomutov, S.Y.: Magnetic observations at Geophysical Observatory "Paratunka" IKIR FEB RAS: tasks, possibilities and future prospects. In: E3S Web of Conferences, vol. 20. no. 02002, pp. 1–18(2017). https://doi.org/e3sconf/20172002002
7. Sapunov, V., Denisov, A., Denisova, O., Saveliev, D.: Proton and Overhauser magnetometers metrology. Contrib. Geophys. Geod. **31**(1), 119–124 (2001)
8. GEM Systems: dIdD Instruction Manual, 7.5 edn. Canada (2010). GEM Systems
9. Ochsner, T.E., Baker, J.M.: In situ monitoring of soil thermal properties and heat flux during freezing and thawing. Soil Sci. Soc. Am. J. **72**(4), 1025–1032 (2008). https://doi.org/10.2136/sssaj2007.0283
10. Bartington Instruments Ltd.: Operation Manual for MS3 Magnetic Susceptibility Meter, OM3227/2 edn. GB. Bartington Instruments Ltd. http://bartingtondownloads.com/wp-content/uploads/OM3227.pdf
11. Parkinson, W.D.: Introduction to Geomagnetism. Elsevier Science Ltd, Edinburgh (1983)
12. Anil, I., Deshmukh, A.T., Veenadhari, B.: Long-term drifts in baselines of ground magnetic observatories. Measurement **102**, 33–41 (2017). https://doi.org/10.1016/j.measurement.2017.01.047
13. He, Z., Hu, X., Teng, Y., Liu, G., Zhang, X., Shen, X.: Characterization of the effects of temperature and instrument drift in long-term comparative geomagnetic vector observations. Atmosphere **13**(3), 449 (2022). https://doi.org/10.3390/atmos13030449
14. Phani Chandrasekhar, N., et al.: One second vector and scalar magnetic measurements at the low-latitude choutuppal (CPL) magnetic observatory. Geosci. Instrum. Method. Data Syst. **6**(2), 547–560 (2017). https://doi.org/10.5194/gi-6-547-2017
15. Utada, H., Neki, M., Kagiyama, T.: A study of annual variations in the geomagnetic total intensity with special attention to detecting volcanomagnetic signals. Earth Planets Space **52**, 91–103 (2000). https://doi.org/10.1186/BF03351617
16. Mishima, T., et al.: Relevance of magnetic properties of soil in the magnetic observatories to geomagnetic observation. Earth Planets Space **65**(4), 337–342 (2013). https://doi.org/10.5194/gi-6-547-2017
17. Helman, D.S.: Earth electricity: a review of mechanisms which cause telluric currents in the lithosphere. Ann. Geophys. **56**(5), 0564 (2014). https://doi.org/10.4401/ag-6184

Conditions for Generating a Chaotic Regime in a Low-Mode $\alpha\Omega$-Dynamo Model with Hereditary α-Quenching by Field Energy

Olga Sheremetyeva(✉)

Institute of Cosmophysical Research and Radio Wave Propagation, Far Eastern Branch of the Russian Academy of Sciences, Mirnaya, Paratunka, 684034 Kamchatka, Russia

sheremeteva@ikir.ru

Abstract. We consider the conditions under which the chaotic mode of magnetic field generation is modeled in the low-mode approximation of the large-scale $\alpha\Omega$-dynamo model. The intensities of the α- and Ω-generators are regulated by the Lorentz force. α-quenching is determined by the influence of the Lorentz force, which is included into the magneto-hydrodynamic system (MHD-system) on a small scale through a process with hereditarity properties (finite 'memory'). The process is defined by the functional. The behavior of the process is determined by the kernel of the functional, which sets the damped oscillations, and the strength of process action depends on the energy of the magnetic field. The variable parameters of the MHD-system are the Reynolds number, the measure of the α-effect and the parameters of the functional kernel. Within the framework of this work, the solutions of the MHD-system are investigated for stability according to the Lyapunov criterion in the vicinity of the rest point for each set of parameters.

Based on the results of the numerical experiment, the limitations of these parameters for modeling the chaotic regime of the magnetic field are determined. A comparison was made with the previously obtained results, where the characteristics of the quenching process depended on the exponential kernel.

Keywords: $\alpha\Omega$-dynamo · hereditarity · α-quenching · chaotic regime · reversals

MSC Classification: 34D08 · 76W05

1 Introduction

In the article [1], a large-scale model of an $\alpha\Omega$-dynamo in a low-mode approximation was considered and the conditions for generating a magnetic field were investigated. In this article, we will focus in more detail on the study of the

A. Dmitriev et al. (Eds.): STRPEP 2023, SPEES, pp. 235–242, 2023.
https://doi.org/10.1007/978-3-031-50248-4_24

chaotic regime while remaining within the framework of the model proposed in the paper [1]. As is known, the characteristic properties of an Earth-type dynamo system are chaotic reversals without a significant restructuring of the nature of the motion of a turbulent conducting medium [2] and a strong differential rotation of the object [3–6]. The relevance of the problem of studying approximations simulating a field close to the observed geomagnetic one remains [7]. In the $\alpha\Omega$-dynamo model used, the intensity of a large- and a small-scale (turbulent) generators depends on the Lorentz force. Small-scale turbulence generates the α-effect, therefore the α-effect intensity is changed by the Lorentz force, which is included in the dynamic $\alpha\Omega$-dynamo model through the process $Z(t)$ with the hereditary properties [1,8]

$$Z(t) = \int_0^t \left[e^{-b(t-\tau)} \cos(a(t-\tau))\right] \mathbf{B}^2(\tau) d\tau. \tag{1}$$

In the context of the study, we believe that characteristics of the behavior and the impact of the process $Z(t)$ at the moment of time t depends on the previous history on a time interval $[0, t]$ and the influence of history decreases with distance from the instant of time t. An alternating kernel $J(t) = e^{-b(t)} \cos a(t)$ of the functional (1) satisfies the above conditions. The study uses a rescaled and dimensionless MHD-system [9]. As a unit of rescaled and dimensionless time, we take the damping time of the magnetic field in the absence of external influence. The use of such a time scale makes it possible to simulate the main dynamic characteristics of the field using low-mode spatial approximations.

The purpose of this work is to find regions of unstable solutions with oscillations in which chaotic regimes are modeled as an action result of the quenching process $Z(t)$, and the model parameters values corresponding to these cases. Comparison of the numerical experiment results with the conclusions of the studies [1,9].

2 Stability Conditions of the Solution According to Lyapunov

The full description of the model is presented in the articles [1,9]. Here we will give only the equations necessary to understand the research carried out in this paper. We investigate the stability of the solution of a system of differential equations [1] obtained by applying the Galerkin method to a dimensionless and rescaled MHD-system [7]

$$\begin{aligned}
\frac{\partial u(t)}{\partial t} &= -P_m u(t) \sum_s \alpha_s^2 \lambda_s + f_{out} + \sum_{s,\, i,\, j} \alpha_s L_{sij} B_j(t) B_j(t), \\
\frac{\partial B_i(t)}{\partial t} &= Re_m u(t) \sum_{s,\, j} \alpha_s W_{isj} B_j(t) - \mu_i B_i(t) + (R_\alpha - Z) \sum_j W_{ij}^\alpha B_j(t), \\
\frac{\partial Z(t)}{\partial t} &= \sum_j B_j^2(t) - bZ(t) - aZ_s(t), \quad \frac{\partial Z_s(t)}{\partial t} = aZ(t) - bZ_s(t), \\
& i,\, j = \{1,\, 2,\, 3\},\ s = \{1,\, 2,\, 3,\, 11,\, 13\},
\end{aligned} \tag{2}$$

where the velocity field amplitude is $u(t)$, the i-th component amplitude of the magnetic field induction is $B_i(t)$; Reynolds number Re_m, measure of α-effect R_α, damping coefficient b and frequency a are variable parameters of the system (2) that take values from the intervals $Re_m \in [10^{-1}, 10^3]$, $R_\alpha \in [10^{-1}, 10^2]$, $a, b \in [0.1, 10]$.

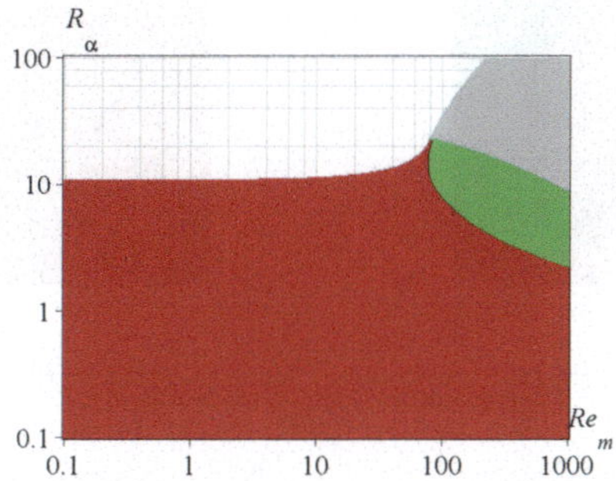

Fig. 1. Distribution of solutions of a linearized system (2) on the phase plane (Re_m, R_α) (on a double logarithmic scale) [1]. Regions of stable solutions are red – without oscillations and green – with oscillations. Regions of unstable solution are white – without oscillations and light-gray – with oscillations.

The study of the system (2) for the stability of the solution is carried out numerically by the predictor-corrector method with initial conditions $u(0) = 1$, $B_2^T(0) = 0$, $B_1^P(0) = 1$, $B_3^P(0) = 0$, $Z(0) = 0$, $Z_s(0) = 0$ [1,9]. In the paper [1], the results were obtained in the study of the stability of the solution of the linearized system (2) when the velocity field is constant ($u = 1$) and without effect on the turbulent generator (without the α-quenching), which are shown in the Fig. 1. Based on the research objectives, we are interested in the region of unstable solutions with oscillations (light-gray region). A divergent magnetic field is obtained in it for a linearized system. Let's consider how the distribution will change if we take into account the α-quenching.

We use the same algorithm as in the article [1]. Namely, at each step of the numerical scheme, considering the control parameters Re_m and R_α fixed and taking the values $u(t)$ and $Z(t)$ as constant, the stability of the solution according Lyapunov was calculated. The number of steps, i.e. the volume of the resulting sample, is 1584620 values. The solution of the system (2) is accepted as stable if the relative frequency ω of obtaining such a solution in the sample exceeds the threshold value 0.5. The frequency ω is considered as a probabilistic characteristic of the solution stability of the system (2) at fixed values of control parameters.

In the current study, additional conditions have been introduced. Namely, the range of values $\omega \in [0, 1]$ is divided into four intervals. If the solution is unstable (with/without oscillations), i.e. $\omega < 0.5$, then three cases are considered: $\omega = 0$ (light-gray/blue region), $\omega \in (0, 0.25]$ (gray/light-blue region), $\omega \in (0.25, 0.5)$

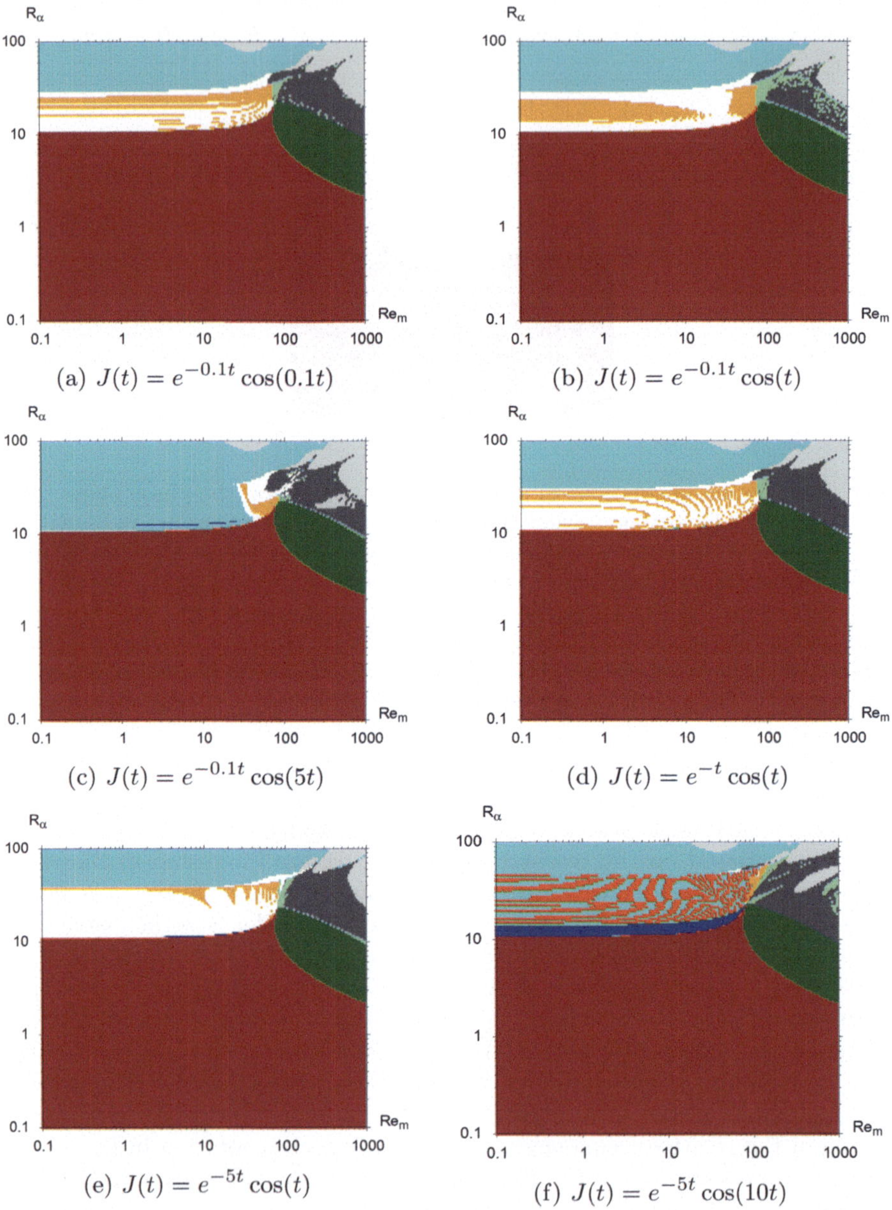

Fig. 2. Distribution of stable and unstable solutions of the system (2) according to the Lyapunov criterion on the phase plane (Re_m, R_α) (on a double logarithmic scale).

(dark-gray/white region). If the solution is stable (with/without oscillations), i.e. $\omega \geq 0.5$, then similarly: $\omega \in [0.5, 0.75)$ (light-green/orange region), $\omega \in [0.75, 1)$ (green/red region), $\omega = 1$ (dark-green/dark-red region).

The results of numerical calculation of the stability characteristic ω of the system of differential equations (2) for some values of a and b parameters of the functional (1) are shown in Fig. 2. In accordance with the research purposes the regions of unstable solutions with oscillations are of interest. In comparison with the picture on the phase plane in Fig. 1, the region of unstable solutions with oscillations (light-gray region, Fig. 1) was divided into two subdomains (gray and dark-gray regions, Fig. 2). In these regions, a chaotic regime may appear as a result of α-quenching. Consequently, we obtain primary constraints on the values of the control parameters $Re_m \geq 100$ and $R_\alpha \geq 10$, determined by the regions of unstable solutions with oscillations. There is note that when using the quenching process $Z(t)$, there are no regions with a characteristic $\omega = 0$ for the solutions with oscillations at the specified limitations of the control parameters. In regions with $\omega \in (0, 0.25)$ (the gray area in the Fig. 2), there are no significant changes when changing the values of the parameters a and b. Significant changes occur in the regions where the stability characteristic belongs to the interval $\omega \in [0.25, 0.5)$ (dark-gray region on the Fig. 2). Increasing the value of the parameter a to one reduces the region due to the appearance of regions of stable solutions with $\omega \in [0.5, 0.75)$ (light-green regions) (Fig. 2a, b). A further increase in the values of a has the opposite tendency (Fig. 2b, c). If we increase the value of the parameter b (Fig. 2b, d, e), then the regions of stable solutions with $\omega \in [0.5, 0.75)$ (light-green regions) disappear inside of the considered region of unstable solutions (dark-gray). The obtained changes in the region of unstable solutions with oscillations allow us to declare that the divergence of the magnetic field is suppressed by the process $Z(t)$, provided that the stability characteristic $\omega \in [0.25, 0.5)$ and the values $R_\alpha \leq 50$.

We investigate the simulated regimes of the field obtained in the region under consideration. The magnetic field generation regime is classified as chaotic according to the Benettin algorithm (positivity of Lyapunov characteristic exponents) [10]. Numerical simulation gives only a divergent magnetic field in the region with the characteristic $\omega \in (0, 0.25)$ (gray region in the Fig. 2). In a region with a stability characteristic $\omega \in [0.25, 0.5]$ for a kernel function with parameters $a = b = 0.1$, the divergence regime of the magnetic field and a dynamo-burst one were obtaned by numerical experiment. Increasing the values of the parameter a to one leads to the appearance of either a chaotic regime, which gradually turns into a steady regime, with small amplitude of oscillations in the velocity field (Fig. 3a), or a chaotic regime with large amplitude of oscillations (Fig. 3b). A further increase in the a values leads only to an increase in the oscillations amplitude of the velocity field. Increasing the parameter b values adds a chaotic regime with minor oscillations amplitude of the velocity field at $b > 5$ (Fig. 3d).

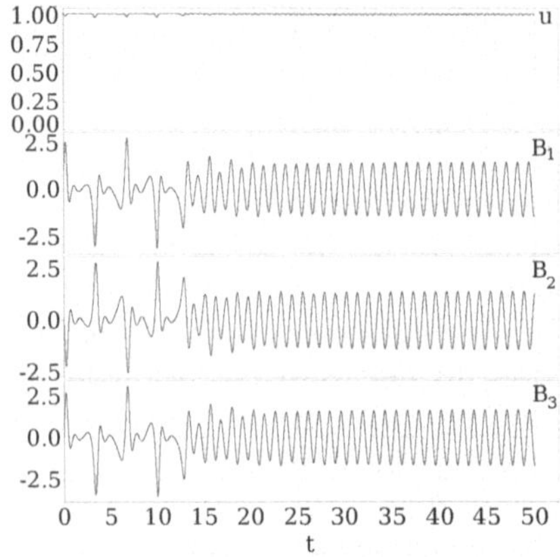

(a) $J(t) = e^{-0.1t}\cos t,\ Re_m = 87,\ R_\alpha = 23$

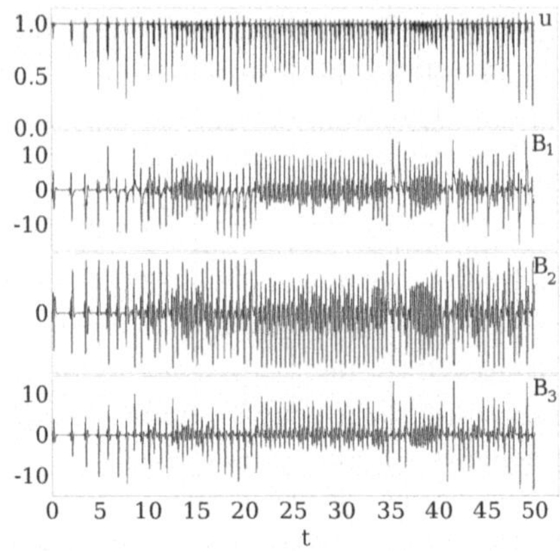

(b) $J(t) = e^{-0.1t}\cos t, Re_m = 398, R_\alpha = 25$

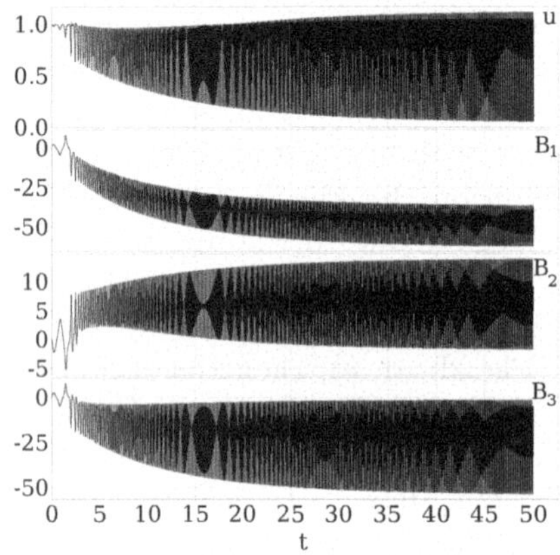

(c) $J(t) = e^{-0.1t}\cos 5t, Re_m = 87, R_\alpha = 23$

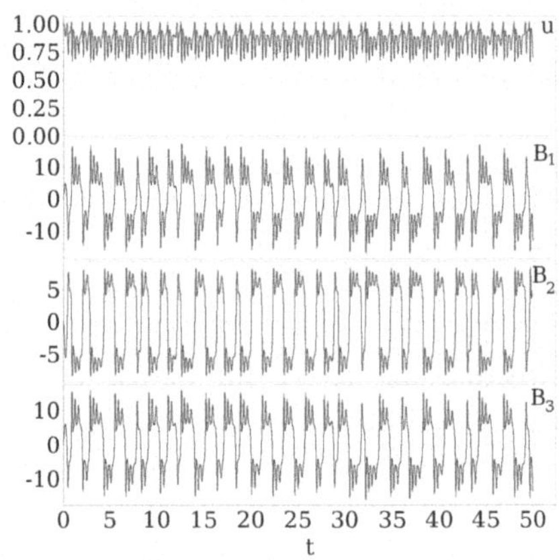

(d) $J(t) = e^{-5t}\cos 10t,\ Re_m = 101,\ R_\alpha = 30$

Fig. 3. Some type of magnetic field regimes.

3 Discussion

Investigation of the solution stability of the system (2) divides the region of unstable solutions with oscillations (light-gray region, Fig. 1) into two subdomains (gray and dark-gray regions, Fig. 2). In the subdomain with $\omega \in (0, 0.25)$, the divergence of the magnetic field remains. Chaotic and regular regimes of magnetic field generation arise in the subdomain with the characteristic $\omega \in [0.25, 0.5)$. Therefore, this subdomain can be considered transitional between the regions of stable $\omega = 1$ and unstable $\omega \in (0, 0.25)$ solutions with oscillations.

The picture on the phase plane vary depending on both the control parameters values and the kernel parameters. An increase in any of the parameters entails an increase in the amplitude of the oscillations in the consideration fields (Fig. 3). If the a parameter value is close to zero, then the pictures on the phase planes that obtained in the case of choosing as the kernel one of the function $J(t) = e^{-bt}\cos(at)$ [1] or $J(t) = e^{-bt}$ [9], differ little. The increase in the a parameter value from the interval [0.1, 10] leads to an increase in the oscillation frequency, and at $a > 1$ also the oscillation amplitude increases (for comparison Fig. 3a, c). An increase in the impact of large- or small-scale generators, expressed in an increase in the values of control parameters, is also accompanied by an increase in the oscillations amplitude in the velocity field (for comparison Fig. 3a, b).

Varying the parameters allowed us to obtain a chaotic regime in this model with small changes in the amplitude of the velocity field (Fig. 3d). In some cases, the initially chaotic behavior changes to a regular one, as, for example, in Fig. 3a. For parameters a and b values close to zero with Reynolds number values of the order of 100, the initially chaotic behavior of the field is changed by a regular regime. An increase in the Reynolds number leads to the predominance of the chaotic regime and the transition to the divergence of the field. Increasing the values of the α-effect measure to 10–30 supports the simulation of regular regimes (Fig. 3a), but a further increase leads either to a chaotic regime (Fig. 3b) or to a divergence of the field. This trend persists for all values of parameters a and b.

In the model under consideration, the simulation of a chaotic regime is often accompanied by the occurrence of a large amplitude oscillation in the velocity field at parameters values $a > 1$ and $b > 1$, but without changing the sign of $u(t)$. Consequently, in the model under consideration, reversals are realized in the absence of changes in the structure of convection.

4 Conclusion

The including into the MHD-system of the process $Z(t)$, which is set by the functional (1) of the alternating kernel, allowed within the framework of the $\alpha\Omega$-dynamo model to quench the turbulent α-effect.

The use of the Lyapunov criterion bounded the region of possible localization of the chaotic regime to the values of the stability characteristic ω from the interval $[0.25, 0.5)$ (dark-gray region in Fig. 2). The boundaries of this region set following limits on the values of the control parameters $Re_m \in [70, 1000]$ and $R_\alpha \in [10, 50]$. Chaotic regimes were simulated in this region (Fig. 3b), including those with insignificant oscillations in the velocity field (Fig. 3d), compared with the previously obtained results [9].

Acknowledgments. The work was carried out within the framework of realization of the State task № AAAA-A21-121011290003-0.

References

1. Sheremetyeva, O.: Magnetic field dynamical regimes in a large-scale low-mode $\alpha\omega$-dynamo model with hereditary α-quenching by field energy. Mathematics **11**(10), 2297 (2023). https://doi.org/10.3390/math11102297

2. Merril, R.T., McElhinny, M.W., McFadden, P.L.: The Magnetic Field of the Earth: Paleomagnetism, the Core, and the Deep Mantle. Academic Press, London (1996)
3. Parker, E.N.: Hydromagnetic dynamo models. Astrophys. J. **122**, 293–314 (1955)
4. Steenbek, M., Krause, F.: Zur dynamotheorie stellarer und planetarer magnetfelder i. berechnung sonnenähnlicher wechselfeldgeneratoren. Astron. Nachr. (291), 49–84 (1969)
5. Zeldovich, Y.B., Rusmaikin, A.A., Sokoloff, D.D.: Magnetic Fields in Astrophysics. The Fluid Mechanics of Astrophysic and Geophysics. Gordon and Breach, New York (1983)
6. Krause, F., Rädler, K.H.: Mean-Filed Magnetohydrodynamics and Dynamo Theory. Pergamon Press, Oxford (1980)
7. Gledzer, E.B., Dolzhanskiy, F.V., Obukhov, A.M.: Hydrodynamic Type Systems and Their Application. Nauka, Moscow (1981)
8. Godomskaya, A.N., Sheremetyeva, O.V.: The modes of magnetic field generation in a low-mode model of $\alpha\omega$-dynamo with α-generator varying intensity regulated by a function with an alternating kernel. In: EPJ Web of Conferences, vol. 254, p. 02015 (2021). https://doi.org/10.1051/epjconf/202125402015
9. Sheremetyeva, O.V., Godomskaya, A.N.: Modelling the magnetic field generation modes in the low-mode model of the $\alpha\omega$-dynamo with varying intensity of the α-effect. Bull. South Ural State Univ. Ser. "Math. Model. Program. Comput. Softw." **14**(2), 27–38 (2021). https://doi.org/10.14529/mmp210203
10. Benettin, G., Galgani, L., Giorgilli, A., Strelcyn, J.-M.: Lyapunov characteristic exponents for smooth dynamical systems and for Hamiltonian systems: a method for computing all of them. Part 1: theory. Meccanica **1**(15), 9–20 (1980). https://doi.org/10.1007/BF02128236

Physics of Earthquake Precursors

Disturbances in E and F Layers of the Ionosphere Preceding Earthquakes in the Kamchatka Region

Vadim Bogdanov and Aleksey Pavlov(✉)

Institute of Cosmophysical Research and Radio Wave Propagation FEB RAS, Paratunka, Russia
pavlov@ikir.ru

Abstract. Anomalous changes in the parameters characterizing the state of the ionospheric layers E and F, observed before the onset of seismic events, can be considered as possible ionospheric precursors of these earthquakes. In this work, according to the data obtained at the ionospheric station of vertical radio sounding, located in Paratunka (52.97° N, 158.24° E), an analysis of ionospheric disturbances recorded before the onset of earthquakes with magnitude $M \geq 5.0$ that occurred in the Kamchatka region was carried out. Anomalous values of the limiting reflection frequency $foEs$, the screening frequency $fbEs$ and the virtual height $h'Es$ of the sporadic Es layer, as well as the critical frequency $foF2$ of the $F2$ layer and the virtual height $h'F$ of the F layer in the absence of geomagnetic disturbances were considered as possible ionospheric precursors of earthquakes. The predictive efficiency of the anomalous behavior of the complex of considered ionospheric parameters for earthquakes with $M \geq 5.0$ is estimated.

Keywords: ionosphere · earthquake precursor · earthquake

1 Introduction

The development of methods for short-term (hours-days) forecasting of strong earthquakes is important scientific task. Searches for short-term precursors of earthquakes in the atmosphere and ionosphere may be useful for solving this problem. According to numerous studies, anomalous variations in ionospheric parameters were observed both before the onset of earthquakes [1–3] and after them [4–6]. The works [7–9] described the possible physical causes of the formation of disturbances in the ionosphere that precede the onset of earthquakes. However, despite the large amount of work carried out on the study of lithospheric-ionospheric relations, the question of the existence of ionospheric precursors of earthquakes is still far from its final answer. This is largely due to the difficulties of unambiguous interpretation of ionospheric data in each specific case. Seismic-ionospheric effects, as a rule, are small in amplitude, and the ionosphere itself is highly variable and subject to the influence of various heliogeophysical factors and especially magnetic disturbances. Thus, precursors in

A. Dmitriev et al. (Eds.): STRPEP 2023, SPEES, pp. 245–254, 2023.
https://doi.org/10.1007/978-3-031-50248-4_25

the ionosphere can be identified only with some probability, because a significant proportion of ionospheric anomalies, even in calm geomagnetic conditions, are not accompanied by earthquakes [10]. A more reliable way to identify a possible ionospheric precursor of an earthquake is to analyze the anomalous values of a set of ionospheric parameters. The purpose of this work is to identify ionospheric disturbances in the E and F regions, compare them to the Kamchatka earthquakes with magnitudes $M \geq 5.0$ for 2014–2022, and evaluate the predictive efficiency of these ionospheric disturbances.

2 Data Analysis Methodology

As noted in [11,12], in the sporadic Es layer before earthquakes, in addition to an increase in the limiting reflection frequency, a number of specific effects are observed: sharp increase in the screening frequency, an increase in the semi transparency range, the appearance of diffuse reflections. But none of these effects is observed consistently for all earthquakes of a certain class. Typically, the formation of a sporadic Es layer in mid-latitudes is observed at altitudes of 100–110 km [13]. However, according to the results obtained, presented in [14,15], several days before the onset of earthquakes, the formation of sporadic Es layers can be observed at significantly higher altitudes, exceeding the median values by tens of kilometers. In addition, high-lying sporadic layers are often accompanied by an increase in the values of the frequency parameters of the Es and $F2$ layers lasting several hours, which are observed during the same daily interval. Thus, these anomalous variations in ionospheric parameters can be attributed to the supposed precursors of upcoming earthquakes.

In this paper, anomalous temporal variations of the following ionospheric parameters will be considered as possible earthquake precursors:

- $h'Es$ is the smallest virtual height of the sporadic Es-layer for an ordinary wave;
- $foEs$ is the limiting frequency of the ordinary wave of the sporadic Es-layer of the ionosphere;
- $fbEs$ is the screening frequency of the ordinary wave of the sporadic Es-layer of the ionosphere;
- $foF2$ is the critical frequency of the ordinary wave of the $F2$-layer of the ionosphere;
- $h'F$ is the minimum virtual height of the ordinary wave reflection trail for the F region of the ionosphere.

The presented study uses hourly values of ionospheric parameters obtained from radiophysical observations at a vertical radio sounding station of the ionosphere located in the village Paratunka ($\varphi = 52.97°$ N, $\lambda = 158.24°$ E). Accounting for the level of geomagnetic activity was carried out according to the values of K-indices measured at the complex geophysical observatory (GPhO) "Paratunka" ($\varphi = 52.97°$ N, $\lambda = 158.25°$ E), which is part of the International Network magnetic observatories INTERMAGNET.

To determine seismoionospheric effects in diurnal variations of ionospheric parameters, it is necessary to obtain background distributions. Also, when analyzing variations in the ionosphere, an important role is played by the presence or absence in the considered period of time of significant geomagnetic disturbances that can affect the ionosphere.

The technique for identifying seismoionospheric anomalies used in this work is as follows. For each moment of the day of each ionospheric parameter $X(t_i)$ on the previous interval of duration $T_{\text{med}} = 30$ days, the median $X_{\text{med}}(t_i)$ was calculated. At the same time, magnetically quiet days were chosen as the background, and the median was calculated only for days in which the values of the geomagnetic index were $K \leq 2$ in all three hourly intervals. To determine the measure of deviation from a quiet background distribution, the parameter of interquartile range IQR was used [6,16,17]:

$$Y_{\pm} = X_{\text{med}}(t_i) \pm 1.5IQR \tag{1}$$

where $X_{\text{med}}(t_i)$ is the median calculated for selected magnetically quiet days, and $IQR = Q_3 - Q_1$ is the difference between the third (upper) Q_3 and the first (lower) Q_1 quartiles, which is also calculated for days with low geomagnetic activity. Then the values of $Y_{\pm}$ limit the range of values of $X(t_i)$, which appear in this range with some probability due to the influence of a complex of random factors. If the distribution of $X(t_i)$ values corresponds to a normal distribution, then the value of $1.5IQR$ will be approximately equal to two standard deviations [18]. Thus, with a 95% probability, the values of $X(t_i)$ will appear within the boundaries of $Y_{\pm}$ due to the influence of various random factors, and outside the boundaries of $Y_{\pm}$ the probability of occurrence of values of $X(t_i)$ will be 5%. Values $X(t_i)$ outside the "noise" bands were referred to as anomalous values if the duration of such a disturbance in time was at least one and a half hours (taking into account the averaging over three points).

To assess the efficiency of predicting possible ionospheric precursors of earthquakes, the following parameters were calculated: the reliability of the precursor R, the veracity of the precursor V, the efficiency of the precursor J_G according to the method of A.A. Gusev, the effectiveness of the precursor of J_M according to the method of G.M. Molchan. These parameters have already been used by the authors of the article to assess the predictive effectiveness of other possible earthquake precursors in the Kamchatka region [19–21].

The reliability of the precursor R is defined as the ratio of the number of earthquakes N_+ for which the precursor was identified to the number of all earthquakes N [22]:

$$R = N_+/N \tag{2}$$

The veracity of a precursor V is defined as the ratio of the number of precursor anomalies $n(A_E)$ to the total number of identified anomalies $n(A)$ [22]:

$$V = n(A_E)/n(A) \tag{3}$$

The efficiency of earthquake prediction according to the method of A.A. Gusev is calculated for a specific spatial area and a certain energy range of earthquakes according to the formula [23]:

$$J_G = \frac{N_+/T_{\text{alarm}}}{N/T} \tag{4}$$

where T is the total time of seismic monitoring; N_+ is the number of earthquakes corresponding to a successful forecast over time T; N is the total number of earthquakes (having spatio-temporal characteristics similar to those predicted) that occurred during time T; T_{alarm} is the total alarm time (the total duration of all time intervals in which the forecast was valid according to the estimated method during the total monitoring time). The J_G efficiency shows how many times the flow velocity of predicted earthquakes exceeds the average over the observation time T, i.e. is the ratio of the probability of an earthquake at the time of the alarm to their average probability. In the absence of a "precursor-earthquake" connection, i.e. with random guessing, the efficiency of J_G is 1.

The efficiency of earthquake prediction by the method of G.M. Molchan [24] is determined by the formula:

$$J_M = 1 - \nu - \tau \tag{5}$$

where $\tau = T_{\text{alarm}}/T$ – alarm measure; $\nu = 1 - N_+/N$ – the share of the misses of the target. For a random prediction $J_M = 0$, and for the ideal (without a miss target and with zero alarm time) – $J_M = 1$. A diagonal $\tau + \nu = 1$ is built on the error diagrams, which connects the points (0;1) and (1;0) and corresponds to a random forecast. For this diagonal, confidence intervals are constructed for different significance levels α. The forecast on the diagram is marked with a dot with coordinates (τ, ν). If the point lies under the lower boundary of the confidence interval, then this can be interpreted as a high degree of reliability of the revealed relationship between the considered precursor and earthquakes of the considered energy range.

3 Results of the Analysis of the Predictive Efficiency of Ionospheric Parameters

When analyzing the predictive efficiency of the anomalies of each ionospheric parameter separately, they were considered as a possible ionospheric precursor of earthquakes, provided that their values go beyond the limits of $K_\pm$. In addition, within a daily time interval (±12 h relative to the moment of occurrence of anomalous values), the levels of three-hour values of the geomagnetic activity index were $K \leq 2$. The results of calculations of the parameters V, R, J_G and J_M for earthquakes with magnitudes $M \geq 5.0$ that occurred over the time interval T =2014–2022 at depths up to 100 km and at distances up to $r = 1000$ km from the location of the vertical radio sounding station of the ionosphere (geographic coordinates $\varphi = 52.97°$ N, $\lambda = 158.25°$ E) in Kamchatka are presented in Table 1. This table also contains estimates for the following quantities: the expecting time for earthquakes T_{exp} from the moment of identification of anomalous values of ionospheric parameters; the number of predicted earthquakes N_+ that occurred during T; the total number of earthquakes with $M \geq 5.0$ that occurred during time T; the number of anomalies of the ionospheric parameter $n(A_E)$, after which, during the expecting time T_{exp}, earthquakes with $M \geq 5.0$ occurred; the total number $n(A)$ of identified anomalies of the ionospheric parameter. The catalogue of earthquakes compiled by the Kamchatka branch of the Geophysical Service of the Russian Academy of Sciences (KB GS RAS) [25] was used in this study.

Table 1. Predictive efficiency of anomalous values of ionospheric parameters for earthquakes with magnitude $M \geq 5.0$.

Ion. param	$h'Es$	$foEs$	$fbEs$	$foF2$	$h'F$
T_{exp} (day)	1.99 ± 1.97	2.31 ± 2.26	1.9 ± 2.09	2.74 ± 2.36	2.79 ± 2.63
N_+	320	279	316	253	247
N	482	482	482	482	482
$n(A_E)$	230	204	233	169	162
$n(A)$	1088	895	1213	600	638
R	0.66	0.57	0.65	0.52	0.51
V	0.21	0.23	0.19	0.28	0.25
J_G	1.28	1.2	1.24	1.26	1.26
J_M	0.14	0.098	0.127	0.11	0.1

According to the results obtained (Table 1), the values of the predictive efficiency J_G for the studied ionospheric parameters are greater than 1 and are in the range from 1.2 to 1.28. This indicates a connection between these prognostic parameters and earthquakes with $M \geq 5.0$. It follows from this that the forecast made on the basis of these ionospheric parameters is not accidental. The forecast reliability R (the relative number of predicted earthquakes) according to these parameters is in the range from 0.51 to 0.66. At the same time, these prognostic anomalies do not have a sufficiently high veracity V (the relative number of anomalies after which an earthquake occurred), which takes values from 0.19 to 0.28. Therefore, to increase the efficiency and reliability of the forecast, as well as to reduce the number of false alarms, it is necessary to use a set of prognostic parameters.

The method for identifying a possible ionospheric precursor of earthquakes with $M \geq 5.0$ based on a set of prognostic parameters was built according to the following procedure:

1. On the time interval T, anomalous values $h'Es$ with a duration of $\Delta t \geq 1.5$ hours are identified, which go beyond the background values of $Y_\pm$.
2. Deviations with a duration $\Delta t \geq 1.5$ hours of additional ionospheric parameters $foEs$, $fbEs$, $foF2$ and $h'F$ from the boundaries of their background $Y_\pm$ ranges are determined within the daily time interval ΔT (± 12 hours relative to the moment of occurrence of anomalous $h'Es$ values).

3. The group of ionospheric parameters was considered as a possible short-term ionospheric precursor of earthquakes with magnitude $M \geq 5.0$ if low geomagnetic activity was observed in the daily time interval ΔT (all three-hour values of the geomagnetic index were $K \leq 2$) and at least three of the four additional parameters were anomalous deviations of their values from the boundaries of the background range $Y_{\pm}$ were found.

As an example, Figs. 1-2 show the changes in the values of the time series of the ionospheric parameters under consideration, their median values and the boundaries of the range of background $Y_{\pm}$ values that preceded the onset of earthquakes.

Figure 1 shows that against the background of low geomagnetic activity (from 08.08.2018 to 10.08.2018 the three-hour K-index did not exceed 2), anomalous values of the parameters $h'Es$, $foEs$, $fbEs$, $foF2$ and $h'F$ were identified before the earthquake, occurred on 10.08.2018 18:12:03 (UT) with epicenter coordinates $\varphi = 48.15°$ N, $\lambda = 155.27°$ E, magnitude $M = 6.5$, hypocenter depth $h = 60$ km, and epicentral distance $r = 575$ km to the location of the station for vertical radio sounding of the ionosphere.

Figure 2 shows the anomalous values of ionospheric parameters identified on 29.07.2022 before the earthquake that occurred on the same day, 29.07.2022 23:05:43 (UT) with epicenter coordinates $\varphi = 51.2°$ N, $\lambda = 161.03°$ E, magnitude $M = 5.9$, hypocenter depth $h = 68.9$ km and epicentral distance $r = 273$ km to the location of the ionospheric station. At the same time, the values of the three-hour index of geomagnetic activity during the day on 29.07.2022 were $K \leq 2$.

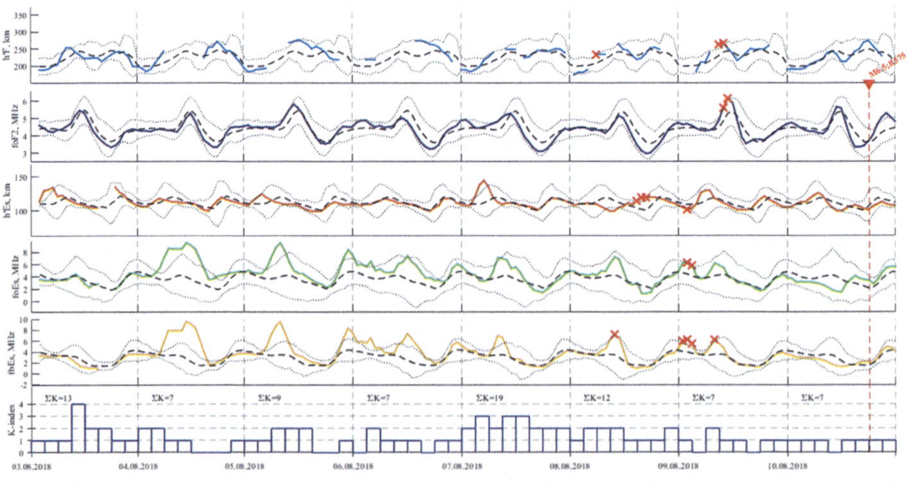

Fig. 1. Time series of values of ionospheric parameters $foEs$, $fbEs$, $h'Es$, $foF2$, $h'F$, their median values (dashed lines) and boundaries of the background range of values (dotted lines) for the time interval 03.08.2018–10.08.2018. Identified anomalies are marked with a red 'x' marker. The earthquake with magnitude $M = 6.5$ that occurred on 10.08.2018 is marked on the time axis with a red triangle (Color figure online).

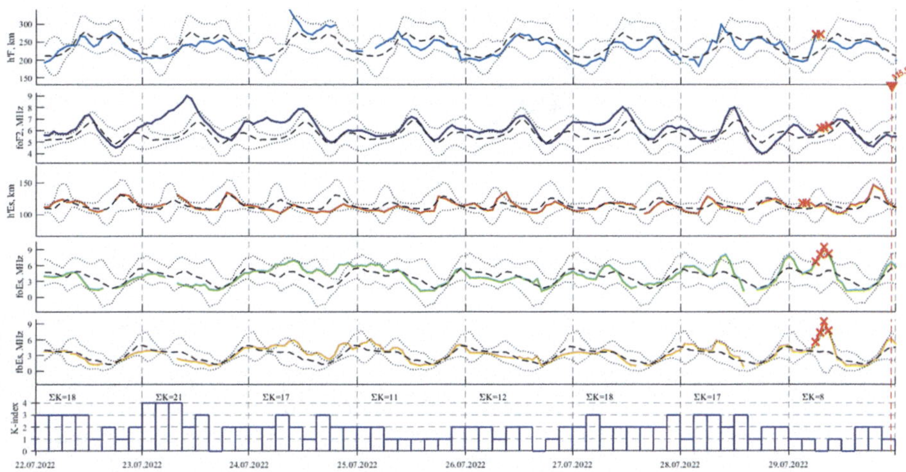

Fig. 2. Time series of values of ionospheric parameters $foEs$, $fbEs$, $h'Es$, $foF2$, $h'F$, their median values (dashed lines) and boundaries of the background range of values (dotted lines) for the time interval 22.07.2022–29.07.2022. Identified anomalies are marked with a red 'x' marker. The earthquake with magnitude $M = 5.9$ that occurred on 29.07.2022 is marked on the time axis with a red triangle (Color figure online).

The results of a retrospective estimation of the predictive efficiency of the method for identifying a possible ionospheric precursor using a set of ionospheric parameters for earthquakes with magnitudes $M \geq 5.0$ that occurred at depths of up to 100 km and at epicentral distances of up to 1000 km from the ionospheric observation point in Kamchatka are presented in Table 2 and Fig. 3.

Table 2. Predictive efficiency of the complex of ionospheric precursors.

T_{exp} (day)	N_+	N	$n(A_E)$	$n(A)$	R	V	J_G	τ	ν	J_M
2.9 ± 2.3	156	482	100	214	0.32	0.47	1.49	0.22	0.68	0.106

According to the results obtained (Table 2), when predicting earthquakes with a magnitude of $M \geq 5.0$ using a complex of ionospheric anomalies, the reliability of the forecast R was 0.32 (i.e., 32% of earthquakes with $M \geq 5.0$ were preceded by a complex of ionospheric disturbances), and the veracity of the forecast was $V = 0.47$ (i.e. after 47% of the detected anomalies during T_{exp}, earthquakes with $M \geq 5.0$ occurred). The expecting time for an earthquake with $M \geq 5.0$ when identifying a complex of ionospheric anomalies was $T_{\text{exp}} = 2.9 \pm 2.3$ days. The efficiency value $J_G = 1.49$ indicates that the prediction by this method is statistically significant and differs from random guessing. Thus, the probability of an earthquake with $M \geq 5.0$ occurring during the waiting period T_{exp} after the appearance of a complex of ionospheric anomalies is 1.49

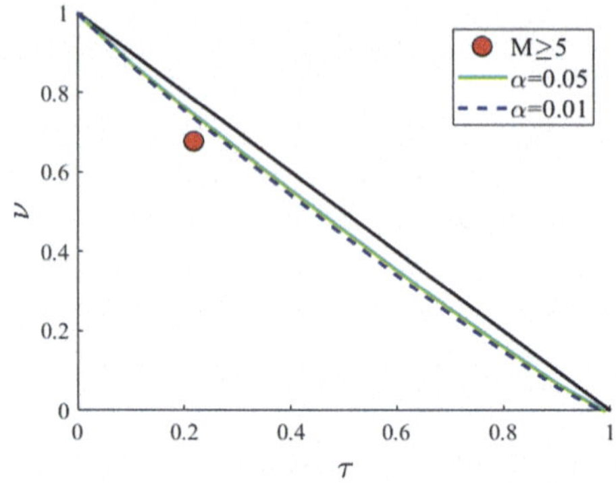

Fig. 3. Diagram of errors for the method of identifying a possible precursor of earthquakes with magnitude $M \geq 5.0$ based on a complex of ionospheric predictive parameters. Confidence intervals are constructed for significance levels $\alpha = 0.01$ and $\alpha = 0.05$.

times higher than during the rest of the time. The total alarm time τ in the forecast using the method under consideration was 22% of the total observation time T =2014–2022. On the error diagram (Fig. 3), the forecast point lies under the lower boundary of the 99% confidence interval, which can be interpreted as a fairly reliable connection between the identified complex of ionospheric anomalies and earthquakes with magnitude $M \geq 5.0$ that occurred at epicentral distances up to 1000 km from the ionospheric observation point.

4 Conclusions

Anomalies in the ionospheric parameters $h'Es$, $foEs$, $fbEs$, $foF2$ and $h'F$, which can be considered as possible earthquake precursors, have been identified. The predictive efficiency of each parameter was evaluated separately for earthquakes with magnitude $M \geq 5.0$ that occurred in the Kamchatka region for the period 2014–2022 at epicentral distances up to 1000 km from the ionospheric observation point. It was found that the predictive efficiency $J_G > 1$ (i.e. differs from random guessing), but the veracity V of each prognostic parameter is not very high and takes values from 0.19 to 0.28.

A retrospective analysis of the reliability and veracity of the methodology for predicting earthquakes with $M \geq 5.0$ based on a complex of ionospheric parameters showed that 32% of earthquakes were preceded by a complex of ionospheric disturbances, while seismic events of the considered energy range occurred within 2.9 ± 2.3 days after 47% of the detected anomalies. An analysis of the predictive efficiency of complex of ionospheric parameters showed that the prediction of earthquakes with a magnitude of $M \geq 5.0$ differs from random guessing. The use of a complex of ionospheric prognostic parameters makes it possible to increase the efficiency of the J_G forecast and reduce the number of false alarms, but does not always increase the reliability of the forecast R.

Further work related to the identification of ionospheric precursors and improving the reliability, veracity and efficiency of short-term earthquake prediction with $M \geq 5.0$ in the Kamchatka region on the basis of the proposed approach can be carried out both by introducing additional criteria for the prognostic parameters, and by involving other ionospheric parameters, which could be considered as the precursors of earthquakes.

Acknowledgments. The work was carried out within the framework of the State task on the topic (2021–2023) "Physical processes in the system of near space and geospheres under solar and lithospheric impact", registration number AAAA-A21-121011290003-0.

The work was realized by the means of the Common Use Center "North-Eastern Heliogeophysical Center" CKP_558279, USU 351757.

References

1. Davies, K., Baker, D.M.: Ionospheric effects observed around the time of the Alaskan earthquake of march 28, 1964. J. Geophys. Res. **70**(9), 2251–2253 (1965)
2. Choosakul, N., Saito, A., Iyemori, T., Hashizume, M.: Excitation of 4-min periodic ionospheric variations following the great Sumatra-Andaman earthquake in 2004. J. Geophys. Res. **114**, A10313 (2009). https://doi.org/10.1029/2008JA013915
3. Hegai, V.V., Legen'ka, A.D., Kim, V.P., Georgieva, K.: Wave like perturbations in the ionospheric F2-layer observed after the Ms8.1 Samoa earthquake of september 29, 2009. Adv. Space Res. **47**(11), 1979–1982 (2011)
4. Hegai, V.V., Legen'ka, A.D., Kim, V.P., Georgieva, K.: Ionospheric precursor of a destructive earthquake that occured on April 6, 2009 at L'Aquila (Italy). Geomag. Aeron. **51**(1), 71–77 (2011)
5. Hegai, V.V., Legen'ka, A.D., Kim, V.P.: Anomalous increase in foF2 critical frequency prior to the Spanish earthquake of may 11, 2011. Geomag. Aeron. **54**(1), 82–86 (2014)
6. Pulinets, S.A., Legen'ka, A.D., Hegai, V.V., Kim, V.P., Korsunova, L.P.: Ionosphere disturbances preceding earthquakes according to the data of ground based station of vertical ionospheric sounding WAKKANAI. Geomag. Aeron. **58**(5), 686–692 (2018)
7. Liperovskiy, V.A., Pokhotelov, O.A., Shalimov, S.L.: Ionosfernye predvestniki zemletryaseniy. Nauka, Moscow (1992). (in Russian)
8. Pulinets, S.A., Boyarchuk, K.A.: Ionospheric Precursors of Earthquakes. Springer, Berlin (2004). https://doi.org/10.1007/b137616
9. Pulinets, S.A., Uzunov, D.P., Davidenko, D.V., Dudkin, S.A., Tsadikovskiy, E.I.: Prognoz zemletryaseniy vozmozhen?! Trovant, Moscow (2014). (in Russian)
10. Perrone, L., Korsunova, L.P., Mikhailov, A.V.: Ionospheric precursors for crustal earthquakes in Italy. Ann. Geophysicae. **28**(4), 941–950 (2010)
11. Liperovskaya, E.V., Pokhotelov, O.A., Oleynik, M.A., Alimov, O.A., Pavlova, S.S., Khakimova, M.: Nekotorye effekty v sporadicheskom sloe E ionosfery pered zemletryaseniem. Fizika Zemli. **11**, 86–88 (1994). (in Russian)
12. Ondoh, T.: Anomalous spradic E layers observed before M7.2 Hyogo-ken Nanbu earthquake; terrestrial gas emanation model. Adv. Polar Upper Atmos. Res. **17**, 96–108 (2003)
13. Chavdarov, S.S., Chasovitin, Y.K., Chernysheva, S.N., Sheftel', V.N.: Sredneshirotnyy sporadicheskiy sloy E ionosfery. Nauka, Moscow (1975). (in Russian)

14. Korsunova, L.P., Khegay, V.V.: Seysmoionosfernye effekty sil'nykh korovykh zemletryaseniy v Tikhookeanskom regione. Geomagnetizm i aeronomiya. **45**(5), 706–711 (2005). (in Russian)
15. Korsunova, L.P., Khegai, V.V.: Medium term ionospheric precursors to strong earthquakes. Int. J. Geomagn. Aeron. **6**, G13005 (2006). https://doi.org/10.1029/2005GI000122
16. Liu, J.Y., Chen, Y.I., Chuo, Y.J., Chen, C.S.: A statistical investigation of pre-earthquake ionospheric anomaly. J. Geophys. Res. **111**, A05304 (2006). https://doi.org/10.1029/2005JA011333
17. Bychkov, V.V., Smirnov, S.E., Korsunova, L.P., Hegai, V.V.: Atmospheric anomalies and anomalies of electricity in the near-surface atmosphere before the Kamchatka earthquake of January 30, 2016, based on the data from the Paratunka observatory. Geomag. Aeron. **57**(4), 491–499 (2017)
18. Klotz, S., Johnson, N.L.: Encyclopedia of Statistical Sciences. John Wiley, Hoboken (1983)
19. Bogdanov, V., Pavlov, A.: Analysis of the efficiency of earthquake prediction based on the anomalous behavior of ionospheric parameters on the eve of earthquakes in the Kamchatka region. In: E3S WEB of Conferences, vol. 62, no. 03001 (2018) https://doi.org/10.1051/e3sconf/2018620
20. Bogdanov, V.V., Pavlov, A.V.: A methodology to estimate a region and waiting period for strong Kamchatka earthquakes based on seismic and ionospheric predictive signs. Vestnik KRAUNC. Fiz.-Mat. Nauki. **30**(1), 59-78 (2020) https://doi.org/10.26117/2079-6641-2020-30-1-59-78
21. Bogdanov, V., Gavrilov, V., Pulinets, S., Ouzounov, D.: Responses to the preparation of strong Kamchatka earthquakes in the lithosphere-atmosphere-ionosphere system, based on new data from integrated ground and ionospheric monitoring. In: E3S WEB of Conferences, vol. 196, no. 03005 (2020). https://doi.org/10.1051/e3sconf/202019603005
22. Saltykov, V.A.: On the possibility of using the tidal modulation of seismic waves for forecasting earthquakes. Izvestiya, Phys. Solid Earth. **53**(2), 250–261 (2017)
23. Gusev, A.A.: Prognoz zemletryaseniy po statistike seysmichnosti. In: Seysmichnost' i seysmicheskiy prognoz, svoystva verkhney mantii i ikh svyaz' s vulkanizmom na Kamchatke, pp. 109-119. Nauka, Novosibirsk (1974). (in Russian)
24. Molchan, G.M.: Strategies in strong earthquake prediction. Phys. Earth Planet. Inter. **61**, 84–98 (1990)
25. Chebrova, A.Yu., Chemarev, A.S., Matveenko, E.A., Chebrov, D.V.: Seismological data information system in Kamchatka branch of GS RAS: organization principles, main elements and key functions. Geophys. Res. **21**(3), 66-91 (2020) https://doi.org/10.21455/gr2020.3-5. (in Russian)

Nonextensive Analysis of Natural and Technogenic Seismicity of Sakhalin Island

V. N. Sychev[1(✉)], L. M. Bogomolov[1], D. V. Kostylev[1,2], and N. V. Kostyleva[1]

[1] The Institute of Marine Geology and Geophysics of the far Eastern branch of the Russian Academy of Sciences, Nauki St. 1B, 693022 Yuzhno-Sakhalinsk, Russia
koitash@mail.ru

[2] Sakhalin Branch, Geophysical Survey, Russian Academy of Sciences, Yuzhno-Sakhalinsk Russia, Tikhookeanskaya St, 2-a, Yuzhno-Sakhalinsk 693010, Russia

Abstract. The Sakhalin Island earthquake catalog is considered from the standpoint of non-extensive statistical physics (NESP). The analysis is based on the concept of entropy, which was introduced in 1988 by Constantino Tsallis as a generalization of the standard Boltzmann–Gibbs entropy. To describe the distribution functions of earthquakes, a modified stick-slip earthquake source model was used - stick-slip of two plates relative to each other along a fault in the presence of friction and filling fragments between the fault surfaces and the Tsallis entropy maximum principle. It is shown that the earthquake flux is a system with memory and long-range spatial correlations, and the calculated values of the Tsallis parameter $q \sim 1.5$ almost coincide with the values obtained for the magnitudes of the catalogs of various seismically active regions. At the same time, when analyzing the flow of events related to technogenic impacts (explosions), it was determined that the value of the Tsallis parameter q is lower than that calculated from the earthquake catalog. Such areas where blasting is carried out are characterized by a low value of the Tsallis parameter q.

Keywords: earthquake · industrial explosion · non-extensive statistical physics · energy distribution function of earthquakes · long-range correlations · Tsallis parameter

1 Introduction

The paper [1] outlines the stages in the development of regional seismology and the results obtained. In particular, it was found that in the region of the Kuril Islands and the Sea of Okhotsk, earthquakes are concentrated mainly in the inclined seismic focal zone (SFZ) with a thickness of about 70 km, extending under the mainland to depths of about 650 km. Here, seismic activity reaches almost the maximum level on Earth. In the last 100 years alone, about a dozen earthquakes with M ~ 8 or more have occurred on the Kuril Islands. The upper edge of the SFZ comes out on the day surface on the continental slope near the deep-sea trench, 60–70 km east of the islands. The width of the epicentral region in the area of the outlet of the SFZ is about 100 km. The vast

A. Dmitriev et al. (Eds.): STRPEP 2023, SPEES, pp. 255–265, 2023.
https://doi.org/10.1007/978-3-031-50248-4_26

majority of earthquakes that cause 7–9 magnitude shaking on the islands are located at depths up to 200 km. About 55% of earthquakes in the zone occur in the depth interval of 30–50 km. The hypocenters of the strongest earthquakes are located at a distance of 60–160 km from the Kuril Islands, which weakens the effect of shaking from them to 8–9 points (instead of 10–11 points at the epicenter). Deep-focus earthquakes make up only about 10% of the total number and release less than 1% of the energy in the sources of all earthquakes in the zone [1].

In the territory under consideration, technogenic seismicity associated with the extraction of minerals is also observed. At present, coal mining in the region is carried out in an open way in several areas. The most active mining is carried out at the Solntsevsky coal mine (SCM) of the Eastern Mining Company [2]. A large number of explosions can serve as a tool for seismic exploration, as well as act as a trigger mechanism for earthquakes or perform the function of stress relief [3, 4]. At the same time, there is the issue of identifying explosions and separating the natural seismic process from technogenic seismicity [5].

The purpose of this work is to evaluate the application of an earthquake model based on the principles of non-extensive statistical physics to the seismic process occurring on the island. Sakhalin. The task also consists in determining how, within the framework of the model used, the natural seismic process differs from man-made impacts, namely from explosions. To solve this problem, the work also considers the catalog of explosions of the Solntsevo coal mine.

2 Initial Data

The earthquake catalog of the Sakhalin branch of the Federal Research Center "Geophysical Survey of the Russian Academy of Sciences" (FRC EGS RAS) since 1997 was used for the study (for example, [6]. The catalog is supplemented annually on the website of the FRC EGS RAS organization (gsras.ru).

In addition, the catalog of explosions of the Solntsevsky coal mine (SCM) of the Eastern Mining Company [2] for 2020–2022 was used for calculations. Figure 1 shows the epicentral positions of earthquakes (8127 events) that occurred from 1997 to 2022. The inset shows the explosions (560 events) of the Solntsevsky coal mine for 2020–2022.

3 Methods

Time series, which are successive events of recorded earthquakes with magnitude values in a wide range, demonstrate complex behavior, which is associated with different scales of ruptures in the earth's crust. In turn, small seismic movements occur much more often than large earthquakes. To describe the statistics of the distribution of earthquakes by energy, the cumulative distribution *N(>M)* is widely used, which indicates the number of earthquakes with a magnitude equal to or greater than *M*, has an exponential behavior and is known as the Gutenberg-Richter law [7]:

$$\lg N(> M) = a - bM, \tag{1}$$

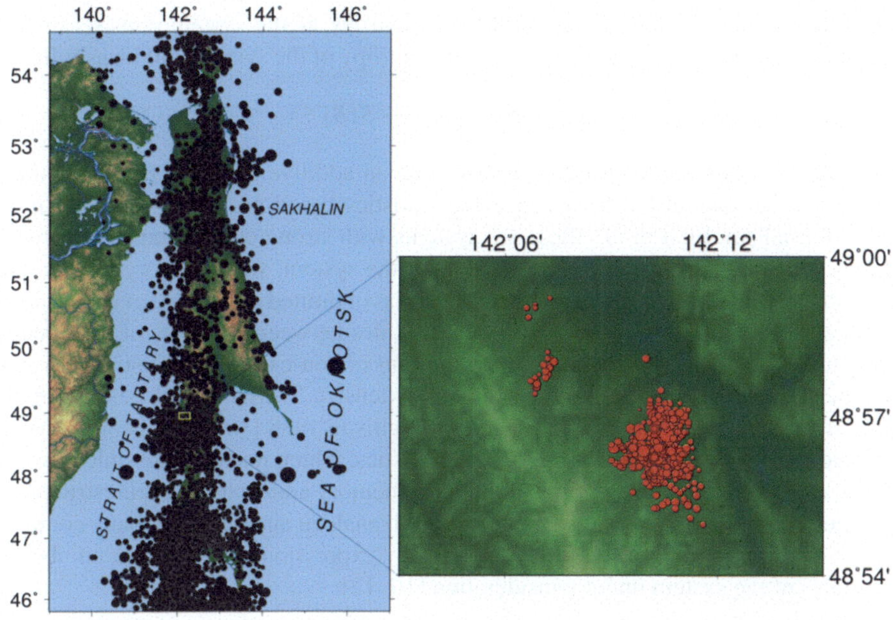

Fig. 1. Epicentral position of earthquakes (8127 events) that occurred from 1997 to 2022. The inset shows the explosions (560 events) of of the Solntsevsky coal mine for 2020–2022.

where $N(>M)$ is the number of earthquakes with magnitudes (or classes) of at least M, a and b are the equation constants. The parameter a (a-value) means seismic activity at $M = 0$, and the slope of the linear part of the graph b determines the rate of decrease in the relative number of events with increasing magnitude. For most regions of the earth, the angular coefficient of the descending linear section of the distribution of earthquakes by energy is $b \approx 0.9$ [8].

It should be noted that dependence (1) is empirical and was obtained without using the positions of equilibrium (classical) thermodynamics, where the state of the system is characterized by entropy.

If p_i is the probability that the system is in state number i ($i = 1, \ldots, N$, $\sum_{i=1}^{N} p_i = 1$), then the entropy of the Boltzmann-Gibbs thermodynamic system is calculated as [9]:

$$S = -k \sum_{i=1}^{N} p_i \ln p_i \tag{2}$$

where k is the Boltzmann constant ($k = 1.38 \cdot 0\text{–}23$ J/K), N is the number of possible states of the system. Boltzmann entropy (2) is characterized by additivity, i.e. the entropy of an equilibrium system is equal to the sum of the entropies of its individual parts, and the change in the entropy of the entire system is equal to the sum of the changes in its parts.

In the case of using the Boltzmann-Gibbs entropy, if the system consists of two independent subsystems A and B, then the total entropy of the system will be as follows:

$$S(A + B) = S(A) + S(B) \tag{3}$$

In statistical physics, entropy is shown to be an additive quantity. However, there are systems that cannot be described using statistics based on the Boltzmann-Gibbs entropy formalism. First of all, these are systems with strong correlations, with strong and long-range interactions between all parts of the system, with memory effects. For such systems, Boltzmann-Gibbs statistical physics is of limited use, and its generalization was necessary to consider correlations over all scales between elements of the system, leading to power-law distributions. The foundations of non-extensive statistical physics, sometimes referred to in the literature as non-extensive Tsallis q-statistics or simply Tsallis entropy, were originally introduced by Tsallis in 1988 [10]. The theory is based on a generalization of the Boltzmann-Gibbs statistics, which makes it possible to study systems with long-range interactions, long-term memory and/or multifractal structures. Non-extensive statistical physics is widely used to analyze and describe many complex dynamic systems, based on the following entropy expression on a discrete number of microstates of the system under consideration [10–12]:

$$S_q = \frac{k}{q-1}\left(1 - \sum_{i=1}^{N} p_i^q\right) \tag{4}$$

where pi is the probability that the system is in the i-state ($\sum_{i=1}^{N} p_i = 1$), N is the number of system states, k is some positive constant that determines the unit of entropy and in physical formulas is used to connect dimensions, such as the Boltzmann constant. q is the so-called entropy index. The entropy proposed by Constantino Tsallis for a composite system satisfies the property of non-additivity. In this case, for a system consisting of two subsystems, the total entropy will be determined by the following expression:

$$\mathrm{S}_q(A + B) = \mathrm{S}_q(A) + \mathrm{S}_q(B) + \frac{(q-1)}{k}\mathrm{S}_q(A)\mathrm{S}_q(B) \tag{5}$$

The parameter q, in this case, is a measure of the non-extensiveness of the system under consideration: $q < 1$ corresponds to superadditivity (superextensiveness), for $q >$ 1—subadditivity (subextensiveness), and the value $q = 1$ corresponds to the extensive case. When $q = 1$, the last term on the right side of the equation disappears, and the additivity (extensiveness) property is restored [13].

To test the hypothesis of the application of non-extensive statistical physics, at the first stage we use the q-Gaussian to estimate the series of differences in the magnitudes of earthquakes adjacent in time. The probability density distribution function can be described by a q-Gaussian [12, 14] as:

$$G_q(\beta, x) = \frac{\sqrt{\beta}}{C_q} e_q^{-\beta x^2} \tag{6}$$

$$\text{for } 1 < q < 3 \; C_q = \frac{\sqrt{\pi}\,\Gamma\left(\frac{3-q}{2(q-1)}\right)}{\sqrt{(q-1)}\,\Gamma\left(\frac{1}{q-1}\right)}$$

Expression (6) uses expressions of the so-called q-algebra, in which the following functions are introduced: q-exponent

$$e_q^x = \left[1 + (1-q)x\right]^{\frac{1}{1-q}}, \tag{7}$$

and q-logarithm

$$ln_q(x) = \frac{x^{1-q} - 1}{1-q}. \tag{8}$$

The shape parameter $q < 3$ controls the degree of non-extensiveness of the system, and distribution (6) reduces to the Gaussian distribution at $q = 1$ [15].

Sotolongo-Costa and Posadas in their work [16] proposed to modify the stick-slip earthquake model - "stick-slip" of two plates over each other along a fault in the presence of friction [17]. The model (Fragment-Asperity Interaction Model for Earthquakes, [17]), based on the principles of non-extensive statistical physics, considers the interaction of fault walls and rigid fragments filling the gaps between the walls. In this model, the released seismic energy ε is related to the size of the fragments that fill the space between the fault blocks.

In [18], the proposed model was improved using the volume dependence between seismic energy and fragment size instead of linear. Subsequently, in works [19] and [20] they continued further refinement of the model by introducing the relationship between the earthquake magnitude and relative energy, according to [21]: $M \propto \frac{2}{3}\log(\varepsilon)$.

According to these studies, if *N(M > Mth)* is the cumulative distribution of the number of earthquakes *N* with a magnitude greater than *Mth*, then:

$$\log\left(\frac{N(M > M_{th})}{N}\right) = \left(\frac{2-q}{1-q}\right)\log\left[\frac{1 - \left(\frac{1-q}{2-q}\right)\left(\frac{10^{M_{th}}}{a^{2/3}}\right)}{1 - \left(\frac{1-q}{2-q}\right)\left(\frac{10^{M_0}}{a^{2/3}}\right)}\right] \tag{9}$$

where M_0 is the threshold magnitude in the catalog, a is the coefficient of proportionality between the earthquake energy E and the size of the block fragment r^3 between faults, q is the Tsallis parameter from the expression for the Tsallis entropy (4). Equation (9) generalizes the Gutenberg-Richter relation in a wide range of values [22] and demonstrates good agreement for various earthquake catalogs [23, 24 and 25, etc.].

The values of the Tsallis parameter q obtained for the seismic catalogs of different regions of the world [26, 27] are $q \approx 1.5$–1.7, which indicates the universality of this parameter.

Quantitative estimates of the distribution of seismic activity involve the calculation of the average number of earthquakes in a certain volume for some time, as well as the calculation of the intensity of seismotectonic deformations (STD) [28].

4 Results and Discussion

The purpose of this work is to determine how, in the light of non-extensive statistical physics, the seismic process differs from technogenic impacts, namely from explosions. Therefore, it was decided to use three data sets: the FRC EGS RAS earthquake catalog, the SCM explosion catalog, and a summary catalog that combines both data sets. The latter variant will be used to determine how the seismicity pattern under consideration will change when the catalog of explosions is "mixed" into the catalog of earthquakes.

It was already shown in [29] that the probability density distribution of the energy difference between successive earthquakes in real data and in numerical models corresponds to the q-Gaussian distribution given by Eq. (6). Let us construct the probability density distribution function of the magnitude difference between successive earthquakes, and then approximate the normal and q-Gaussian ones using expression (6). The result is shown in Fig. 2.

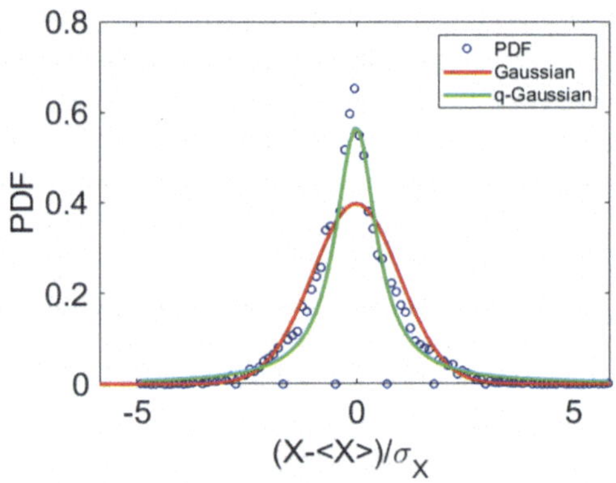

Fig. 2. Probability density distribution function of magnitude difference between successive earthquakes and its approximation by Gaussian and q-Gaussian with parameters $q = 2.0119 \pm 0.03$, $\beta = 3.1173 \pm 0.325$.

It can be seen that the approximation of the distribution function by the q-Gaussian has a more realistic form than by the normal distribution. Moreover, the q parameter indicates that we are dealing with a non-extensive system in which memory and cross-correlations are present. Therefore, for data analysis, methods of non-extensive statistical physics based on the entropy formalism proposed by K. Tsallis are applicable.

Let us construct cumulative distributions of the number of earthquakes by magnitude and approximate them using expression (9). We use three catalogs: a catalog of earthquakes for 1997 - 2022, a catalog of earthquakes + a catalog of explosions, and a separate catalog of explosions. The result is shown in Fig. 3.

Adding information about explosions to the catalog of earthquakes does not significantly change the picture. The Tsallis q parameters found by the method of nonlinear regression for the earthquake catalog and the mixed catalog ($q = 1.557 \pm 0.003$ and $q = 1.554 \pm 0.003$) characterize the region under study as seismically active and are close to the values obtained for other seismically active regions $q \approx 1.5$ presented in the papers.

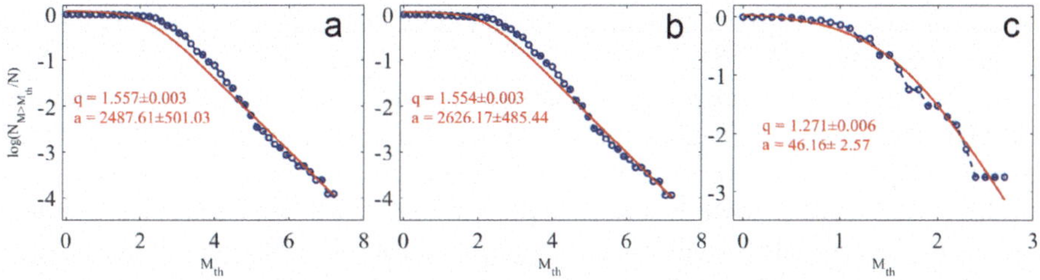

Fig. 3. Distribution of the relative number of earthquakes with magnitude $M > M_{th}$ depending on the magnitude. a – earthquake catalog, b – earthquake catalog + explosion catalog, c – explosion catalog.

Interesting, in this case, is the value of the Tsallis parameter q obtained for the catalog of explosions. It can be assumed that the work related to the explosions at the SCM is carried out according to a predetermined schedule. However, despite the fact that the process described by the catalog of explosions is deterministic rather than stochastic, the obtained value of the Tsallis parameter $q = 1.271 \pm 0.006$ indicates the opposite. That is, even if there is some mutual correlation between successive explosion events, it is not as strong as in the case of earthquakes.

We define the Tsallis parameter q in another way. In many works, in addition to energy characteristics, distances or time between successive events are considered [29–32, etc.]. Let us construct cumulative time distributions between successive events. We still use three catalogs: the catalog of earthquakes for 1997–2022, the catalog of earthquakes + the catalog of explosions, and the separate catalog of explosions. We approximate the obtained distributions by the q-exponent of Tsallis, expression (7). The parameters of the equation will be determined by the method of non-linear regression. The result is shown in Fig. 4.

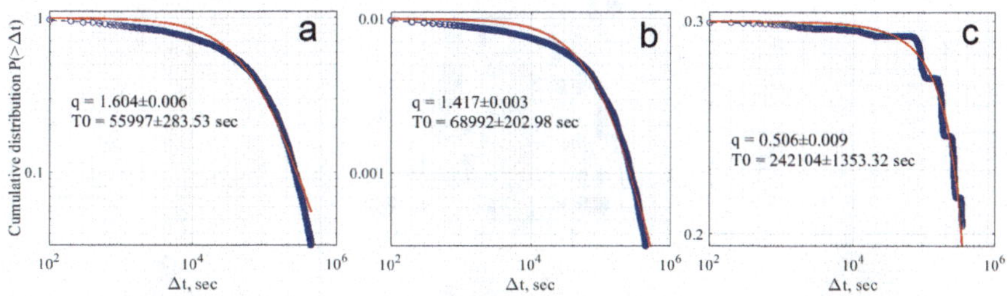

Fig. 4. Cumulative distributions of time intervals between successive events:. a – catalog of earthquakes, b – catalog of explosions.

The obtained values of the Tsallis parameter q for the earthquake catalog ($q = 1.604 \pm 0.006$) differ significantly from the value obtained for the explosion catalog of the Solntsevo coal mine ($q = 0.506 \pm 0.009$). In contrast to the natural seismic process, according to the catalog of explosions, it can be argued that there is no mutual correlation between individual acts of event-explosions, and individual acts of explosions are in no way connected with each other.

The calculation of the quantitative distribution of seismic events, the distribution of the STD intensity and the Tsallis parameter q over the territory was carried out in cells sized 0.5 × 0.5° with a step of 0.25° with the obtained value tied to the center of the cell. The results of constructions are presented in Fig. 5. The resulting distribution of the number of events varies from 0 to 5. A significant part of the number of events refers to the southern part of the island. Sakhalin, north and central part. The intensity of STD varies from 0 to $2{\cdot}10^{-6}$ yr^{-1}. Most of the events of moderate magnitude are located in cells where $\lg I\Sigma$ is 10^{-8} yr^{-1}. The intensity maxima are also observed in the southern, central and northern parts of the island. Sakhalin. The areas of increased seismicity that stand out in Fig. 6 (a) correspond to the maps of the seismic activity of Sakhalin [33]. The values of the Tsallis parameter q vary over the territory of the island. Sakhalin and are in the range from 1.3 to 1.9.

On the whole, the pattern of the distribution of the Tsallis parameter q coincides with the distributions of the number of events and the intensity of the STD. The areas of increased values of the number of events and the intensity of STDs are accompanied by increased values of the Tsallis parameter q. This can be explained by the fact that the seismogenic zones are in a metastable state, and an active seismic process is taking place in areas with an increased value of the parameter q.

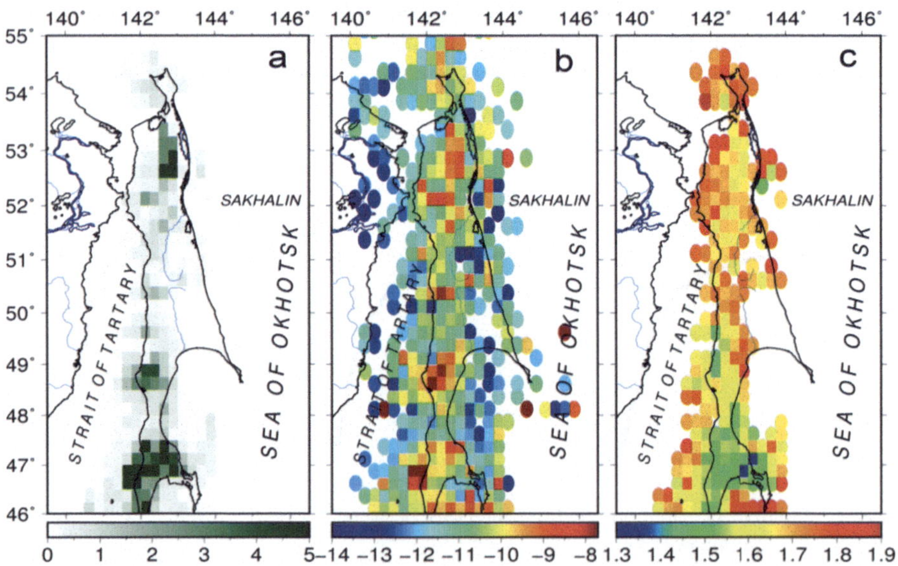

Fig. 5. Distribution of the number of earthquakes (a), the logarithm of the STD intensity (b), and the Tsallis parameter q(c) over the territory.

Despite the fact that the number of events and the intensity of STDs demonstrate an active seismic process, the Tsallis parameter q has lower values in this region. The distribution of seismicity characteristics used in the work shows the same effect as in

Fig. 4 and 5. Despite the determinism and the presence of periodicities in the catalog of explosions, individual acts of event-explosions themselves weakly or even do not correlate with each other, and even more so with the catalog of earthquakes.

5 Conclusion

The paper considers the catalog of earthquakes of the Sakhalin branch of the Federal Research Center "Unified Geophysical Service of the Russian Academy of Sciences", 8127 events for the period from 1997 to 2022, and the catalog of explosions of the Solntsevsky coal mine, 560 events for 2020–2023. When analyzing the data, the principles of non-extensive statistical physics were used, which is based on the entropy formalism proposed by K. Tsallis. For calculations, a modified stick-slip earthquake model was used - "stick-slip" of two plates over each other along a fault in the presence of friction, built on the principles laid down in the description of the Tsallis entropy. It is shown that the natural seismic process is well described by the proposed model, the obtained model parameters coincide with similar values for other seismically active regions, and characterize the island. Sakhalin as seismically active. The set of earthquake catalog events is a process with memory and cross-correlations, where individual events are connected not only with each other, but with the entire system as a whole. Increased values of the Tsallis q parameter are observed in areas with an active seismic process and coincide with areas of increased seismicity, which correspond to the maps of seismic activity of Sakhalin.

It should be noted that, unlike the natural seismic process, which looks like a stochastic one, explosions occur according to the rules, schedule, and are closer to a deterministic process. It could be assumed that the Tsallis parameter q, which is an indicator of the degree of correlation, will be higher than the value calculated for the seismic process. According to the results of data processing on the explosions of the Solntsevo coal mine, it can be noted that the catalog of explosions did not reveal such strong correlations between individual explosion events. Areas where blasting is carried out are marked by lower values of the Tsallis parameter q.

In addition, it can be assumed that when observing the dynamics of changes in the Tsallis parameter q in mining areas, an increase in this parameter can increase the likelihood of an underground shock.

Funding and Thanks. The work was carried out within the framework of the state task of the Institute of marine Geology and Geophysics of the far Eastern branch of the Russian Academy of Sciences.

References

1. Tarakanov, R.Z., Tikhonov, I.N.: Sakhalin seismology: development and some results. Bull. Far East. Branch Russ. Acad. Sci. **6**, 34–42 (2011)
2. The Solntsevsky open pit is the flagship of the Far Eastern coal mining. Coal. **3(1116),** 36–39 (2019)
3. Sadovsky, M.A.: Selected works. Geophysics and physics of explosion. Science, Moscow (2004)

4. Sharmanov, B.F., Sharmanova, L.N.: On the use of industrial explosions in seismic exploration. Geol. Geophys. **21**(7), 116–122 (1980)
5. Dobrynina, A.A., German, V.I., Sankov, V.A.: Recognition of industrial explosions and weak natural earthquakes. Coal. **S12**(1162), 23–29 (2022) https://doi.org/10.18796/0041-5790-2022-S12-23-29
6. Sokhatyuk, A.S., Decik, I.V., Boginskaya, N.V., Parshina, I.A., Fercheva, V.N.: Sakhalin (M $\geq$ 2.8). Earthquakes in Russia in 2014. Obninsk: Federal Research Center EGS RAS. 127–131 (2016)
7. Gutenberg, B., Richter, C.F.: Frequency of earthquakes in California. Bull. Seismol. Soc. Am. **34**, 185–188 (1944)
8. Kasahara, K.: Earthquake Mechanics. Mir, Moscow ((1985)
9. Gibbs, J. W.: Thermodynamics. Statistical mechanics. Nauka, Moscow (1982)
10. Tsallis, C.: Possible generalization of Boltzmann-Gibbs statistics. J. Stat. Phys. **52**(1–2), 479–487 (1988). https://doi.org/10.1007/bf01016429
11. Tsallis, C., Brigatti, E.: Nonextensive statistical mechanics: a brief introduction. Continuum Mech. Thermodyn. **16**, 223–235 (2004)
12. Tsallis, C.: Introduction to Nonextensive Statistical Mechanics: Approaching a Complex World. Springer, New York (2009)
13. Ramírez-Rojas, A., Sigalotti, L.Di G., Márquez, E.L.F., Rendón, O.: Time Series Analysis in Seismology: Practical Applications. Elsevier, vol. 397 (2019)
14. Iliopoulos, A.C., Pavlos, G.P., Papadimitriou, E.E., Sfiris, D.S., Athanasiou, M.A., Tsoutsouras, V.G.: Chaos, self organized criticality, intermittent turbulence and nonextensivity revealed from seismogenesis in North Aegean area. Int. J. Bifurcat. Chaos **22**(09), 1250224 (2012). https://doi.org/10.1142/S0218127412502240
15. Witkovský, V.: Characteristic function of the Tsallis q-Gaussian and its applications in measurement and metrology. Metrology. **3**(2), 222–236 (2023)
16. Sotolongo-Costa, O., Posadas, A.: Fragment-asperity interaction model for earthquakes. Phys. Rev. Lett. **92**(4), 048501 (2004). https://doi.org/10.1103/PhysRevLett.92.048501
17. Brown, S.R., Scholz, C.H., Rundle, J.B.: A simplified spring-block model of earthquakes. Geophys. Res. Lett. **18**(2), 215–218 (1991)
18. Silva, R., Franca, G.S., Vilar, C.S., Alcaniz, J.S.: Nonextensive models for earthquakes. Phys. Rev. E **7**, 026102 (2006). https://doi.org/10.1103/PhysRevE.73.026102
19. Darooneh, A., Mehri, A.: A nonextensive modification of the Gutenberg-Richter law: q-stretched exponential form. Physica A **389**, 509–514 (2010). https://doi.org/10.1016/j.physa.2009.10.006
20. Telesca, L.: Maximum likelihood estimation of the nonextensive parameters of the earthquake cumulative magnitude distribution. Bull. Seismol. Soc. Am. **102**(2), 886–891 (2012). https://doi.org/10.1785/0120110093
21. Kanamori, H.: Quantification of earthquakes. Nature **271**, 411–414 (1978)
22. Vallianatos, F, Michas, G, Papadakis, G.: A description of seismicity based on non-extensive statistical physics: a review. In: D'Amico S, (ed.) Earthquakes and their impact on society. Cham: Springer Natural Hazards, pp. 1–41. Springer, Cham (2015) https://doi.org/10.1007/978-3-319-21753-6
23. Varotsos, P.A., Sarlis, N.V., Skordas, E.S.: Tsallis entropy index q and the complexity measure of seismicity in natural time under time reversal before the M9 Tohoku earthquake in 2011. Entropy **20**, 757–773 (2018). https://doi.org/10.3390/e20100757
24. Michas, G.: Generalized statistical mechanics description of fault and earthquake populations in corinth rift (greece) (Ph.D. thesis). University College London.(2016)
25. Sychev, V.N., Sycheva, N.A.: Nonextensive analysis of aftershocks following moderate earthquakes in Tien Shan and North Pamir. J. Volcanol. Seismolog. **15**, 58–711 (2021). https://doi.org/10.1134/S0742046321010127

26. Sarlis, N.V., Skordas, E.S., Varotsos, P.A.: Nonextensivity and natural time: the case of seismicity. Phys Rev E. **82,** 021110 (2010). https://doi.org/10.1103/PhysRevE.82.021110
27. Sardeli, E., Michas, G., Pavlou, K., Vallianatos, F., Karakonstantis, A., Chatzopoulos, G.: Complexity of recent earthquake swarms in Greece in terms of non-extensive statistical physics. Entropy **25**(4), 667 (2023). https://doi.org/10.3390/e25040667
28. Yunga, S.L.: Methods and results of studying seismotectonic deformations. Nauka, Moscow (1990)
29. Vallianatos, F., Michas, G., Papadakis, G., Tzanis, A.: Evidence of non-extensivity in the seismicity observed during the 2011–2012 unrest at the Santorini volcanic complex, Greece. Nat. Hazard. **13**, 177–185 (2013)
30. Abe, S., Tirnakli, U., Varotsos, P.A.: Complexity of seismicity and nonextensive statistics. Europhys. News **36**, 206–208 (2005)
31. Avgerinou, S.E., Anyfadi, E.A., Michas, G., Vallianatos, F.: A non-extensive statistical physics view of the temporal properties of the recent aftershock sequences of strong earthquakes in Greece. Appl. Sci. **13**, 1995 (2023). https://doi.org/10.3390/app13031995
32. Sigalotti, L.D.G., Ramírez-Rojas, A., Vargas, C.A.: Tsallis q-statistics in seismology. Entropy **25**, 408 (2023). https://doi.org/10.3390/e25030408
33. Oskorbin, L.S.: Seismicity of Sakhalin. In: Seismic zoning of Sakhalin. Vladivostok, pp. 3–23 (1977)

Power-Law Compound and Fractional Poisson Process in the Theory of Anomalous Phenomena

Boris Shevtsov(✉) and Olga Sheremetyeva

Institute of Cosmophysical Research and Radio Wave Propagation, Far Eastern Branch of the Russian Academy of Sciences, Mirnaya, Paratunka, 684034 Kamchatka, Russia
{bshev,sheremeteva}@ikir.ru

Abstract. The power-law compound and time-fractional Poisson process is considered as a statistical model of anomalous phenomena in the hereditarian theory of criticality. This model can be useful in studies of energy-active zones. Regardless of their nature, anomalous phenomena have universal statistical properties, among which, first of all, it should be noted scale invariance. In the proposed model, the special role of scaling in the properties of anomalous phenomena is shown taking into account the heredite effects, the physical meaning of which is explained by examples of analogies between anomalous phenomena of different nature. Critical process modes and exceptional values of critical indexes are determined. The structural instability of the process caused by scaling and catastrophes in its statistical characteristics are discussed. The obtained results are used to study seismic data and determine the critical indices of the deformation process.

Keywords: hereditarian theory of criticality · anomalous phenomena · scaling of random event flows of a compound fractional Poisson process · critical indices · critical regimes · coherent effects · deformation theory

MSC Classification: 33E12 · 60G22 · 60G55

1 Introduction

The signs of critical phenomena arising in the form of anomalous modes of the wave process are scaling and power-law divergence of process characteristics near critical points. The hereditarity of the process affects the values of critical indices, slows down the dynamics and causes an increase in fluctuations, the nature of which indicates coherent effects.

To describe the anomalous phenomenon, consider the critical modes of the compound Poisson process [1,2] in its fractional representation [3–5] and with event flow scaling defined by the Guttenberg-Richter law [6,7]. It will be shown

O. Sheremetyeva—These authors contributed equally to this work.

A. Dmitriev et al. (Eds.): STRPEP 2023, SPEES, pp. 266–275, 2023.
https://doi.org/10.1007/978-3-031-50248-4_27

below that such a statistically scale-invariant and hereditarian process under critical ratios of scaling and hereditarity parameters goes into a mode characterized by a catastrophic increase in statistical characteristics.

Due to the hereditarity of the process, there is a statistical dependence and power-law correlations of random events. At the same time, relaxation slows down and the role of fluctuations increases. If scaling generates divergences at critical points, then the memory of the process determines the features of its dynamics.

The applicability of the hereditarian theory of criticality will be considered on the example of seismicity. The critical indices of anomalous modes of the geodeformation process will be determined by the peculiarities in the ratios of scaling and hereditarity parameters.

2 Compound Fractional Poisson Process

Equations of Compound Fractional Poisson Process (CFPP) of order k with integer random state changes by $r = 1,\ 2, ...,\ k$ can be represented as [3]:

$$\begin{cases} \dfrac{dp_0^\nu(t)}{dt^\nu} = -\Lambda p_0^\nu(t), & \Lambda = \sum\limits_{r=1}^{k} \lambda_r, \\ \dfrac{dp_j^\nu(t)}{dt^\nu} = \sum\limits_{r=1}^{j} \lambda_r p_{j-r}^\nu(t) - \Lambda p_j^\nu(t), & j = 1,\ 2, ..., k-1, \\ \dfrac{dp_j^\nu(t)}{dt^\nu} = \sum\limits_{r=1}^{k} \lambda_r p_{j-r}^\nu(t) - \Lambda p_j^\nu(t), & j = k,\ k+1, ..., \end{cases} \tag{1}$$

with initial conditions

$$p_j^\nu(0) = \begin{cases} 1, & j = 0, \\ 0, & j \geq 1, \end{cases}$$

where ν is the exponent of the fractional derivative [3], $0 < \nu \leq 1$.

3 Series of Repeatability Frequencies of Events

The power-law distribution of dislocation changes in the deformation process can be determined using the Gutenberg-Richter law [6]

$$N = 10^{a-bM}, \tag{2}$$

where $N = N(m \geq M)$ is the number of events with magnitudes $m \geq M$, M is fixed magnitude, m is magnitudes of events, a and b are constants, $b \approx 1$, $0.5 < b < 1.5$, $1 < M < 9.5$, $N_{total} = 10^a$ is total number of events.

The Kanamori magnitude is defined as $M = (2/3)(\lg M_0 - C)$ [7], where $M_0 = \mu S u$ is the seismic moment, μ is the rock shear modulus, about 30 GPa, S is the area of the geological fault, u is the average displacement along the fault, $C = \lg m_0$ is the normalization constant, the logarithm of the seismic moment m_0

of the reference (calibration) event, for example, a controlled explosion. Seismic moment – the work of elastic forces expended on the destruction of rocks, the displacement of the sides of the fault and seismic vibrations. Only a small part of the released elastic energy (on the order of a percent) is converted into an earthquake.

There are other definitions of magnitude, which essentially come down to renormalization of the energy of an event, but the power-law distribution of events over energies, which is characteristic of nonequilibrium systems, remains unchanged under any normalization.

Taking into account that $S = L^2$ and $u = \epsilon L$, where L is the dislocation size and ϵ is the relative strain, we get $M_0 = \mu\epsilon L^3$.

Then for the dislocation probability distribution function, we obtain $P(L) = 1 - N/N_{total} = 1 - 10^{-bM} = 1 - 10^{-(2b/3)(\lg M_0 - C)} = 1 - 10^{-(2b/3)(\lg \mu\epsilon L^3 - C)}$.

However, if we introduce the minimum dislocation size L_{min} using the relation $C = \lg \mu\epsilon(L_{min})^3$, then the probability distribution can be represented as: $P(L) = 1 - (L/L_{min})^{-2b}$. The event with the minimum dislocation size is chosen as a reference (calibration).

It is noteworthy that the definition of the magnitude in terms of the dislocation size $M = 2lg(L/L_{min})$ in Bells gives the probability distribution of events without the medium parameters μ and ϵ, i.e. in this case, the deformation process is represented exclusively by geometric characteristics, and in such a way that there is no dependence on the units of their measurement. The fact that the resulting distribution is related to the energy of the event remained in the two in front of the logarithm.

For the probability density, we get $p(L) = dP(L)/dL = 2b(L/L_{min})^{-2b-1}/L_{min}$. This function is defined on the interval $L_{min} < L < L_{max}$, where L_{max} is the maximum size of dislocations, limited by the polygon size (hundreds of kilometers).

From the relations $M_{max} = 2lg(L_{max}/L_{min}) = 9.5$ and $L_{max} \approx 10^5$ meters we get $L_{min} \approx 1$ meter. Note that $M = 0$ at $L = L_{min}$, however, the magnitude interval $0 < M < 1$ is not considered in seismology, this is the region of geoacoustic emission, but since it makes a well-defined high-frequency contribution to the energy of events, we will consider the extended (consolidated) interval of magnitudes $0 < M < 9.5$ and the corresponding interval of dislocation sizes $L_{min} < L < L_{max}$.

The frequency dependence of the propagation of acoustic oscillations in inhomogeneous media leads to the effects of renormalization of the source of oscillations in the range of magnitudes $0 < M < 1$ and significant deviations from the Gutenberg-Richter law. These effects at the left end of the $1 < M < 9.5$ magnitude range are called roll-offs.

The energy spectrum of seismoacoustic oscillations is non-stationary and very dynamic, as it should be for a turbulent process, so the Gutenberg-Richter law should be considered as an approximation of a real spectral characteristic. We will use this power law in the magnitude range $0 < M < 9.5$ as the average value of the energy spectrum of seismoacoustic oscillations. Deviations from the mean in narrower spectral ranges are treated as non-stationary [8,9]. Below, the conditions for the instability of the power-law distribution will be determined.

Data in the seismic catalogs are given with a magnitude step of $\Delta M = 0.1$, i.e. in deciBells, but the division of the region $L_{min} < L < L_{max}$ into intervals ΔL will not be equidistant. On the contrary, we will make the partitioning in L equidistant with the step $\Delta L = L_{min}$ in order to form a generalized harmonic series of event recurrence frequencies. With such a partition, we get the number of intervals $k = L_{max}/L_{min} = 10^{4.75} \approx 10^5$. The number of events in each interval with number r will be $\Delta N_r = N_{total}\, p(rL_{min})L_{min} = 2b \cdot 10^a \cdot r^{-2b-1}$. Dividing this by the observation period T, we obtain a series of event recurrence frequencies ω_r that fall within the ΔL_r interval,

$$\omega_r = 2b\Omega \cdot r^{-2b-1}, \ \sum_{r=1}^{k} \omega_r = \Omega = \frac{N_{total}}{T}, \tag{3}$$

where Ω is the total frequency of events. In our case of large k we can put $k \to \infty$.

Equations (1) are supplemented by a power-law distribution (3) of frequency ω_r of random events occurrence with jump amplitude $r = 1, 2, ..., k$, which are related to the parameters λ_r of Eq. (1) (fractional event rates) as follows:

$$\lambda_r = \omega_r^\nu = \left(2b\Omega \cdot r^{-2b-1}\right)^\nu, \tag{4}$$

$$\sum_{r=1}^{k} \lambda_r = \Lambda = (2b\Omega)^\nu \sum_{r=1}^{k} r^{-(2b+1)\nu}. \tag{5}$$

4 Critical Indexes and Process Instability

Let us consider the role of the nonlocality of the process in time, which is determined by the fractional parameter ν. The sum of fractional event frequencies $\Lambda = \sum_{r=1}^{k} \lambda_r = (2b\Omega)^\nu \sum_{r=1}^{k} r^{-(2b+1)\nu}$ characterizes the rate of the process. According to Eq. (1), $\Lambda^{1/\nu} = 2b\Omega\left(\sum_{r=1}^{k} r^{-(2b+1)\nu}\right)^{1/\nu}$ – rate of decay of the initial and all subsequent states. The dimension and value of this expression is determined by the total frequency of occurrence of random events Ω, see (3). The probability distributions of the waiting times for the first movement for each scale r are represented by expressions

$$P(t) = \lambda_r t^\nu E_{\nu,\nu+1}(-\lambda_r t^\nu), \ t \geq 0, \ r = 1, 2, \ldots, k, \tag{6}$$

where $E_{\nu,\nu+1}(x)$ is the Mittag-Leffler function [3]. Distributions (6) define correlations in event streams. The fact that the correlations are represented by a fractional exponent indicates the non-locality of the process in time, memory effects, statistical dependence of events, and delayed relaxation that arise as a result of hardening of the medium.

Three types of process instability are determined by three types of divergences in its statistical characteristics. The first one is related to the divergence of the partial sum of the generalized harmonic series $S_k = \sum_{r=1}^{k} r^{-(2b+1)\nu}$ in expression (5), for which the following is satisfied $S_\infty = \lim_{k\to\infty} S_k = \zeta((2b+1)\nu)$, where $\zeta(x)$ is the Riemann zeta function. For $(2b+1)\nu = 1$, the zeta function goes to infinity, which gives a critical relation between the parameters of scaling b and the time nonlocality ν of the process. If $b \approx 1$, then the critical value is $\nu \approx 0.33$. If ν is greater than this value, then the process is in subcritical mode, but there may be divergences in other characteristics of the process.

Consider the first and second moments of distribution (5), which determine the mean and variance of the CFPP. Note that the sum of the fractional frequency of occurrence of events Λ is the zero moment. If it is finite and far from the critical value, then we find the conditions for the existence of the other two moments. Generating moment functions $\mathbb{E}(t,s) = E_{\nu,1}\left(\sum_{r=1}^{k} \lambda_r (e^{r\cdot s} - 1)t^\nu\right)$, $t \geq 0$, $s \in R$ [3], gives for the mean $\mathbf{E}(t)$ and variance $\mathbf{Var}(t)$ of the CFPP [3]:

$$
\begin{aligned}
\mathbf{E}(t) &= S_{k,1}(2b\Omega \cdot t)^\nu / \Gamma(\nu+1), \quad t \geq 0, \\
\mathbf{Var}(t) &= S_{k,2}(2b\Omega \cdot t)^\nu / \Gamma(\nu+1) + \left(S_{k,1}(2b\Omega \cdot t)^\nu\right)^2 Z(\nu), \quad t \geq 0, \\
S_{k,p} &= \sum_{r=1}^{k} r^{-(2b+1)\nu+p}, \; p = 1,2, \\
Z(\nu) &:= \frac{1}{\nu}\left(\frac{1}{\Gamma(2\nu)} - \frac{1}{\nu\Gamma^2(\nu)}\right),
\end{aligned}
\tag{7}
$$

where $\Gamma(x)$ is the gamma function. The first term of the dispersion $\mathbf{Var}(t)$ is proportional to the average as in the usual Poisson process, and the second is proportional to the square of the average, which can be considered a manifestation of the coherence of random events, known, for example, as superluminescence in laser physics. This is a direct consequence of the non-locality (memory) of the process.

In this case, the limit $S_{\infty,p} = \lim_{k\to\infty} S_{k,p} = \zeta((2b+1)\nu - p)$ tends to infinity when $(2b+1)\nu - p = 1$, which gives the critical relation $\nu = (1+p)/(2b+1)$ between the scaling parameters b and the time nonlocality ν of the process. Critical values ν (critical indices) can be defined as $\nu_p = (1+p)/(2b+1)$. To combine this result with the one obtained earlier for the zero moment of the power-law distribution of the frequencies of random events, we must set $p = 0, 1, 2$.

If the zero moment determines the decay rate of the states, then the first moment determines the rate of growth of deformations, and the second determines the rate of energy release (power) of the process. Which of them are finite for a given power-law distribution of event frequencies is determined by critical indices. For example, if $b \approx 1$, then the critical values $\nu_p \approx 1/3,\ 2/3,\ 1$, for $p = 0, 1, 2$. Using the value of the hereditarity parameter ν, we determine at what statistical moments instability can occur.

5 Analysis of Seismic Data

When obtaining the values of critical indices, it is not necessary to proceed to a geometric description of the process, you can operate directly with the class or magnitude, in tenths of its fractions (in decibels), as is customary in seismic catalogs [10]. By the nonlinear regression method we will find the process parameters b and ν averaged over the intervals of classes or magnitudes of the catalog in which the power-low distribution of the frequency of recurrence of random events is performed. According to the real value of the parameter b and in accordance with the hereditary model discussed above, critical indices ν_p will be found, by the value of which and in comparison with ν the state of the seismic process will be determined.

5.1 Determination of the Parameters of the Distribution of Repeatability Frequencies

Distribution of the frequencies ω_r $[day^{-1}]$ of events repeatability (3) is defined by the parameters a and b of the Gutenberg-Richter law (2), the logarithmic form of which is as follows

$$\lg N = a - bM. \tag{8}$$

The earthquake catalog of the Kamchatka Branch of the Geophysical Survey of Russian Academy of Sciences for the period from 1 January 1962 to 31 December 2002 for the Kuril-Kamchatka island arc subduction zone was used to determine the necessary parameters (area 46°–62° N, 158°–174° E) [10]. The size of the catalog $n = 79282$ earthquakes, the catalog contains events of 4.1–16.1 energy classes.

The distribution of the number of earthquakes obtained from the catalog data, depending on their energy, showed that the sample for earthquakes with energy less than class 8.3 is not representative. The sample size of earthquakes of energy classes $K \in [8.3,\ 16.1]$ is $n = 46917$. At this interval, we will find the best approximation by the exponential function (2) of the empirical Gutenberg-Richter law (Fig. 1a). In logarithmic form, we obtain an approximation by the function (8) of the linear part of the empirical Gutenberg-Richter law (Fig. 1b).

The least squares method (LSM) was used to approximate the empirical Gutenberg-Richter law by the exponential function (2) (nonlinear regression). Suitable intervals of changing the parameters a and b were determined. The values of the parameters of the approximating function (2) were found by iterating over the values from these intervals in increments of 0.001 based on minimizing the approximation error ε with the accepted constraint $1\% < \varepsilon < 10\%$ with sequential exclusion of classes from the beginning and end interval $K \in [8.3,\ 16.1]$ ($M \in [2.33,\ 7.53]$) and the largest correlation index R. At each interval obtained in this way, statistical characteristics for the logarithmic Gutenberg-Richter law were also calculated. If the approximation error ε of the logarithmic law (2) on the considered interval of magnitudes satisfied the minimization condition and the constraints of $1\% \le \varepsilon \le 2\%$, then this interval was chosen as the approximation interval.

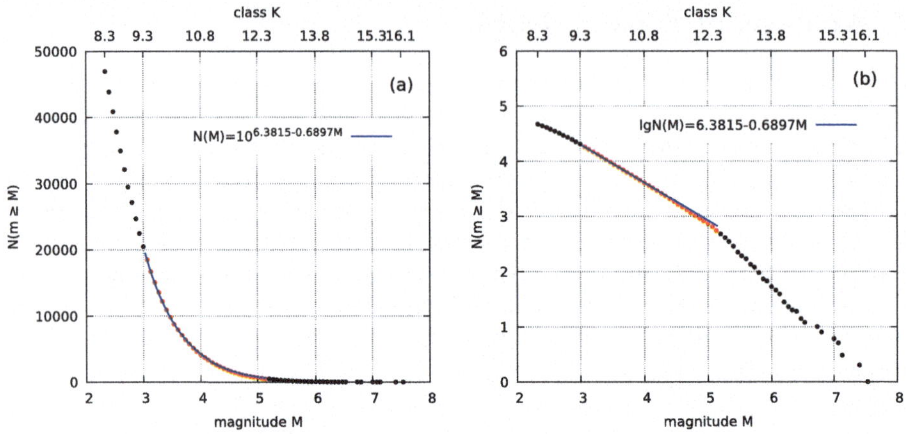

Fig. 1. The empirical Gutenberg-Richter law: (a) initial, (b) logarithmic. The blue graph – nonlinear regression, the dot graph – empirical law, where the red dots – approximation interval.

According to the results of calculations, all the conditions formulated above are satisfied by the approximation of the empirical Gutenberg-Richter law by an exponential function on the interval $K \in [9.2,\ 12.9]$ ($M \in [2.93,\ 5.40]$) (Table 1, first row). At the selected approximation interval, the closeness of the nonlinear regression is estimated using the correlation index $R = 0.9857$, the proximity of which to one indicates a close relationship. The statistical significance of the nonlinear regression equation as a whole is estimated using the F-criterion $F(\alpha, k_1, k_2)$ at the significance level $\alpha = 0.05$ with degrees of freedom $k_1 = m-1$ and $k_2 = n_K - m$, where n_K is the number of classes in the approximation interval, m – number of regression parameters ($m = 2$). The empirical value of the statistics F exceeds the critical value $\tilde{F} = F(0.05, 1, 42) = 4.08$, therefore, at a given level α, we recognize the statistical significance of the nonlinear regression equation as a whole. Note that the statistical characteristics of the logarithmic empirical Gutenberg-Richter law also allow us to conclude about the statistical significance of the linear equation ($F > \tilde{F}$) as a whole (Table 1, second row).

Table 1. Parameters of the Gutenberg-Richter law and statistical characteristics of approximating functions

Function of approximation	$[K_1,\ K_2]$	$[M_1,\ M_2]$	n_K	N_{total}	a	b	R	F	$\varepsilon, \%$
$Y = 10^{a-bX}$	[9.2, 12.9]	[2.93, 5.40]	38	22230	6.3815	0.6897	0.9857	1233	1.658
$\lg Y = a - bX$							0.8567	99	1.687

5.2 Distributions of the First-Passage Times

For each of the classes that fall within the approximation interval $K \in [9.2,\ 12.9]$ ($M \in [2.93,\ 5.40]$), we find the distribution of times between pairs of consecutive events (distribution of inter-event times) – distribution of the first-passage times. The calculation procedure is as follows. We select events of the class K_r from the catalog, calculate the lengths of time intervals between neighboring events and find the length of the largest interval T_{max}. We find an empirical cumulative distribution function (eCDF) of the inter-event times by the values of their length. The time step is one day. The eCDF is approximated by the probability function $P(t)$ (6) taking into account the expression (4). The LSM approximation was carried out in two ways. In the first case, the frequency of repeatability ω_r $[day^{-1}]$ events of the class K_r were calculated based on the empirical distribution of earthquakes by energy (classes), and the parameter $\nu = \nu_r$ of the approximation function $P(t)$ was found from a suitable interval by iterating with increment of 0.001, based on the condition of minimizing the approximation error ε. As an example, some results are presented in the Table 2 (col. 4–7). In the second case, a two-parameter approximation was used, where both parameters ω_r $[day^{-1}]$ and $\nu = \nu_r$ of the function $P(t)$ were found from suitable intervals by iterating with increments of 0.001, provided the approximation error ε was minimized (Table 2, col. 8–11). It should be noted that the two-parameter approximation of the eCDF of the first-passage times is more accurate. In this case the error ε for classes from the approximation interval does not exceed 7 % (Table 2, col. 11), whereas using the empirical value of the frequency ω_r leads to errors exceeding 10 % for classes $K \geq 11.7$ (Table 2, col. 7). Therefore, the results of two-parameter approximation were used for further calculations.

Table 2. Parameters of first-passage times distributions

K_r	M_r	n_r[a]	Approximation by a function $P(t)$							
			one-parameter				two-parameter			
			RSS	$\omega_r, [day^{-1}]$	ν_r	$\varepsilon, \%$	RSS	$\omega_r, [day^{-1}]$	ν_r	$\varepsilon, \%$
1	2	3	4	5	6	7	8	9	10	11
9.2	2.93	57	0.102	0.135	0.961	4.71	0.025	0.182	0.891	2.33
11.0	4.13	113	0.331	0.025	0.922	7.19	0.041	0.033	0.865	3.76
11.7	4.6	95	0.505	0.011	0.825	11.18	0.023	0.021	0.784	4.07
12.9	5.4	50	0.428	0.004	0.868	14.85	0.032	0.006	0.883	5.32

[a] n_r – number of inter-event time intervals.

5.3 Results

Based on the results of processing experimental data and their regularities, the parameters characterizing a CFPP (1), (4) of the order of k with integer random state changes by a value $r = 1, \ldots, k$ [3] are calculated on the selected approximation interval. The intensity of this process is determined by a parameter $\Lambda\,[day^{-\nu}]$ that was calculated using the formula (5) taking into account the equality (4). As a result of two-parameter approximation of the distributions of the first-passage times (Table 2, col. 9–10), the value of the intensity of the process is obtained $\Lambda = 1.8654\ day^{-0.7849} = (2.2129/day)^{0.7849}$, where the average of the exponent of the fractional derivative $\nu = 0.7849$. Then the process stability parameter (parameter b is taken from the Table 1) takes the value $(2b+1)\nu \approx 1,8675$.

6 Discussion

From the value $b = 0.6897$, the critical values $\nu_p \approx 0.4, 0.8, 1.2$ are found for $p = 0, 1, 2$. And according to the value of the hereditarity parameter $\nu = 0.7849$, it is determined that the process is in the subcritical mode by the zero moment, and in the supercritical mode by the first and second moments. And this means that instabilities can arise in the accumulated deformations and energy, and will manifest in strong fluctuations.

The nature of these fluctuations can be explained using $\mathbf{Var}(t)$ in (7). In this energy characteristic of the process (dispersion), the second term is proportional to the square of the mean $\mathbf{E}(t)$ (7), and this indicates the presence of coherent effects in the deformations arising due to the nonlocality of the process in time. In other words, the memory of the process contributes to the consolidation of dislocation changes, as a result of which not the energies of events are added, but the amplitudes of dislocations. In quantum optics, this effect is known as superluminescence, and in acoustics, on this principle, they tried to create a generator of coherent radiation using processes in a superheated liquid. This idea arose in connection with the analogy between the physics of boiling and lasers.

In a nonequilibrium medium, as a result of the emerging instability, the process sporadically passes into a nonstationary regime, due to which intense fluctuations are formed, the description of which using the nonstationary Poisson process was considered in [8,9]. Coherence in anomalous phenomena is a self-organized criticality due to the scaling and hereditary nature of the process. Collective and time-nonlocal effects arise in systems that are far from thermodynamic equilibrium.

For $\nu = 1$, due to the property of the gamma function $\Gamma(z+1) = z\Gamma(z)$, the second term of $\mathbf{Var}(t)$ in (7) vanishes, and coherent effects disappear. The variance of the process becomes proportional to its mean, as in the usual Poisson process.

7 Conclusion

The considered model has three special properties: complexity, power-law scaling, and hereditarity. Each of these properties contributes to the statistical features of the model.

Complexity (compositeness) can take different forms, but a special case gives power-law scaling, which causes instability and non-stationarity of the process. In seismology, these are foreshocks, mainshock and aftershocks. In laser physics, this is a gigantic impulse. In condensed matter physics, this is explosive boiling. In the infectious sciences, this is an epidemic. In investments, this is a financial disaster. Scaling through scale consolidation generates divergences in the statistical characteristics of the process.

But the main feature of the model is its heredity, which generates the consolidation of random events in time with an infinite correlation radius, according to distribution (6). The process slows down, and due to this, the collectivization of events occurs.

As has been shown, scaling and hereditarity produce a multiplier effect of the consolidation of events, which results in a catastrophic rather than a gigantic phenomenon. The second term of the process dispersion is responsible for the coherence.

Acknowledgments. The work was carried out within the framework of realization of the State task № AAAA-A21-121011290003-0.

References

1. Janossy, L., Renyi, A., Aczel, J.: On composed Poisson distributions. I. Acta Math. Acad. Sci. Hungar. **1**, 209–224 (1950)
2. Adelson, R.M.: Compound Poisson distributions. Oper. Res. Quart. **17**, 73–75 (1966)
3. Di Crescenzo, A., Martinucci, B., Meoli, A.: A fractional counting process and its connection with the Poisson process. ALEA Lat. Am. J. Probab. Math. Stat. **13**(1), 291–307 (2016). https://doi.org/10.30757/ALEA.v13-12
4. Beghin, L., Macci, C.: Multivariate fractional Poisson processes and compound sums. Adv. Appl. Probab. **48**(3), 961–711 (2016). https://doi.org/10.1017/apr.2016.23
5. Khandakar, M., Kataria, K.K.: Some compound fractional Poisson processes. Fractal Fract. **7**(1), 15 (2023). https://doi.org/10.3390/fractalfract7010015
6. Gutenberg, B., Richter, C.F.: Frequency of earthquakes in California. Bull. Seismol. Soc. Am. **34**, 185–188 (1944)
7. Kanamori, H.: The energy release in great earthquakes. J. Geophys. Res. **82**(20), 2981–2987 (1977)
8. Shevtsov, B., Sheremetyeva, O.: Fractional models of seismoacoustic and electromagnetic activity. In: E3S Web of Conferences, vol. 20, p. 02013 (2017). https://doi.org/10.1051/e3sconf/20172002013
9. Sheremetyeva, O., Shevtsov, B.: Fractional model of the deformation process. Fractal Fract. **6**(7), 372 (2022). https://doi.org/10.3390/fractalfract6070372
10. FRC: The Geophysical Service of the Russian Academy of Sciences. Figshare (2022). http://www.gsras.ru/new/eng/catalog/

Operational Precursors of the Earthquake on March 16, 2021 with Mw = 6.6, Kamchatka

Evgeniy Makarov(✉) and Rinat Akbashev

Kamchatka Branch Geophysical Survey, Russian Academy of Sciences, 9 Piip Boulevard, Petropavlovsk-Kamchatsky 683006, Kamchatka region, Russian Federation
cice@yandex.ru, arr@emsd.ru

Abstract. The paper provides new information about the occurrence of anomalous variations in the flow of gamma and beta pulses, as well as in the atmospheric electric field before the earthquake that occurred off the coast of Kamchatka on March 16, 2021 with Mw = 6.6. It is assumed that deformation processes before this earthquake led to increased radon runoff, which affected the ionization balance of the surface layer of the atmosphere. The detected short-term anomalous disturbances that precede an earthquake can be considered as its operational precursors and evidence of the impact of the upper layer of the earth's crust on the surface atmosphere during its preparation and implementation.

Keywords: radon · atmospheric electric field · precursors · ionization · earthquake

Introduction

Radon (222Rn), formed as a result of the decay of radium and available for continuous registration in the subsurface air, is very sensitive to changes in the geodynamic state of the medium. This makes it possible to consider it as an indicator of changes in the structure of the studied area of the Earth's crust, porosity, permeability of gas migration channels, as well as a response of the geomedia to external influences. Anomalies in the radon field, as a herald of an approaching earthquake, have been widely studied in recent decades. In order to search for precursors of strong earthquakes in many seismically active regions of the world, since the 60 s of the last century, the volumetric activity of radon dissolved in water and in the air of the subsurface has been recorded. Information about the informativeness of the radon method for searching for earthquake precursors is given in a number of papers. The first brief description of radon precursors was published in [1]. Another article [2] provides a more detailed description of the geochemical precursors of strong earthquakes, including radon ones, recorded

E. Makarov and R. Akbashev—Equal Contribution.

A. Dmitriev et al. (Eds.): STRPEP 2023, SPEES, pp. 276–284, 2023.
https://doi.org/10.1007/978-3-031-50248-4_28

before 2009. According to these data, with the involvement of observations of the dynamics of subsurface radon before earthquakes with $M \geq 5.0$ on the Kamchatka Peninsula, in the work [3], the dependences of the parameters of radon precursors on magnitude and distance were calculated. The theoretical substantiation of the possibility of the occurrence of harbinger anomalies is quite fully described in [4]. Short-term precursors of earthquakes with magnitude $M \geq 4.5$ in the field of subsoil radon with a lead time of up to 15 days have been reported in many parts of the world [3,5]. Observations of the behavior of the radon volume activity (RVA) over a long period at the Petropavlovsk-Kamchatsky geodynamic polygon (PKGP) suggest that there is a definite connection between the RVA and strong earthquakes with a magnitude $M_w \geq 5.5$ occurring in the Kamchatka region [3]. The impact of the lithosphere on the atmosphere is an integral part of the interaction of solid and gaseous geospheric shells. It is determined by the dynamics of lithospheric processes and occurs intensively at the boundary of the contact of geospheres, where radon emanation into the atmosphere occurs. At the same time, disturbances in the electric field of the atmosphere are observed. The lithospheric-atmospheric effect is most strongly manifested in seismically active regions at the final stage of earthquake preparation, when the deformation of rocks increases. In the works of [6,7] shows the relationship between high-frequency geoacoustic emission and the electric field of the atmosphere. These results indicate the interaction of geospheric shells, including due to changes in the radon flux over a large area into the near-surface layer of the atmosphere [8].

Methodology and Equipment

The network of subsurface gas monitoring points has been operating in one configuration or another since 1998. During the operation of the network in the radon field, two types of precursor anomalies for subduction earthquakes with $M \geq 5.5$ have been identified. Type A - is recorded at several points in the form of in-phase bays lasting from 3 to 12 days and reflects the large-scale manifestation of geodeformation processes on the last stage of earthquake preparation. It is this type of anomalies that precede earthquakes that is supposed to be associated with the passage of solitary deformation waves in the geomedium that arise due to the quasi-plastic or cataclastic flow of rock masses at the last stage of earthquake preparation. Type B - is registered in a single observation point and is associated with a special state of the hydrogeological system of the registration point. As shown in the work [3], the mechanism of the type B precursor is in good agreement with the theoretical model of Rn transport in an aqueous medium with complete transverse mixing. Currently, the subsurface gas monitoring network includes 6 points (Fig. 1) and is equipped with modern hardware and software for collecting, processing and storing the received time series with data transmission in a mode close to real time. It is distinguished by uniformity in the conditions of the creation of points and the location of sensors: all points are located in alluvial-deluvial deposits of river valleys; two sensors in the aeration zone at depths of 1 and 2 m from the daytime surface; a large number of recorded parameters, including atmospheric values. The analysis of the obtained data and comparison with the seismicity of the region continue continuously, which makes it possible to identify abnormal changes in the flow of subsurface radon that precede earthquakes. The network of points is described in detail in [3]. The data obtained by the subsurface gas monitoring network and the atmospheric electric field potential gradient (PG) data are used in the work.

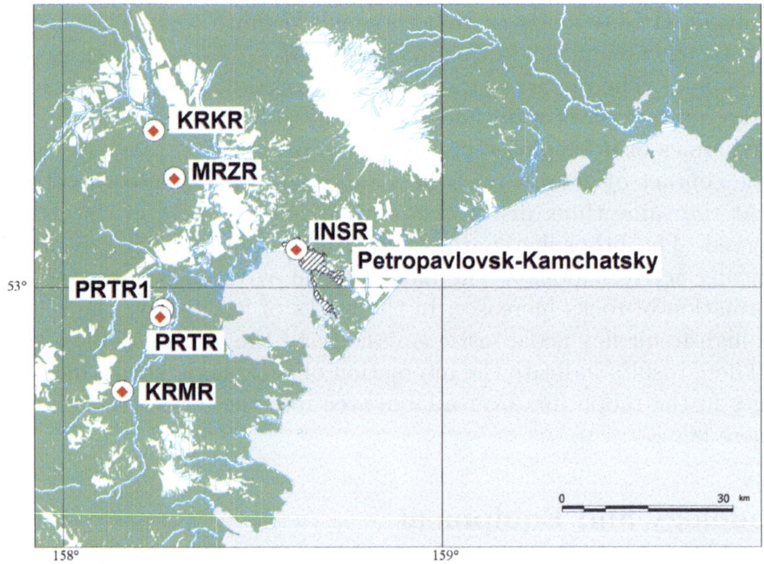

Fig. 1. Layout of the registration points for the concentration of subsurface gases at the Petropavlovsk-Kamchatsky geodynamic polygon.

A network of PG observation points has been implemented in Kamchatka [9]. For data collection, storage and preprocessing, as well as monitoring the operability of the equipment, a hardware and software complex based on an electrostatic sensor of the "EF-4" type has been developed and implemented, this complex allows transmitting data close to real time [9]. In one of the points of the radon observation network (PRTR1) located in the valley of the Paratunka River, in addition to the RVA sensors, β and γ radiation is recorded in the air at altitudes of 2.5 and 5 m (Fig. 1, 2). This point is equipped with a complex for recording the concentration of subsurface gases [3].

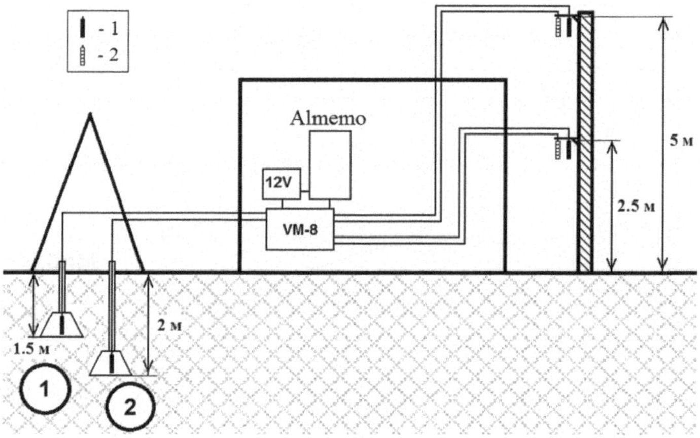

Fig. 2. Layout of sensors for monitoring the concentration of soil gases at the point PRTR1. 1 - gas-discharge counter of β- radiation; 2 - gas-discharge counter of γ - radiation; VM-8 - pulse generator for recording β radiation as part of the complex [3].

Earthquake on March 16, 2021 with $M_W = 6.6$, Kamchatka

An earthquake with $M_W = 6.6$ occurred in the Pacific Ocean on the traverse of the Kronotsky Peninsula on March 16, 2021 at a distance of 350 km from INSR at a depth of 65 km (Fig. 3). The RVA anomalies that preceded this earthquake appeared at several points of the network, had a bay-shaped form of positive and negative polarity and a duration of up to 18 days. Details about the development of anomalies of subsurface gases (radon and hydrogen) and deformations of the Earth's crust recorded by tilt meters before this earthquake are described in [10].

In this work, it is concluded that the anomalous variations of radon recorded in Kamchatka before the earthquake on March 16, 2021 with $M_W = 6.6$ are associated with the impact on the geomedium of several deformation pulses that passed through all the registration points, causing changes in the permeability of the soil, which led to a change in the flow of radon in the sensor installation sites. The probable area of generation of such pulses is associated with the spatial location of the future focus, and the occurrence of deformations in it, similar to the processes of inelastic deformation (crip) indicates the final stage of earthquake preparation. Sufficiently long anomaly times (up to 18 days) and anticipations also indicate that the processes of the final stage of preparation for this earthquake were stretched over time and are probably related to the creep of rocks. In addition to the identified anomalous variations of the RVA, which were mentioned above, preliminary analysis and comparison of PG data with variations of subsurface radon revealed coincident anomalous variations in parameters that occurred $\sim$ 13 h before the earthquake on March 16, 2021. Figure 4 shows the RVA curves of MRZR and PRTR1, β, γ-radiation points at PRTR1, the gradient of the potential of the electric field of the atmosphere at PETT, and meteorological parameters for the period March 12–19, 2021. Anomalous variations with

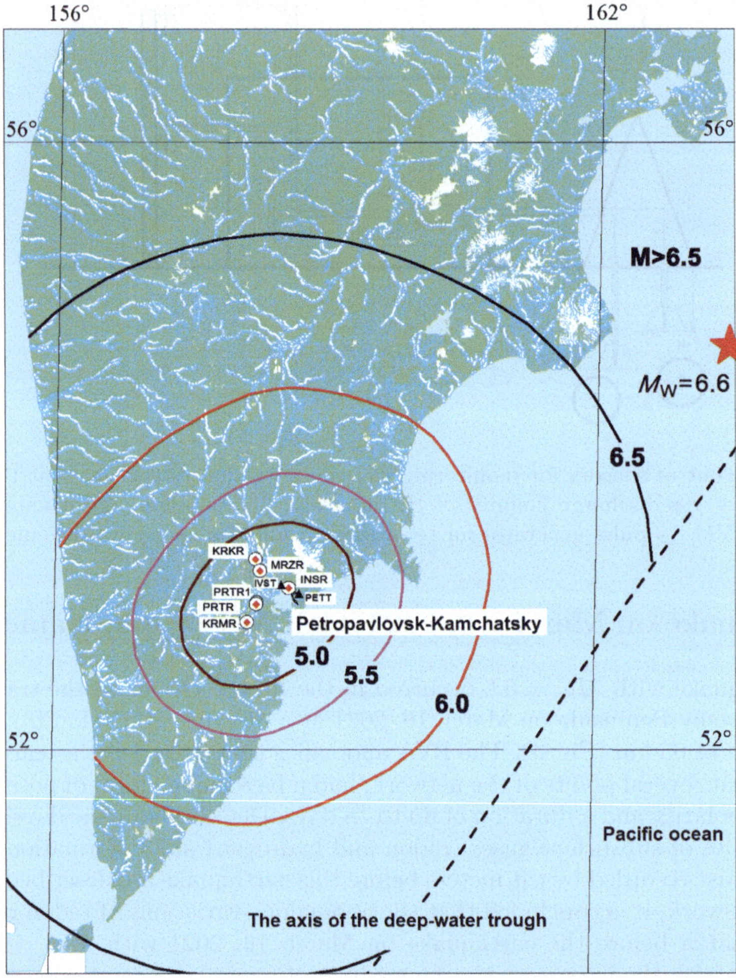

Fig. 3. Map of the boundaries of areas within which the preparation process for earthquakes of corresponding magnitudes can cause anomalies in the field of subsoil radon with a relative amplitude $\delta_{min} \geq 20$ minus the lower boundary of the 95% confidence corridor [3]. The star shows the epicenter of the earthquake on March 16, 2021 with $M_W = 6.6$.

a duration of ~ 30 h in RVA (MRZR, PRTR1), ~ 24 h in β, γ radiation are clearly visually distinguished on the curves.

Attention is drawn to the increase in the concentration of carbon dioxide (Fig. 4) in the INSTR point, organized on the basis of the NIS-1 well. Earlier, anomalous postseismic effects were observed at this point after the Zhupanov earthquake with $M = 7.2$ [3]. Probably, minor changes in the permeability of the upper layer of loose sediments in the aeration zone where the sensor is located, resulting from the impact of seismic waves, led to a change in the flow

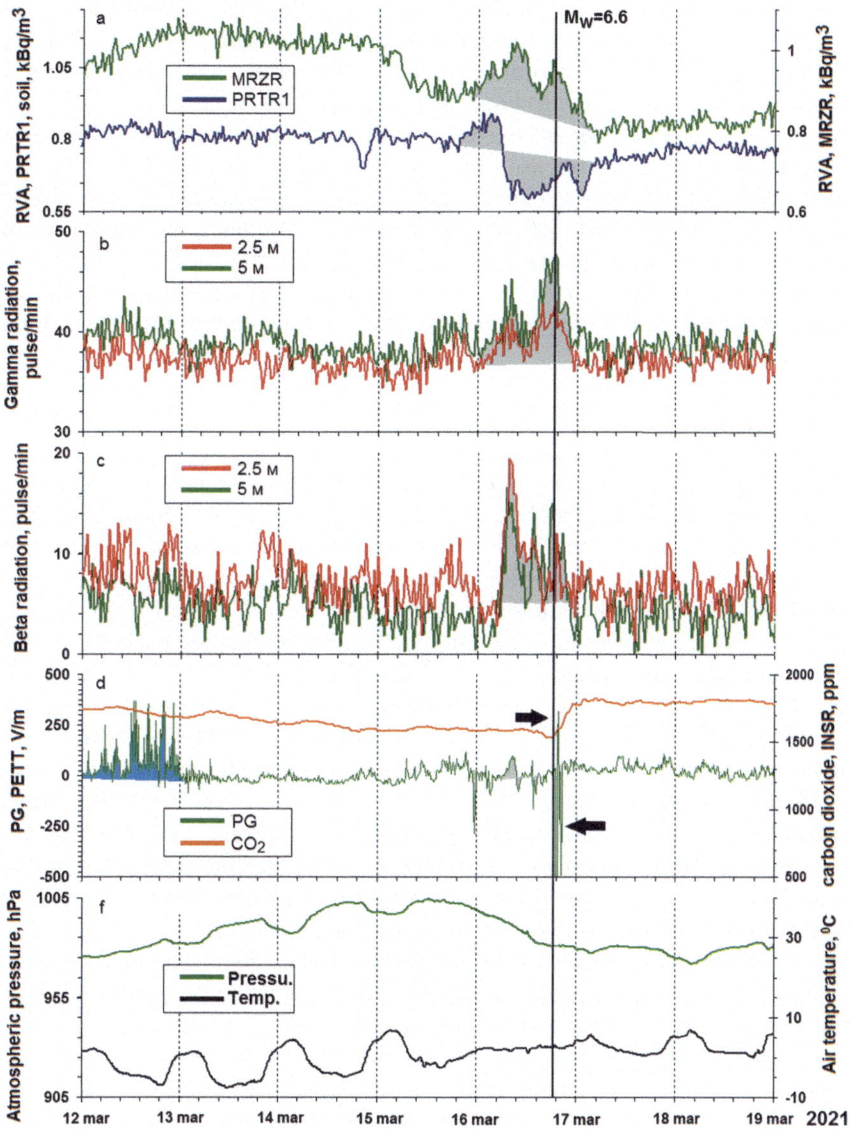

Fig. 4. Dynamics of the RVA at points MRZR, PRTR1 (a), β and γ radiation at point PRTR1 (b, c), carbon dioxide at point INSR, PG at point PETT (d), variations in atmospheric pressure and temperature (f) for the period March 12–19, 2021. The moment of the earthquake is shown by a black vertical line, anomalies highlighted in light gray. Post-seismic variations in the concentration of carbon dioxide and PG are shown by black arrows. The blue color on the PG curve shows the response to cyclonic activity. (Color figure online)

of carbon dioxide, which was recorded. The strain impulse of stretching could also lead to such a result. Similarly, changes in permeability could lead to the release of some excess radon over a large area in the area of the PETT point where PG is recorded. Since radon sensors are located in accumulation chambers and, as a result, register local changes in the flow, as well as are inertial, the appearance of some insignificant excess volume of radon was not reflected in their data. However, this radon released over a large area of the earth's surface, due to ionization, could affect the electric field of the atmosphere, as demonstrated by the PG data in the form of a short abnormal bipolar burst that appeared almost immediately after the earthquake (Fig. 4). Similar anomalies in curves of RVA were observed on the PKGP before the first of the Simushir earthquakes doublet, with $M = 8.3$, which occurred in the area of the middle Kuril Islands on November 15, 2006 [3].

Conclusions

Based on the weekly analysis of data received from the network of observation points on March 11, 2021, a standard conclusion on seismic hazard for the Kamchatka Territory was submitted, made on the basis of the emanation method described in [3]. According to the conclusion, during the next week it was possible to have earthquakes with magnitudes greater than those indicated in the corresponding areas marked in Fig. 3, or earthquakes with $M > 6.5$ in the band bounded by latitude $49.5°N$ - $56°N$ and the axis of the deep-water trough. The submitted forecast was based on an expert assessment of the development of the identified anomaly and was additionally justified using the method of "aigenoscopy" [3]. The earthquake that occurred in the area of the Kronotsky Peninsula on March 16, 2021 with $M_W = 6.6$ is considered as predicted. The effect of radon and its daughter products on the ionization balance of the surface layer of the atmosphere is determined by the intensity of exhalation of subsurface radon and meteorological conditions (convection, turbulence and stratification of the surface layer of the atmosphere), which was shown on long time series in [8]. The search for short-period variations of the electric field of the atmosphere associated with the activation of deformation processes may be complicated by weather conditions. There is no effective methodology for assessing or compensating the influence of meteorological processes on the recording of variations in the electric field of the atmosphere. However, in some individual cases, the identification of anomalies, in comparison with the data of radon observations, becomes possible. These cases, as a rule, are associated with good weather conditions, under which it is considered to be such weather conditions at the observation point, in which the effect of local sources of formation of volumetric electric charges is minimal (wind speed less than $6m/s$, lack of clouds and strong magnetic disturbances). As noted in [8], the choice of days with good weather for the Kamchatka region, as a rule, corresponds to the nature of anticyclonal atmospheric circulation, characterized by cloudless weather and distance from the registration point of active generators of the electric field of the atmosphere due to the vertical transfer of air masses to the Earth's surface. Such conditions provide the greatest effect of radon exhalation from the Earth's surface on the electric field of the atmosphere. The detected short-term anomalous disturbances that precede an earthquake can be considered as its operational precursors and evidence of the impact of the upper layer of the Earth's crust on the surface atmosphere during its preparation and implementation. The ongoing analysis of continuously received data and comparison with the seismicity of the region make it possible to identify anomalous changes in the RVA that precede earthquakes. Detection of in-phase variations at several registration points is extremely important for expert assessment and identification of anomalies in the dynamics of subsurface gases preceding earthquakes against the background of regularly occurring changes associated with other processes. Equally important is the search for optimal techniques that help reduce the influence of meteorological factors and ensure reliable network operation without data loss during long-term continuous registration. The development of an emanation method for earthquake prediction requires modern digital equipment that allows obtaining time series of various parameters in time close to real time. The hardware base of the Kamchatka network of subsurface gas registration points is constantly being improved and developed in order to provide detailed continuous qualitative data series for research and meet international standards.

Acknowledgments. The work was carried out according to the RSF project 22-17-00125 "Physical analysis of seismoelectromagnetic phenomena at the Kamchatka geodynamic test site: modernization of the observation system and theoretical modeling".

References

1. Hauksson, E.: Radon content of groundwater as an earthquake precursor: evaluation of worldwide data and physical basis. Journ. Geophys. Res. **86**, 9397–9410 (1981)
2. Cicerone, R.D., Ebel, J.E., Beitton, J.A.: Systematic compilation of earthquake precursors. Tectonophysics **476**, 371–396 (2009)
3. Firstov, P.P., Makarov, E.O.: Dynamics of Subsurface Radon in Kamchatka and Strong Earthquakes. Kamchatka State University named after Vitus Bering, Petropaulowsk-Kamchatskiy (2018). (in Russian)
4. Giuseppina, I., Morelli, D.: Radon as earthquake precursor. In: D'Amico, S. (ed.) Earthquake Research and Analysis - Statistical Studies, Observations and Planning, pp. 143–160 (2012)
5. Inan, S., et al.: Geochemical monitoring in the marmara region (NW Turkey): a search for precursors of seismic activity. J. Geophys. Research: Solid Earth **113**(B3) (2008). https://doi.org/10.1029/2007JB005206
6. Marapulets, Y.V., Rulenko, O.P., Mishchenko, M.A., Shevtsov, B.M.: Relationship of high-frequency geoacoustic emission and electric field in the atmosphere in seismotectonic process. Dokl. Earth Sci. **431**(1), 361–364 (2010). https://doi.org/10.1134/S1028334X10030207
7. Rulenko, O.P., Marapulets, Y.V., Mishchenko, M.A.: An analysis of the relationship between high-frequency geoacoustic emission and the electrical field in the atmosphere near the ground surface. Volcanol. Seismol. 8(3), 183–193 (2014) https://doi.org/10.7868/S0203030614030055

8. Firstov, P.P., Cherneva, N.V., Ponomarev, E.A., Buzevich, A.V.: Ubsoil radon and electric field intensity in atmosphere in case of petropavlovsk-kamchatsky geodinamic chain in 1998–2003 (in russian). Bull. KRAESC. Earth Sci. **7**(1), 102–109 (2006)
9. Akbashev, R.R., Firstov, P.P., Budilov, D.I., Zavodevkin, I.A.: Monitoring the potential gradient of the electric field the atmosphere on the kamchatka peninsula and on the paramushir island (kuril islands). In: AIP Conference Proceedings, vol. 2467, no. 1, p. 080013 (2022). https://doi.org/10.1063/5.0092738
10. Firstov, P.P., Makarov, E.O.: Reaction in the field of subsoil gases to the preparation of the earthquake on march 16, 2021 with mw = 6.6 (kamchatka, russia). In: Journal of Physics: Conference Series, vol. 2094, no. 5, p. 052026 (2021). https://doi.org/10.1088/1742-6596/2094/5/052026

Author Index

A. Dmitriev et al. (Eds.): STRPEP 2023, SPEES, pp. 285–286, 2023.
https://doi.org/10.1007/978-3-031-50248-4

SPRINGER NATURE

GPSR Compliance

The European Union's (EU) General Product Safety Regulation (GPSR) is a set of rules that requires consumer products to be safe and our obligations to ensure this.

If you have any concerns about our products, you can contact us on ProductSafety@springernature.com

In case Publisher is established outside the EU, the EU authorized representative is:

Springer Nature Customer Service Center GmbH
Europaplatz 3
69115 Heidelberg, Germany

The manufacturer's authorised representative in the EU is Springer Nature Customer Service Centre GmbH, Europaplatz 3, 69115 Heidelberg, Germany. If you have any concerns regarding our products, please contact ProductSafety@springernature.com

Printed and bound by CPI Group (UK) Ltd, Croydon, CR0 4YY
01/10/2026
02225892-0007